D0374912

MODERN OPTICAL ENGINEERING

Warren J. Smith

Director of Development,
Electro-Optics Division
Infrared Industries, Inc.
Santa Barbara, California

Modern
Optical
Engineering

The Design of Optical Systems

McGRAW-HILL BOOK COMPANY

New York   San Francisco   Toronto   London   Sydney

to Mary, David and Barbara

Preface

This book is directed to the practicing engineer or scientist who requires effective practical technical information on optical systems and their design. The recent increase in the utilization of optical devices in such fields as alignment, metrology, automation, and space/defense applications has brought about a need for technical people conversant with the optical field. Thus, many individuals whose basic training is in electronics, mechanics, physics, or mathematics, find themselves in positions requiring a relatively advanced competence in optical engineering. It is the author's hope that this volume will enable them to undertake their practice of optics soundly and with confidence.

Although the reader is presumed to be at least familiar with the optical material contained in a first-year physics course, the book begins with a general orientation chapter dealing with electromagnetic waves, Snell's law, interference, diffraction, and the photoelectric effect. The second chapter goes quite deeply into image formation at the first-order (Gaussian) level, and includes several numerical examples. The departures from first-order imagery represented by the aberrations are discussed in the third chapter. Prisms and mirrors are covered in both general and specific terms, in such a way that the reader can independently proceed beyond the standard systems. A chapter on the eye (as the basic "detector" involved in the vast majority of optical systems) follows, and includes a brief summary of the specification of colors.

The chapter on stops and apertures covers the usual aperture, field and glare stops and integrates the diffraction and resolution effects of apertures. The seventh chapter discusses optical materials and optical coatings, including the computation of the reflectance and transmittance of interference films.

The chapter on radiation and photometry introduces the basic radiation concepts, which are so necessary to a complete understanding of the relationship between the optics of a larger "system" and its performance. Chapter 9 discusses the basic "tools" of optics, the devices such as telescopes, microscopes, radiometers, variable focus lenses, and the like, from which complete systems and instruments are built.

Chapters 10 through 13 are fairly advanced and contain sufficient material to permit the reader to undertake the complete design of an optical system. The chapter on optical computation covers ray-tracing through spherical and aspheric surfaces and includes techniques for determining the third-order aberrations. Image evaluation is discussed at length, from both a geometrical and physical optics basis; the concept of the Optical Transfer Function is introduced and computing techniques are demonstrated. Design procedures, both specific and generalized, are presented, and the individual design characteristics of a wide range of optical systems are discussed. Chapter 13 also includes a number of equations and charts which are of great value in preliminary engineering and proposal work and which permit a very rapid estimation of performance level for many basic optical systems (with emphasis on those used in space/defense applications).

The final chapter of the book includes discussions of optical manufacturing processes and the specification of optics for the shop, as well as brief discussions of optical-mechanics and laboratory practice.

An extensive bibliography and technical-journal reference list are provided for those who may wish to pursue their specific interests further.

The general approach throughout has been to emphasize the application of basic optical principles to practice. Many numerical examples are included for the purpose of guiding the reader through typical engineering problems. Most chapters are followed by a set of exercises (and answers), designed to provide the reader with a close approximation to practical experience. The mathematical level required has been deliberately kept low; derivations are limited and are designed primarily to demonstrate either the technique of manipulation of optical quantities or the application of the relationships previously presented. The notation used is basically that of Conrady with modifications, since this is probably the most widely known and used system.

Probably every author owes a considerable debt to his family, teachers, colleagues, and employers; these I acknowledge with pleasure. More particularly, the help of my friend and close associate, Max J. Riedl, who read the manuscript, and that of Wanda Wilburn, who patiently typed and retyped, is gratefully acknowledged.

W. J. S.

Contents

MODERN OPTICAL ENGINEERING

1

General Principles

1.1 The Electromagnetic Spectrum

This book deals with certain phenomena associated with a relatively narrow slice of the electromagnetic spectrum. Optics is often defined as being concerned with radiation visible to the human eye; however, in view of the recent expansion of optical applications in the regions of the spectrum on either side of the visible region, it seems not only prudent, but necessary, to include certain aspects of the infrared and ultraviolet regions in our discussions.

The known electromagnetic spectrum is diagrammed in Fig. 1.1 and ranges from cosmic rays to radio waves. All the electromagnetic radiations transport energy and all have a common

FREQUENCY
IN CYCLES PER SECOND

WAVELENGTH
IN MICRONS

Frequency	Region	Wavelength	Unit
10^{23}	COSMIC RAYS	10^{-9}	
10^{22}		10^{-8}	
10^{21}	GAMMA RAYS	10^{-7}	X-UNIT
10^{20}		10^{-6}	
10^{19}		10^{-5}	
10^{18}	X-RAYS	10^{-4}	ANGSTROM UNIT (°A)
10^{17}		10^{-3}	MILLIMICRON (mμ)
10^{16}		10^{-2}	
10^{15}	ULTRAVIOLET	10^{-1}	
10^{14}	VISIBLE	1	MICRON (μ)
10^{13}	INFRARED	10	
10^{12}		10^{2}	
10^{11} EHF		10^{3}	MILLIMETER (mm)
10^{10} SHF	MICROWAVE	10^{4}	CENTIMETER (cm)
10^{9} UHF		10^{5}	
10^{8} VHF	F.M. RADIO	10^{6}	METER (m)
10^{7} HF	TELEVISION	10^{7}	
10^{6} MF	A.M. RADIO	10^{8}	
10^{5} LF		10^{9}	KILOMETER (km)
10^{4}		10^{10}	
10^{3}		10^{11}	
10^{2}		10^{12}	
10		10^{13}	
1		10^{14}	

FIG. 1.1. The Electromagnetic Spectrum.

velocity in vacuum of $c = 2.998 \times 10^{10}$ cm/sec. In other respects, however, the nature of the radiation varies widely, as might be expected from the tremendous range of wavelengths represented. At the short end of the spectrum we find gamma radiation with wavelengths extending below a billionth of a micron (one micron = $1 \mu = 10^{-6}$ m) and at the long end, radio waves with wavelengths measurable in miles. At the short end of the spectrum electromagnetic radiation tends to be quite particle-like in its behavior, whereas toward the long wavelength end the behavior is mostly wavelike. Since the optical portion of the spectrum occupies an intermediate position, it is not surprising that optical radiation exhibits both wave and particle behavior.

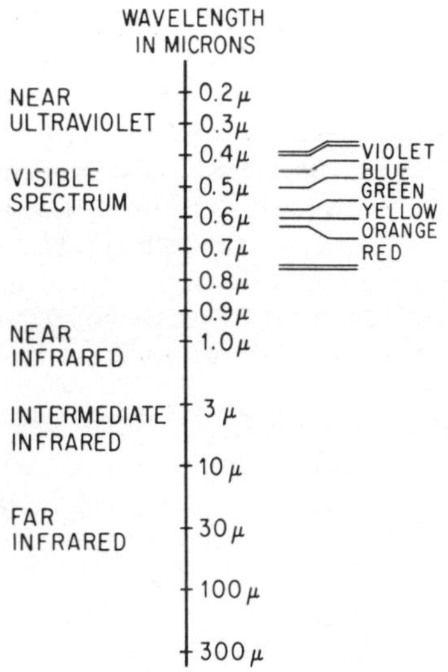

WAVELENGTH
IN MICRONS

NEAR ULTRAVIOLET   0.2 μ
                    0.3 μ
                    0.4 μ ═══ VIOLET
VISIBLE   0.5 μ ─── BLUE
SPECTRUM   ─── GREEN
                    0.6 μ ═══ YELLOW
                          ─── ORANGE
                    0.7 μ   RED
                    0.8 μ ═══
                    0.9 μ
NEAR   1.0 μ
INFRARED

INTERMEDIATE   3 μ
INFRARED
                    10 μ

FAR   30 μ
INFRARED
                    100 μ

                    300 μ

FIG. 1.2. The "Optical" portion of the Electromagnetic Spectrum.

The visible portion of this spectrum (Fig. 1.2) takes up less than one octave, ranging from violet light with a wavelength of 0.4 μ to red light with a wavelength of 0.76 μ. Beyond the red end of the spectrum lies the infrared region which blends into the microwave region at a wavelength of about one millimeter. The ultraviolet region extends from the lower end of the visible spectrum to a wavelength of about 0.01 μ at the beginning of the X-ray region. The wavelengths associated with the colors seen by the eye are indicated in Fig. 1.2.

The ordinary units of wavelength measure in the optical region are the Angstrom Å, the millicron mμ, and the micron μ. One micron is a millionth of a meter, a millimicron is a thousandth of a micron, and an Angstrom is one ten thousandth of a micron. Thus, $1.0 \text{ Å} = 0.1 \text{ m}\mu = 10^{-4} \mu$.

1.2 Light Wave Propagation

If we consider light waves radiating from a point source in a vacuum as shown in Fig. 1.3, it is apparent that at a given instant each wave front is spherical in shape, with the curvature (reciprocal of the radius) decreasing as the wave front travels away from the point source. At a sufficient distance from the source the radius of the wave front may be regarded as infinite. Such a wave front is called a plane wave.

The distance between successive waves is of course the wavelength of the radiation. The velocity of propagation of light waves in vacuum is approximately 3×10^{10} cm/sec. In other media the velocity is less than in vacuum. In ordinary glass, for example, the velocity is about two-thirds of the velocity in free space. The ratio of the velocity in vacuum to the velocity in a medium is called the index of refraction of that medium, denoted by the letter N.

i.e. a $f(N)$ or $f(\lambda)$

$$\text{Index of refraction } N = \frac{\text{velocity in vacuum}}{\text{velocity in medium}} \qquad (1.1)$$

Ordinary air has an index of refraction of about 1.0003, and since almost all optical work (including measurement of the index of refraction) is carried out in a normal atmosphere, it is a highly convenient convention to express the index of a material relative to that of air (rather than vacuum), which is then assumed to have an index of 1.0.

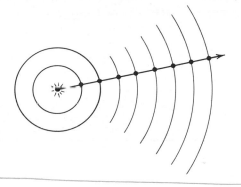

FIG. 1.3. Light waves radiating from a point source take on a spherical form. The path of a point on the wave front is called a ray.

If we trace the path of a hypothetical point on the surface of a wave front as it moves through space, we see that the point progresses as a straight line. The path of the point is thus what is called a ray of light. Such a light ray is an extremely convenient fiction, of great utility in understanding and analyzing the action of optical systems, and we shall devote the greater portion of this volume to the study of light rays.

The preceding discussion of wave fronts has assumed that the light waves were in a vacuum, and of course that the vacuum was isotropic, i.e., of uniform index in all directions. Several optical crystals are anisotropic; in such media wave fronts as sketched in Fig. 1.3 are not spherical. The waves travel at different velocities in different directions, and thus at a given instant a wave in one direction will be further from the source than will a wave traveling in a direction for which the media has a larger index of refraction.

1.3 Snell's Law of Refraction

Let us now consider a plane wave front incident upon a plane surface separating two media, as shown in Fig. 1.4. The light is progressing from the top of the figure downwards and approaches the boundary surface at an angle. The parallel lines represent the positions of a wave front at regular intervals of time. The index of the upper medium we shall call N_1 and that of the lower N_2. From Eq. 1.1, we find that the velocity in the upper medium is given by $v_1 = c/N_1$ (where c is the velocity in vacuum $\approx 3 \times 10^{10}$ cm/sec) and in the lower by $v_2 = c/N_2$. Thus, the velocity in the upper medium is N_2/N_1 times the velocity in the lower and the distance which the wave front travels in a given interval of time in the upper medium will also be N_2/N_1 times that in the lower. In Fig. 1.4 the index of the lower medium is assumed to be larger so that the velocity in the lower medium is less than that in the upper medium.

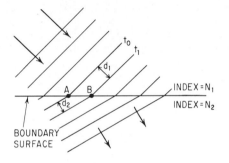

FIG. 1.4. A plane wave front passing through the boundary between two media of differing indices of refraction. $(N_2 > N_1)$

At time t_0 our wave front intersects the boundary at point A; at time $t_1 = t_0 + \Delta t$ it intersects the boundary at B. During this time it has moved a distance

$$d_1 = v_1 \Delta t = \frac{c}{N_1} \Delta t \qquad (1.2a)$$

in the upper media, and a distance

$$d_2 = v_2 \Delta t = \frac{c}{N_2} \Delta t \qquad (1.2b)$$

in the lower media.

In Fig. 1.5 we have added a ray to the wave diagram; this ray is the path of the point on the wave front which passes through point B on the surface, and is normal to the wave front (since we have tacitly assumed isotropic media). If the lines represent the positions of the wave at equal intervals of time, AB and BC, the distances between intersections, must be equal. The angle between the wave front and the surface (I_1 or I_2) is equal to the angle between the ray (which is normal to the wave) and the normal to the surface XX'. Thus we have from Fig. 1.5

$$AB = \frac{d_1}{\sin I_1} = BC = \frac{d_2}{\sin I_2}$$

and if we substitute the values of d_1 and d_2 from Eq. 1.2 we get

$$\frac{c\,\Delta t}{N_1 \sin I_1} = \frac{c\,\Delta t}{N_2 \sin I_2}$$

which, after canceling and rearranging, yields

$$N_1 \sin I_1 = N_2 \sin I_2 \qquad (1.3)$$

This expression is the basic relationship by which the passage of light rays is traced through optical systems. It is called Snell's Law after one of its discoverers.

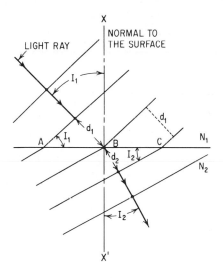

FIG. 1.5.

Since Snell's Law relates the sines of the angles between a light ray and the normal to the surface, it is readily applicable to surfaces other than the plane which we used in the example above; the path of a light ray may be calculated through any surface for which we can determine the point of intersection of the ray and the normal to the surface at that point.

The angle I_1 between the incident ray and surface normal is customarily referred to as the angle of incidence; the angle I_2 is called the angle of refraction.

For all optical media the index of refraction varies with the wavelength of light. In general the index is higher for short wavelengths than for long wavelengths. In the preceding discussion it has been assumed that the light incident on the refracting surface was monochromatic, i.e., composed of only one wavelength of light. Figure 1.6 shows a ray of white light broken into its various component wavelengths by refraction at a surface. Notice that the blue light ray is bent, or refracted, through a greater angle than is the ray of red light. This is because N_2 for blue light is larger than N_2 for red. Since $N_2 \sin I_2 = N_1 \sin I_1 =$ a constant in this case, it is apparent that if N_2 is larger for blue light than red, then I_2 must be smaller for blue than red. This variation in index with wavelength is called dispersion; when used as a differential it is written dN, otherwise dispersion is given by $\Delta N = N_{\lambda_1} - N_{\lambda_2}$ where λ_1 and λ_2 are the wavelengths of the two colors of light for which the dispersion is given. Relative dispersion is given by $\Delta N/(N-1)$ and, in effect, expresses the "spread" of the colors of light as a fraction of the amount that light of a median wavelength is bent.

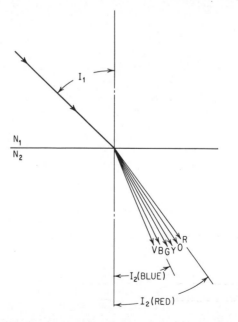

FIG. 1.6. Showing the dispersion of white light into its constituent colors by refraction (exaggerated for clarity).

All of the light incident upon a boundary surface is not transmitted through the surface; some portion is reflected back into the incident medium. A construction similar to that used in Fig. 1.5 can be used to demonstrate that the angle between the surface normal and the reflected ray (the angle of reflection) is equal to the angle of incidence, and that the reflected ray is on the opposite side of the normal from the incident ray (as is the refracted ray). Thus, for reflection, Snell's law takes on the form

$$I_{\text{incident}} = -I_{\text{reflected}} \qquad (1.4)$$

Figure 1.7 shows the relationship between a ray incident on a plane surface and the reflected and refracted rays which result.

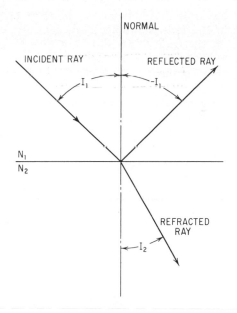

FIG. 1.7. Relationship between a ray incident on a plane surface and the reflected and refracted rays which result.

At this point it should be emphasized that the incident ray, the normal, the reflected ray, and the refracted ray all lie in a common plane, which in Fig. 1.7 is the plane of the paper.

1.4 The Action of Simple Lenses and Prisms on Wave Fronts

In Fig. 1.8 point source P is emitting light; as before, the arcs centered about P represent the successive positions of a wave front at regular intervals of time. The wave front is incident on a lens consisting of two surfaces of rotation bounding a medium of (in this instance) higher index of refraction than the medium in which the source is located. In each interval of time the wave front may be assumed to travel a distance d_1 in the medium of the source; it will travel a lesser distance d_2 in the medium of the lens. (As in the preceding discussion, these distances are related by $N_1 d_1 = N_2 d_2$.) At some instant, the vertex of the wave front will just contact the vertex of the lens surface at point A. In the succeeding interval, the portion of the wave front inside the lens will move a distance d_2, while the portion of the wave front still outside the lens will have moved d_1. As the wave front passes through the lens, this effect is repeated in reverse at the second surface. It can be seen that the wave front has been retarded by the medium of the lens and that this retardation has been greater in the thicker central portion of the lens, causing the curvature of the wave front to be reversed. At the left of the lens the light from P was diverging, and to the right of the lens the light is now converging in the general direction of point P'. If a screen or sheet of paper were placed at P', a concentration of light could be observed at this point. The lens is said to have formed an image of P at P'. A lens of this type is called a converging, or positive, lens.

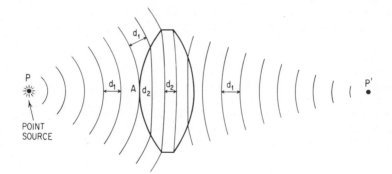

FIG. 1.8. The passage of a wave front through a converging, or positive, lens element.

Figure 1.8 diagrammed the action of a convex lens—that is, a lens which is thicker at its center than at its edges. A convex lens with an index higher than that of the surrounding media is a converging lens, in that it will increase the convergence (or reduce the divergence) of a wave front passing through it.

In Fig. 1.9 the action of a concave lens is sketched. In this case the lens is thicker at the edge and thus retards the wave front more at the edge than at the center and increases the divergence. After passing through the lens, the wave front appears to have originated from the neighborhood of point P', which is the image of point P formed by the lens. In this case, however, it would be futile to place a screen at P' and expect to find a concentration of light; all that would be observed would be the general illumination produced by the light emanating from P. This type of image is called a virtual image to distinguish it from the type of image diagrammed in Fig. 1.8, which is called a real image. Thus a virtual image may be observed directly or may serve as a source to be reimaged by a subsequent lens system, but it cannot be produced on a screen.

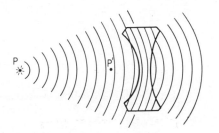

FIG. 1.9. The passage of a wave front through a diverging, or negative, lens element.

The path of a ray of light through the lenses of Figs. 1.8 and 1.9 is the path traced by a point on the wave front. In Fig. 1.10 several ray paths have been drawn for the case of a converging lens. Note that the rays originate at point P and proceed in straight lines (since the media involved are isotropic) to the surface of the lens where they are refracted according to Snell's Law (Eq. 1.3.) After refraction at the second surface the rays converge at the image P'. (In practice the rays will converge exactly at P' only if the lens surfaces are suitably chosen surfaces of rotation, usually nonspherical, whose axes are coincident and pass through P.) This would lead one to expect that the concentration of light at P' would be a perfect point. However, the wave nature of light causes it to

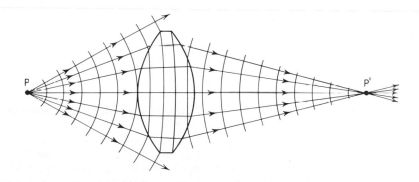

FIG. 1.10. Showing the relationship between light rays and the wave front in passing through a positive lens element.

be diffracted in passing through the limiting aperture of the lens so that the image, even for a "perfect" lens, is spread out into a small disc of light surrounded by faint rings.

In Fig. 1.11 a wave front from a source so far distant that the curvature of the wave front is negligible, is shown approaching a prism, which has two flat polished faces. As it passes through each face of the prism, the light is refracted downwards so that the direction of propagation is deviated. The angle of deviation of the prism is the angle between the incident ray and the emergent ray. Note that the wave front remains plane as it passes through the prism.

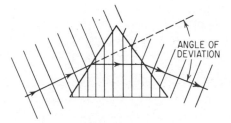

FIG. 1.11. The passage of a plane wave front through a refracting prism.

If the radiation incident on the prism consisted of more than one wavelength, the shorter wavelength radiation would be slowed down more by the medium composing the prism and thus deviated through a greater angle. This is one of the methods used to separate different wavelengths of light and is, of course, the basis for Isaac Newton's classic demonstration of the spectrum.

1.5 Interference and Diffraction

If a stone is dropped into still water, a series of concentric ripples, or waves, are generated and spread outward over the surface of the water. If two stones are dropped some distance apart, a careful observer will notice that where the waves from the two sources meet there are areas with waves twice as large as the original waves and also areas which are almost free of waves. This is because the waves can reinforce or cancel out the action of each other. Thus if the crests (or troughs) of two waves arrive simultaneously at the same point, the crest (or trough) generated is the sum of the two wave actions. However, if the crest of one wave arrives at the same instant as the trough of the other, the result is a cancellation. A more spectacular display of wave reinforcement can often be seen along a sea wall where an ocean wave, which has struck the wall and been reflected back out to sea, will combine with the next incoming wave to produce an eruption where they meet.

Similar phenomena occur when light waves are made to interfere. In general, light from the same point on the source must be made to travel two separate paths and then be recombined, in order to produce optical interference. The familiar colors seen in soap bubbles or in oil films on wet pavements are produced by interference.

Young's Experiment, which is diagrammed schematically in Fig. 1.12, illustrates both diffraction and interference. Light from a source to the left of the figure is caused to pass through a slit or pin hole *s* in an opaque screen. According to *Huygens' principle* the propagation of a wave front may be constructed by considering each point on the wave front as a source of new spherical wavelets; the envelope of these new wavelets indicates the new position of the wave front. Thus *s* may be considered as the center of a new spherical or cylindrical wave (depending on whether *s* is a pinhole or a slit), provided that the size of *s* is sufficiently small. These diffracted wave fronts from *s* travel to a second opaque screen which has two slits (or pin holes), *A* and *B*, from which new wave fronts originate. The wave fronts again spread out by diffraction and fall on an observing screen some distance away.

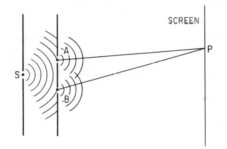

FIG. 1.12. Young's Diffraction Experiment.

Now, considering a specific point *P* on the screen, if the wave fronts arrive simultaneously (or in phase), they will reinforce each other and *P* will be illuminated. However, if the distances *AP* and *BP* are such that the waves arrive exactly out of phase, destructive interference will occur and *P* will be dark.

If we assume that *s*, *A*, and *B* are so arranged that a wave front from *s* arrives simultaneously at *A* and *B* (that is, distance *sA* exactly equals distance *sB*), then new wavelets will start out simultaneously from *A* and *B* toward the screen. Now if distance *AP* exactly equals distance *BP*, or if *AP* differs from *BP* by exactly an integral number of wavelengths, the wave fronts will arrive at *P* in phase and will reinforce. If *AP* and *BP* differ by one-half wavelength, then the wave actions from the two sources will cancel each other.

If the illuminating source is monochromatic, that is, emits but a single wavelength of light, the result will be a series of alternating light and dark bands of gradually changing intensity on the screen (assuming that s, A, and B are slits) and by careful measurement of the geometry of the slits and the separation of the bands, the wavelength of the radiation may be computed. (The distance AB should be less than a millimeter and the distance from the slits to the screen should be to the order of a meter to conduct this experiment.)

With reference to Fig. 1.13, it can be seen that, to a first approximation, the path difference between AP and BP, which we shall represent by Δ, is given by

$$\Delta = \frac{AB \cdot OP}{D}$$

and rearranging this expression we get

$$OP = \frac{\Delta \cdot D}{AB} \tag{1.5}$$

Now as Fig. 1.13 is drawn, it is obvious that the optical paths AO and BO are identical, so that the waves will reinforce at O and produce a bright band. If we set Δ in Eq. 1.5 equal to (plus or minus) one-half wavelength, we shall then get the value of OP for the first dark band

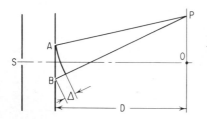

FIG. 1.13. Geometry of Young's Experiment.

$$OP \text{ (1st Dark)} = \frac{\pm \lambda D}{2 AB} \tag{1.6}$$

and if we assume that the distance from slits to screen D is one meter, that the slit separation AB is one-tenth millimeter, and that the illumination is red light of a wavelength of 0.64 μ, we get the following by substitution of these values in Eq. 1.6

$$OP \text{ (1st Dark)} = \frac{\pm \lambda 10^3}{2 \cdot 10^{-1}} = \frac{\pm 10^4 \lambda}{2} = \frac{\pm 10^4 \cdot 0.64 \cdot 10^{-3}}{2} = \pm 3.2 \text{ mm}$$

Thus the first dark band occurs 3.2 mm above and below the axis. Similarly the location of the next light band can be found to be at 6.4 mm by setting Δ equal to one wavelength and so on.

If blue light of wavelength 0.4 μ were used in the experiment, we would find that the first dark band occurs at ± 2 mm and the next bright band at ± 4 mm.

Now if the light source, instead of monochromatic, is white and consists of all wavelengths, it can be seen that each wavelength will produce its own array of light and dark bands of its own particular spacing. Under these conditions the center of the screen will be illuminated by all wavelengths and will be white. As we proceed from the center, the first effect perceptible to the eye will be the dark band for blue light which will occur at a point where the other wavelengths are still illuminating the screen. Similarly, the dark band for red light will occur where blue and other wavelengths are illuminating the screen. Thus a series of colored bands is produced, starting with white on axis and progressing through red, blue, green, orange, red, violet, green, and violet, as the path difference increases. Further from the axis, however, the various light and dark bands from all the visible wavelengths become so "scrambled" that the band structures blend together and disappear.

Newton's Rings are produced by the interference of the light reflected from two surfaces which are close together. Figure 1.14 shows a beam of parallel light incident on a pair of partially reflecting surfaces. At some instant a wave front AA′ strikes the first surface at A. The point on the wave front at A travels through the space between the two surfaces and strikes the second surface at B where it is partially reflected; the reflected wave then travels upward to pass through the first surface again at C. Meanwhile the point on the wave front at A′ has been reflected at point C and the two paths recombine at this point.

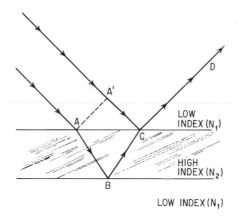

FIG. 1.14.

Now if the waves arrive at C in phase, they will reinforce; if they arrive one-half wavelength out of phase they will cancel. In determining the phase relationship at C we must take into account the index of the material through which the light has traveled and also the phase change which occurs on reflection. This phase change occurs when light traveling through a low index medium is reflected from the surface of a high index medium; the phase is then abruptly changed by 180°, or one-half wavelength. No phase change occurs when the indices are encountered in reverse order. Thus with the relative indices as indicated in Fig. 1.14, there is a phase change at C for the light following the $A'CD$ path, but no phase change at B for the light reflected from the lower surface.

As in the case of Young's Experiment described above, the difference between the optical paths ABC and $A'C$ determines the phase relationship. Since the index of refraction is inversely related to the velocity of light in a medium, it is apparent that the length of time a wave front takes to travel through a thickness d of a material of index N is given by $t = Nd/c$ (where $c \approx 3 \cdot 10^{10}$ cm/sec = velocity of light in vacuum). The constant frequency of electromagnetic radiation is given by c/λ, so that the number of cycles which take place during the time $t = Nd/c$ is given by $(c/\lambda) \cdot (Nd/c)$ or Nd/λ. Thus, if the number of cycles are the same, or differ by an integral number of cycles, over the two paths of light traversed, the two beams of light will arrive at the same phase.

In Fig. 1.14, the number of cycles for the path $A'C$ is given by $1/2 + N_1 A'C/\lambda$ (the one-half cycle is for the reflection phase change) and for the path ABC by $N_2 ABC/\lambda$; if these numbers differ by an integer, the waves will reinforce; if they differ by an integer plus one-half, they will cancel.

The use of cycles in this type of application is inconvenient and it is customary to work in optical path length. It is obvious that if we consider the difference between the two path lengths (arrived at by multiplying the above number of cycles by the wavelength λ), exactly equivalent results are obtained when the difference is an integral number of wavelengths (for reinforcement) or an integral number plus one-half wavelength (for cancellation). Thus, for Fig. 1.14 the *optical path difference* is given by

$$O.P.D. = \frac{\lambda}{2} + N_1 A'C - N_2 ABC$$

when the phase change is taken into account by the $\lambda/2$ term.

The term "Newton's Rings" usually refers to the ring pattern of interference bands formed when two spherical surfaces are placed in intimate contact. Figure 1.15 shows the convex surface of a lens resting on a plane surface. At the point of contact the difference in the optical paths reflected from the upper and lower surfaces is patently zero. The phase change on reflection from the lower surface causes the beams to rejoin exactly out of phase, resulting in complete cancellation and the appearance of the central "Newton's Black Spot." Some distance from the center the surfaces will be separated by exactly one-quarter wavelength, and this path difference of one-half wavelength plus the phase change results in reinforcement, producing a bright ring. A little further from the center, the separation is one-half wavelength, resulting in a dark ring, and so on.

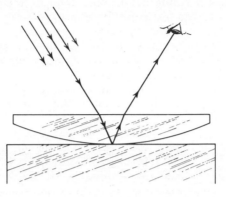

FIG. 1.15.

Just as in Young's experiment, the dark and bright bands for different wavelengths will occur at different distances from the center, resulting in colored circles near the point of contact which fade away toward the edge.

A setup similar to Fig. 1.15 can obviously be used to measure the wavelength of light if the radius of curvature of the lens is known and a careful measurement of the diameters of the light and dark fringes is made. The spacing between the surfaces is the sagittal height of the radius, given by

$$S.H. = R - (R^2 - Y^2)^{\frac{1}{2}} \qquad (1.7)$$

where Y is the semidiameter of the ring measured. $S.H.$ is equal to $\lambda/4$ for the first bright ring, $\lambda/2$ for the first dark ring, $3\lambda/4$ for the second bright ring, and so on.

1.6 The Photoelectric Effect

In the preceding section, the discussion was based upon the assumption that light was wavelike in nature. This assumption provides reasonable explanations for reflection, refraction, interference, diffraction, and dispersion, as well as other effects. The photoelectric effect, however, seems to require for its explanation that light behave as if it consisted of particles.

In brief, when short wavelength light strikes a photoelectric material, it can knock electrons out of the material. As stated, this effect could be explained by the energy of the light waves exciting an electron sufficiently for it to break loose. However, when the nature of the incident radiation is modified, the characteristics of the emitted electrons change in an unexpected way. As the intensity of the light is increased, the number of electrons is increased just as might be expected. If the wavelength is increased, however, the maximum velocity of the electrons emitted is reduced; if the wavelength is increased beyond a certain value (this value is characteristic of the particular photoelectric material used), the maximum velocity drops to zero and no electrons are emitted, regardless of the intensity.

Thus the energy necessary to break loose an electron is not stored up until enough is available (as one would expect of the wavelike behavior of light.) The situation here is more analagous to a shower of particles, some of which have enough energy to break an electron loose from the forces which bind it in place. Thus the particles of shorter wavelength have sufficient energy to release an electron. If the intensity of light is increased, the number of electrons released is increased and their velocity remains unchanged. The longer wavelength particles do not have enough energy to knock electrons loose, and when the intensity of the long wavelength light is increased, the effect is to increase the number of particles striking the surface, but each particle is still insufficiently powerful to release an electron from its bonds.

The apparent contradiction between the wave and particle behavior of light can be resolved by assuming that every "particle" has a wavelength associated with it which is inversely proportional to its momentum. This has proved true experimentally for electrons, protons, ions, atoms, and molecules; for example, an electron accelerated by an electric field of a few hundred volts has a wavelength of a few angstroms ($10^{-4}\mu$) associated with it. Reference to Fig. 1.1 indicates that this wavelength is characteristic of X-rays, and indeed, electrons of this wavelength are diffracted in the same patterns (by crystal lattices) as are X-rays.

2

Image Formation (first-order optics)

2.1

The action of a lens on a wave front was briefly discussed in Section 1.4. Figures 1.8 and 1.9 showed how a lens can modify a wave front to form an image. A wave front is difficult to manipulate mathematically, and for most purposes the concept of a light ray (which is the path described by a point on a wave front) is more convenient. In an isotropic medium, light rays are straight lines normal to the wave front, and the image of a point source is formed where the rays converge (or appear to converge) to a concentration or focus. In a perfect lens the rays converge to a point at the image.

An extended object may be regarded as an array of point sources. The location and size of the image formed by a given optical system can be determined by locating the respective images of the sources making up the object. This can be accomplished by calculating the paths of a number of rays from each object point through the optical system, applying Snell's Law (Eqs. 1.3) at each ray-surface intersection in turn. However, it is possible to locate optical images with considerably less effort by means of simple equations derived from the limiting case of the trigonometrically traced ray. These expressions yield image positions and sizes which would be produced by a perfect optical system.

First-order (or Gaussian) optics is often referred to as the optics of perfect optical systems. The first-order equations are derived by reducing the exact trigonometrical expressions for ray paths to the limit when the angles and ray heights involved approach zero. These equations are completely accurate for an infinitesimal thread-like region about the optical axis, known as the *paraxial* region. The value of first-order expressions lies in the fact that a well-corrected optical system will follow the first-order expressions almost exactly and also that the first-order image positions and sizes provide a convenient reference from which to measure departures from perfection.

We shall begin this chapter by considering the manner in which a "perfect" optical system forms an image, and we will discuss the expressions which allow the location and size of the image to be found when the basic characteristics of the optical system are known. Then we will take up the determination of these basic characteristics from the constructional parameters of an optical system. Finally, methods of image calculation by paraxial ray-tracing will be discussed.

2.2 Cardinal Points of an Optical System

A well-corrected optical system can be treated as a "black box" whose characteristics are defined by its cardinal points, which are its first and second focal points, its first and second principal points, and its first and second nodal points. The focal points are those points at which light rays (from an infinitely distant object) parallel to the optical axis* are brought to a common focus on the axis. If the rays entering the system and those emerging from the system are extended until they intersect, the points of intersection will define a surface, usually referred to as the principal plane. The intersection of this surface with the axis is the principal point. The "second" focal point and the "second" principal plane are those defined by rays approaching the system from the left. The "first" points are those defined by rays from the right.

The effective focal length (efl) of a system is the distance from the principal point to the focal point. The back focal length (bfl), or back focus, is the distance from the vertex of the last surface of the system to the second focal point. The front focal length (ffl) is the distance from the front surface to the first focal point. These are illustrated in Fig. 2.1.

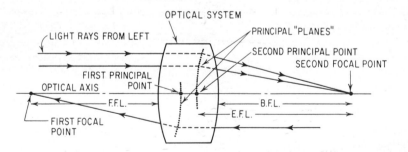

FIG. 2.1. Illustrating the location of the focal points and principal points of a generalized optical system.

The nodal points are two axial points such that a ray directed at the first nodal point appears (after passing through the system)

---

*The optical axis is a line through the centers of curvature of the surfaces which make up the optical system. It is the common axis of rotation for an axially symmetrical optical system.

to emerge from the second nodal point parallel to its original direction. The nodal points of an optical system are illustrated in Fig. 2.2 for an ordinary thick lens element. When an optical system is bounded on both sides by air (as is true in the great majority of applications), the nodal points coincide with the principal points.

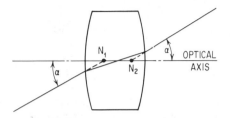

FIG. 2.2. A ray directed at the first nodal point (N_1) of an optical system emerges from the system without angular deviation and appears to come from the second nodal point (N_2).

2.3 Image Position and Size

When the cardinal points of an optical system are known, the location and size of the image formed by the optical system can be readily determined. In Fig. 2.3, the focal points F_1 and F_2 and the principal points P_1 and P_2 of an optical system are shown; the object which the system is to image is shown as the arrow AO. Ray OB, parallel to the system axis, will pass through the second focal point F_2; the refraction will appear to have occurred at the second principal plane. The ray OF_1C passing through the first focal point F_1 will emerge from the system parallel to the axis. (Since the path of light rays is reversible, this is equivalent to starting a ray from the right at O' parallel to the axis; the ray is then refracted through F_1 in accordance with the definition of the first focal point in Section 2.2.)

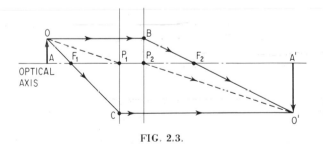

FIG. 2.3.

The intersection of these two rays at point O' locates the image of point O. A similar construction for other points on the object would locate additional image points, which would lie along the indicated arrow $O'A'$.

A third ray could be constructed from O to the first nodal point; this ray would appear to emerge from the second nodal point and would be parallel to the entering ray. If the object and image are both in air, the nodal points coincide with the principal points, and the ray is drawn from O to P_1 and from P_2 to O', as indicated by the dashed line in Fig. 2.3.

At this point, it is necessary to adopt a convention for the algebraic signs given to the various distances involved. The following conventions are used by most workers in the field of optics. There is nothing sacrosanct about these conventions, and many optical workers adopt their own, but the use of some consistent sign convention is a practical necessity.

1. Heights above the optical axis are positive (e.g., OA and P_2B). Heights below the axis are negative (P_1C and $A'O'$).
2. Distances measured to the left of a reference point are negative; to the right, positive. Thus P_1A is negative and P_2A' is positive.
3. The focal length of a converging lens is positive and the focal length of a diverging lens is negative.

Image Position

Figure 2.4 is identical to Fig. 2.3 except that the distances have been given single letters; the heights of the object and image are labeled h and h', the focal lengths are f and f', the object and image distances (from the principal planes) are s and s', and the distances from focal point to object and image are x and x', respectively. According to our sign convention, h, f, f', x', and s' are positive as shown, and x, s, and h' are negative. Note that the primed symbols refer to dimensions associated with the image and the unprimed symbols to those associated with the object.

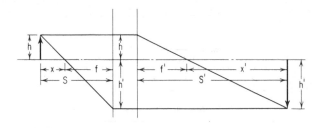

FIG. 2.4.

From similar triangles we can write

$$\frac{h}{(-h')} = \frac{(-x)}{f} \quad \text{and} \quad \frac{h}{(-h')} = \frac{f'}{x'} \qquad (2.1)$$

Setting the right-hand members of each equation equal and clearing fractions, we get

$$ff' = -xx' \tag{2.2}$$

If we assume the optical system to be in air, then f will be equal to f' and

$$f^2 = -xx' \tag{2.3}$$

This is the "Newtonian" form of the image equation and is very useful for calculations where the locations of the focal points are known.

If we substitute $x = s + f$ and $x' = s' - f$ in Eq. 2.3, we can derive another expression for the location of the image, the "Gaussian" form.

$$f^2 = -xx' = -(s + f)(s' - f)$$

$$= -ss' + sf - s'f + f^2$$

Canceling out the f^2 terms and dividing through by $ss'f$, we get

$$\frac{1}{s'} = \frac{1}{f} + \frac{1}{s} \tag{2.4}$$

or alternatively

$$s' = \frac{sf}{(s + f)} \tag{2.5}$$

Image Size

The lateral (or transverse) magnification of an optical system is given by the ratio of image size to object size, h'/h. By rearranging Eq. 2.1, we get for the magnification m,

$$m = \frac{h'}{h} = \frac{f}{x} = \frac{-x'}{f} \tag{2.6}$$

Substituting $x = s + f$ in this expression to get

$$m = \frac{h'}{h} = \frac{f}{(s + f)}$$

and noting (from Eq. 2.5) that $f/(s + f)$ is equal to s'/s, we find that

$$m = \frac{h'}{h} = \frac{s'}{s} \tag{2.7}$$

Note that Eqs. 2.3 through 2.7 assume that both object and image are in air.

Longitudinal magnification is the magnification along the optical axis, that is, the magnification of the longitudinal thickness of the object. If s_1 and s_2 denote the distances to the front and back edges of the object and s_1' and s_2' denote the distances to the front and back edges of the image, then the longitudinal magnification $\overline{m}$ is, by definition,

$$\overline{m} = \frac{s_2' - s_1'}{s_2 - s_1}$$

Substituting Eq. 2.5 for the primed distances and manipulating, we get

$$\overline{m} = \frac{s_1'}{s_1} \cdot \frac{s_2'}{s_2} \qquad (2.8)$$

and noting that $m = s'/s$, we get (if $(s_2' - s_1')$ and $(s_2 - s_1)$ are relatively small),

$$\overline{m} = m^2 \text{ (approximately)} \qquad (2.9)$$

Example A

Given an optical system with a positive focal length of 10 in., find the position and size of the image formed of an object 5 in. high which is located 40 in. to the left of the first focal point of the system.

Using the Newtonian equation, we get, by substituting in Eq. 2.3,

$$f^2 = -xx'$$

$$10^2 = -(-40) x'$$

$$x' = 100/40 = +2.5 \text{ in.}$$

Therefore the image is located 2.5 in. to the right of the second focal point. To find the image height, we use Eq. 2.6.

$$m = \frac{h'}{h} = \frac{f}{x} = \frac{10}{-40} = -0.25$$

$$h' = mh = (-0.25)(5) = -1.25 \text{ in.}$$

Thus if the base of the object were on the optical axis and the top 5 in. above it, the base of the image would also lie on the axis and and the image of the top would lie 1.25 in. below the axis.

The Gaussian equations can be used for this calculation by noting that the distance from the first principal plane to the object is given by $s = x - f = -40 - 10 = -50$; then, by Eq. 2.4,

$$\frac{1}{s'} = \frac{1}{f} + \frac{1}{s} = \frac{1}{10} + \frac{1}{(-50)} = 0.1 - 0.02 = 0.08$$

$$s' = \frac{1}{0.08} = 12.5 \text{ in.}$$

and the image is found to lie 12.5 in. to the right of the second principal plane (or 2.5 in. to the right of the second focal point, in agreement with the previous solution).

The height of the image can now be determined from Eq. 2.7.

$$m = \frac{h'}{h} = \frac{s'}{s} = \frac{+12.5}{-50} = -0.25$$

$$h' = mh = (-0.25)(5) = -1.25 \text{ in.}$$

Example B

If the object of Example A is located 2 in. to the right of the first focal point, where is the image and what is its height? Using Eq. 2.3,

$$f^2 = -xx'$$

$$10^2 = -(2)x'$$

$$x' = -50 \text{ in.}$$

Notice that the image is formed to the *left* of the second focal point; in fact, if the optical system is of moderate thickness, the image is to the left of the optical system and also to the left of the object. From Eq. 2.6 we get the magnification

$$m = \frac{h'}{h} = -\frac{f}{x} = \frac{10}{2} = +5.$$

$$h' = mh = (5)(5) = +25 \text{ in.}$$

The magnification and image height are both positive. In this case the image is a virtual image. A screen placed at the image position will not have an image formed on it, but the image may be observed by viewing through the lens from the right. A positive sign for the lateral magnification of a simple lens indicates that the image formed is virtual; a negative sign for the magnification of a simple lens indicates a real image. Figure 2.5 shows the relationships in this example.

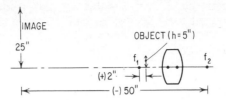

FIG. 2.5. Illustrating the formation of a virtual image. See example B.

Example C

If the object of Example B is 0.1 in. thick, what is the apparent thickness of the image? Since the lateral magnification was found to be 5× in Example B, the longitudinal magnification, by Eq. 2.9, is 5^2 or 25. Thus the apparent image thickness is approximately 25× (0.1 in.) or 2.5 in. If an exact value for the apparent thickness is required, the image position for each surface of the object must be calculated. Assuming that the front of the object was given in Example B as 2 in. to the right of the first focal point, then its rear surface must lie 1.9 in. to the right of f_1. Its image is located at

$$x' = -f^2/x = -100/1.9 = -52.63 \text{ in.}$$

to the left of the second focal point. Thus the distance between the image positions for the front and rear surfaces is 2.63 in., in reasonable agreement with the approximate result of 2.5 in. Had we computed the thickness for the case where the front and back surfaces of the object were 1.95 and 2.05 in. from the focal point, the results from the exact and approximate calculations would have been in even better agreement.

2.4 Refraction of a Light Ray at a Single Surface

As mentioned in Chapter 1, the path of a light ray through an optical system can be calculated from Snell's Law (Eq. 1.3) by the application of a reasonable amount of geometry and trigonometry. Figure 2.6 shows a light ray (GQP) incident on a spherical surface at point Q. The ray is directed toward point P where it would intersect the optical axis at a distance L from the surface if it were extended. At Q the ray is refracted by the surface and intersects the axis at P', a distance L' from the surface. The surface has a radius R with center of curvature at C and separates two media of index N on the left and index N' on the right. The light ray makes an angle U with the axis before refraction, U' after refraction; angle I is the angle between the incident ray and the normal to the surface (HQC) at point Q, and angle I' is the angle between the refracted ray and the normal. Notice that plain or un-primed symbols are used for quantities before refraction at the surface; after refraction, the symbols are primed.→

[handwritten margin note:] SURFACE
OBJECT SURFACE = #1
next SURFACE = #2
image surface = # n

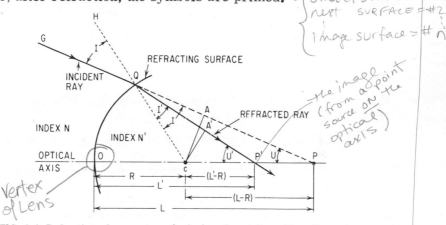

[handwritten note:] the image point (from a source on the optical axis)

[handwritten note:] Vertex of Lens

FIG. 2.6. Refraction of a ray at a spherical surface (all positive diagram).

The sign conventions which we shall observe are as follows:

[handwritten margin note:] Know!!

1. A radius is positive if the center of curvature lies to the right of the surface.
2. As before, distances to the right of the surface are positive; to the left, negative.
3. The angles of incidence and refraction (I and I') are positive if the ray is rotated clockwise to reach the normal.
4. The slope angles (U and U') are positive if the ray is rotated counterclockwise to reach the axis.
5. The light travels from left to right.

(Figure 2.6 is an "all positive" diagram, in that all quantities shown have a positive sign.)

A set of equations which will allow us to trace the path of the ray may be derived as follows. From right triangle PAC,

$$CA = (L - R)\sin U \qquad (2.10)$$

and from right triangle QAC,

$$\sin I = \frac{CA}{R} \qquad (2.11)$$

[handwritten note:] Derivation - see notes Oct 10

Applying Snell's Law (Eq. 1.3), we get the sine of the angle of refraction,

$$\sin I' = \frac{N}{N'} \sin I \qquad (2.12)$$

solve for I'

Then

The exterior angle QCO of triangle PQC is equal to $U + I$, and, as the exterior angle of triangle $P'QC$, it is also equal to $U' + I'$. Thus $U + I = U' + I'$ and

$$U' = U + I - I' \qquad (2.13)$$

substitute
here to
find U'
for eqn 2.16

From right triangle $QA'C$ we get

$$\sin I' = \frac{CA'}{R} \qquad (2.14)$$

and substituting Eqs. 2.11 and 2.14 into Eq. 2.12 gives us

$$CA' = \frac{N}{N'} CA \qquad (2.15)$$

Finally, the location of P' is found by rearranging $CA' = (L' - R) \sin U'$ from right triangle $P'A'C$ into

$$L' = R + \frac{CA'}{\sin U'} \qquad (2.16)$$

= R sin I'

known from 2.12

Thus, beginning with a ray defined by its slope angle U and its intersection with the axis L, we can determine the corresponding data, U' and L', for the ray after refraction by the surface.

2.5 The Paraxial Region

The paraxial region of an optical system is a thin thread-like region about the optical axis which is so small that all the angles made by the rays (i.e., the slope angles and the angles of incidence and refraction) may be set equal to their sines and tangents. At first glance this concept seems utterly useless, since the region is obviously infinitesimal and seemingly of value only as a limiting case. However, calculations of the performance of an optical system based on paraxial relationships are of tremendous utility. Their simplicity makes calculation and manipulation quick and easy. Since most optical systems of practical value form good images, it is apparent that most of the light rays originating at an object point must at least pass reasonably close to the paraxial image point. The paraxial relationships are the limiting relationships of the exact trigonometric relationships derived in the preceding section, and thus give locations for image points which serve as an excellent approximation for the imagery of a well-corrected optical system.

Paradoxically, the paraxial equations are frequently used with relatively large angles and ray heights. This extension of the paraxial region is useful in estimating the necessary diameters of optical elements and in approximating the quality of the image formed by a lens system, as we shall demonstrate in later chapters.

Although paraxial calculations are often used in rough preliminary work on optical systems and in approximate calculations (indeed, the term "paraxial approximation" is often used), the reader should bear in mind that the paraxial equations are perfectly exact for the paraxial region and that as an exact limiting case they are used in aberration determination as a basis of comparison to indicate how far a trigonometrically computed ray departs from its ideal location.

The simplest way of deriving a set of equations for the paraxial region is to substitute the angle itself for its sine in the equations derived in the preceding section. Thus we get

from Eq. 2.10: $\qquad ca = (l - R)u$ $\qquad$ (2.17)

from Eq. 2.11: $\qquad i = ca/R$ $\qquad$ (2.18)

from Eq. 2.12: $\qquad i' = Ni/N'$ $\qquad$ (2.19)

from Eq. 2.13: $\qquad u' - u + i - i'$ $\qquad$ (2.20)

from Eq. 2.15: $\qquad ca' = Nca/N'$ $\qquad$ (2.21)

from Eq. 2.16: $\qquad l' = R + ca'/u'$ $\qquad$ (2.22)

Notice that the paraxial equations are distinguished from the trigonometric equations by the use of lower-case letters for the paraxial values. This is a widespread convention and will be observed throughout this text.

Equations 2.17 through 2.22 may be materially simplified. Indeed, since they apply exactly only to a region in which angles and heights are infinitesimal, we can totally eliminate i, u, and ca from the expressions without any loss of validity. Thus, if we substitute into Eq. 2.22, Eq. 2.21 for ca' and Eq. 2.20 for u', and continue the substitution with Eqs. 2.17, 2.18, and 2.19, the following simple expression for l' is found:

$$l' = \frac{lN'R}{(N' - N)l + NR}$$ $\qquad$ (2.23)

By rearranging we can get an expression which bears a marked similarity to Eq. 2.4 (relating the object and image distances for a complete lens system):

$$\frac{N'}{l'} = \frac{(N' - N)}{R} + \frac{N}{l} \qquad (2.24)$$

These two equations are useful when the quantity of interest is the distance l'.

2.6 Paraxial Raytracing through Several Surfaces

Another form of the paraxial equations is more convenient for use when calculations are to be continued through more than one surface. Figure 2.7 shows a paraxial ray incident on a surface at a height y from the axis, with the ray-axis intersection distances l and l' before and after refraction. The height y in this case is a fictitious extension of the paraxial region, since, as noted, the paraxial region is an infinitesimal one about the axis. However, since all heights and angles cancel out of the paraxial expressions for the intercept distances (as indicated above), the use of finite heights and angles does not affect the accuracy of the expressions. For systems of modest aperture these fictitious heights and angles are a reasonable approximation to the corresponding values obtained by exact trigonometrical calculation.

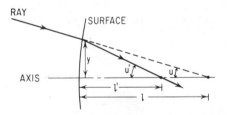

FIG. 2.7. Illustrating the relationship $y = lu = l'u'$ for paraxial rays.

In the paraxial region, every surface approaches a flat plane surface, just as all angles approach their sines and tangents. Thus we can express the slope angles shown in Fig. 2.7 by $u = y/l$ and $u' = y/l'$, or $l = y/u$ and $l' = y/u'$. If we substitute these latter values for l and l' into Eq. 2.24, we get

$$\frac{N'u'}{y} = \frac{N' - N}{R} + \frac{Nu}{y}$$

and multiplying through by y we find

$$N'u' = y \frac{(N' - N)}{R} + Nu \qquad (2.25)$$

It is frequently convenient to express the curvature of a surface as the reciprocal of its radius, $C = 1/R$; making this substitution we have

$$N'u' = y(N' - N)C + Nu \qquad (2.25a)$$

To continue the calculation to the next surface of the system, we require a set of transfer equations. Figure 2.8 shows two surfaces of an optical system separated by an axial distance t. The ray is shown after refraction by surface #1; its slope is the angle u'_1. The intersection heights of the ray at the surfaces are y_1 and y_2, respectively, and since this is a paraxial calculation, the difference between the two heights can be given by tu'_1. Thus, it is apparent that

$$y_2 = y_1 - tu'_1 = y_1 - t \frac{N'_1 u'_1}{N'_1} \qquad (2.26)$$

And if we note that the slope of the ray incident on surface #2 is the same as the slope after refraction by #1, we get the second transfer equation

$$u_2 = u'_1 \quad \text{or} \quad N_2 u_2 = N'_1 u'_1 \qquad (2.27)$$

These equations can now be used to determine the position and size of the image formed by a complete optical system, as illustrated by the following example.

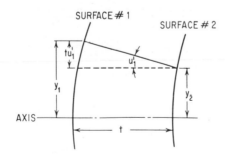

FIG. 2.8. Illustrating the transfer of a paraxial ray from surface to surface by $y_2 = y_1 - tu'_1$.

Example D

Figure 2.9 shows a typical problem. The optical system consists of three surfaces, making a "doublet" lens whose radii, thicknesses, and indices are indicated in the figure. The object is located 300 units to the left of the first surface and extends a height of 20 units above the axis. The lens is immersed in air, so that object and image are in a media of index $N = 1.0$.

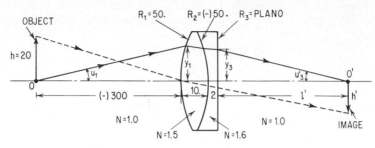

FIG. 2.9. Showing the rays traced in Example D.

The first step is to tabulate the parameters of the problem with the proper signs associated. Following the sign convention given above, we have the following:

$$h = +20$$

$$l_1 = -300$$

$R_1 = +50$	$C_1 = +.02$	
$R_2 = -50$	$C_2 = -.02$	
$R_3 = $ plano	$C_3 = 0$	

$$N_1 = 1.0$$
$$t_1 = 10 \quad N_1' = N_2 = 1.5$$
$$t_2 = 2 \quad N_2' = N_3 = 1.6$$
$$N_3' = 1.0$$

The location of the image can be found by tracing a ray from the point where the object intersects the axis (O in the figure); the image will then be located where the ray re-crosses the axis at O'. We can use any reasonable value for the starting data of this ray. Let us trace the path of the ray starting at O and striking the first surface at a height of ten units above the axis. Thus $y_1 = +10$ and we get the initial slope angle by

$$u_1 = \frac{y_1}{l_1} = \frac{10}{-300} = -.0333$$

and since $N_1 = 1.0$, $N_1 u_1 = -0.0333$. The slope angle after refraction is obtained from Eq. 2.25a.

$$N_1' u_1' = y_1 (N_1' - N_1) C_1 + N_1 u_1$$

$$= 10 (1.5 - 1.0)(+0.02) - 0.0333$$

$$= 0.1 - 0.0333$$

$$N_1' u_1' = +0.0666$$

The ray height at surface #2 is found by Eq. 2.26.

$$y_2 = y_1 - \frac{t_1 N_1' u_1'}{N_1'}$$

$$= 10 - \frac{10(+0.0666)}{1.5}$$

$$= 10 - 0.444$$

$$y_2 = 9.555$$

Noting that $N_2 u_2 = N_1' u_1'$, the refraction at the second surface is carried through by

$$N_2' u_2' = y_2 (N_2' - N_2) C_2 + N_2 u_2$$

$$= 9.555(1.6 - 1.5)(-0.02) + 0.0666$$

$$= -0.019111 + 0.0666$$

$$= +0.047555$$

and the ray height at the third surface is calculated by

$$y_3 = y_2 - \frac{t_2 N_2' u_2'}{N_2'} = 9.555 - 2(+0.04755)/1.6$$

$$= 9.555 - 0.059444 = 9.496111$$

Since the last surface of the system is plane, that is, of infinite radius, its curvature is zero and the product Nu is unchanged at this surface:

$$N_3' u_3' = y_3 (N_3' - N_3) C_3 + N_3 u_3$$

$$= 9.496111(1.0 - 1.6)(0) + 0.047555 - +0.047555$$

and

$$u_3' = \frac{N_3' u_3'}{N_3'} = +0.047555$$

Now the location of the image is given by the final intercept length l', which is determined by

$$l_3' = \frac{y_3}{u_3'} = \frac{9.496111}{+0.047555}$$

$$l_3' = +199.6846$$

The execution of a long chain of calculations such as the preceding is much simplified if the calculation is arranged in a convenient table form. By ruling the paper in squares, a simple arrangement of the constructional parameters at the top of the sheet and the ray data below helps to speed the calculation and eliminate errors. The following table (Fig. 2.10) sets forth the curvatures, thicknesses, and indices of the lens in the first three rows; the next two rows contain the ray heights and index-slope angle products of the calculation worked out above.

	Surface #1		Surface #2		Surface #3	
Curvature	+0.02		−0.02		0.0	
Thickness		10.		2.		
Index	1.0		1.5		1.6	1.0
Ray Height (y)		10.		9.555		9.496111
Nu	−0.0333		+0.0666		+0.047555	+0.047555
y		0.0		−0.444		−0.52888
Nu	+0.0666		+0.0666		+0.067555	+0.067555

FIG. 2.10.

The image height can now be found by tracing a ray from the top of the object and determining the intersection of this ray with the image plane we have just computed. Such a ray is shown by the dashed line in Fig. 2.9. If we elect to trace the ray which strikes the vertex of the first surface, then y_1 will be zero and the initial slope angle will be given by

$$u_1 = \frac{y_1 - h}{l_1} = \frac{0 - 20}{-300} = +0.0666$$

The calculation of this ray is indicated in the sixth and seventh rows of Fig. 2.10 and yields $y_3 = -0.52888 \ldots$ and $n_3' u_3' = +0.067555$.

The height of the image, h' in Fig. 2.9, can be seen to equal the sum of the ray height at surface #3 plus the amount the ray climbs or drops in traveling to the image plane.

$$h' = y_3 - l_3' \frac{N_3' u_3'}{N_3'} = -0.26444 - (199.6846)\frac{(+0.067555)}{1.0} = -14.0187$$

Notice that the expression used to compute h' is analogous to Eq. 2.26; if we regard the image plane as surface #4 and the image distance l_3' as the spacing between surfaces #3 and 4, Eq. 2.26 can be used to calculate y_4, which is h'.

Similarly, Eq. 2.26 can be used to determine the initial slope angle u_1 by regarding the object plane as surface zero and re-arranging the equation to solve for $u_0' = u_1$ as shown below:

$$y_1 = y_0 - t_0 \frac{N_0' u_0'}{N_0'}$$

$$u_0' = u_1 = \frac{y_0 - y_1}{t_0} = \frac{h - y_1}{-l_1} = \frac{y_1 - h}{l_1}$$

2.7 Focal Points and Principal Points of a Thick-Lens Element

The cardinal points of a single-lens element can be readily determined by use of the formulas given in the preceding section. The focal point is the point where the rays from an infinitely distant axial object cross the optical axis at a common focus. This point can be located by tracing a ray with an initial slope (u_1) of zero through the lens and determining the axial intercept.

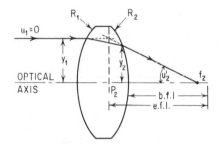

FIG. 2.11. A ray parallel to the axis is traced through an element to determine the effective focal length and back focal length.

Figure 2.11 shows the path of such a ray through a lens element. The principal plane (p_2) is located by the intersection of the extensions of the incident and emergent rays. The effective focal length (efl) or focal length (usually symbolized by f), is the distance from p_2 to f_2 and, for the paraxial region, is given by

$$\text{e.f.l.} = f = \frac{y_1}{u_2'}$$

The back focal length (bfl) can be found from

$$\text{b.f.l.} = \frac{y_2}{u_2'}$$

Owing to the frequency with which these quantities are used, it is worth while to work up a single equation for each of them. If the lens has an index of refraction N and is surrounded by air of index 1.0, then $N_1 = N_2' = 1.0$ and $N_1' = N_2 = N$. The surface radii are R_1 and R_2, and the surface curvatures are c_1 and c_2. The thickness is t. At the first surface, using Eq. 2.25a,

$$N_1' u_1' = N_1 u_1 + (N_1' - N_1) y_1 c_1 = 0 + (N - 1) y_1 c_1$$

The height at the second surface is found from Eq. 2.26:

$$y_2 = y_1 - \frac{t N_1' u_1'}{N_1'} = y_1 - \frac{t(N-1) y_1 c_1}{N} = y_1 \left[1 - \frac{(N-1)}{N} t c_1 \right]$$

And the final slope is found by Eq. 2.25a:

$$N_2' u_2' = N_1' u_1' + y_2 (N_2' - N_2) c_2$$

$$= (N-1) y_1 c_1 + y_1 \left[1 - \frac{(N-1)}{N} t c_1 \right] (1 - N) c_2$$

$$(1.0) u_2' = u_2' = y_1 (N-1) \left[c_1 - c_2 + t c_1 c_2 \frac{(N-1)}{N} \right]$$

Thus the power ϕ (or reciprocal focal length) of the element is expressed as

$$\phi = \frac{1}{f} = \frac{u_2'}{y_1} = (N-1) \left[c_1 - c_2 + t c_1 c_2 \frac{(N-1)}{N} \right] \tag{2.28}$$

or, if we substitute $c = 1/R$,

$$\phi = \frac{1}{f} = (N-1) \left[\frac{1}{R_1} - \frac{1}{R_2} + \frac{t(N-1)}{R_1 R_2 N} \right] \tag{2.28a}$$

The back focal length can be found by dividing y_2 by u_2' to get

$$\text{b.f.l.} = f - \frac{ft(N-1)}{NR_1} \tag{2.29}$$

The distance from the second surface to the second principal point is just the difference between the back focal length and the effective focal length (see Fig. 2.11); this is obviously the second term of Eq. 2.29.

The above procedure has located the second principal point and second focal point of the lens. The "first" points are found simply by substituting R_1 for R_2 and vice versa.

The focal points and principal points for several shapes of el-elements are diagrammed in Fig. 2.12. Notice that the principal points of an equiconvex or equiconcave element are approximately evenly spaced within the element. In the plano forms, one principal point is at the curved surface, the other is about one-third of the way into the lens. In the meniscus forms shown, one of the principal points is completely outside the lens; in extreme meniscus shapes, both the principal points lie outside the lens and their order may be reversed from that shown.

2.8 The Thin Lens

If the thickness of a lens element is small enough so that its effect on the accuracy of the calculation may be neglected, the el-element is called a thin lens. The thin-lens concept is an extremely useful one for the purposes of quick preliminary calculations and analysis.

The focal length of a thin lens can be derived from Eq. 2.28 by setting the thickness equal to zero.

$$\frac{1}{f} = (N - 1)(c_1 - c_2) \tag{2.30}$$

$$\frac{1}{f} = (N - 1)\left(\frac{1}{R_1} - \frac{1}{R_2}\right) \tag{2.30a}$$

Since the thickness is assumed to be zero, the principal points of a "thin lens" are coincident with the location of the lens. Thus, in computing object and image positions, the distances s and s' of Eqs. 2.4, 2.5, 2.7, etc., are measured from the lens itself.

Example E

An object 10 mm high is to be imaged on a screen 50 mm high and 120 mm distant. What are the radii of an equiconvex lens of index 1.5 which will produce an image of the proper size and location?

The first step in the calculation is the determination of the focal length of the lens. Since the image is a real one, the magnification will have a negative sign, and by Eq. 2.7 we have

$$m = \frac{h'}{h} = (-)\frac{50}{10} = \frac{s'}{s} \quad \text{or} \quad s' = -5s$$

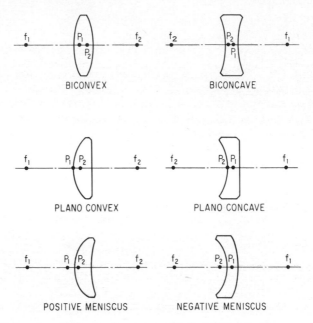

FIG. 2.12. The location of the focal points and principal points for several shapes of converging and diverging elements.

For the object and image to be 120 mm apart,

$$120 = -s + s' = -s - 5s = -6s$$

$$s = -20 \text{ mm}.$$

and $s' = -5s = +100$ mm.

Substituting into Eq. 2.4 and solving for f, we get

$$\frac{1}{100} = \frac{1}{f} + \frac{1}{-20}$$

$$f = 16.67 \text{ mm}.$$

Noting that for an equiconvex lens $R_1 = -R_2$, we use Eq. 2.30a to solve for the radii

$$\frac{1}{f} = +.06 = (N - 1)\left(\frac{1}{R_1} - \frac{1}{R_2}\right) = 0.5 \frac{2}{R_1}$$

$$R_1 = \frac{1}{.06} = 16.67 \text{ mm}.$$

$$R_2 = -R_1 = -16.67 \text{ mm}.$$

2.9 Mirrors

A curved mirror surface has a focal length and is capable of forming images just as a lens does. The equations for paraxial raytracing (Eqs. 2.25, 2.26, and 2.27) can be applied to reflecting surfaces by taking into account two additional sign conventions. The index of refraction of a material was defined in the first chapter as the ratio of the velocity of light in vacuum to that in the material. Since the direction of propagation of light is reversed upon reflection, it is logical that the sign of the velocity should be considered reversed, and the sign of the index reversed as well. Thus the conventions are as follows.

1. The signs of all indices following a reflection are reversed.
2. The signs of all spacings following a reflection are reversed.

Obviously if there are two reflecting surfaces in a system, the signs of the indices and spacings are changed twice, and after the second change revert to the original positive signs, since the direction of propagation is again left to right.

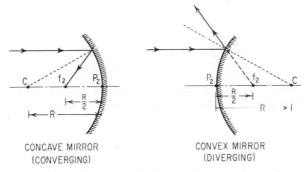

CONCAVE MIRROR
(CONVERGING)

CONVEX MIRROR
(DIVERGING)

FIG. 2.13. The location of the focal points for reflecting surfaces.

Figure 2.13 shows the locations of the focal and principal points of concave and convex mirrors. The ray from the infinitely distant source which defines the focal point can be traced as follows, setting $N = 1.0$ and $N' = -1.0$:

$$Nu = 0 \text{ (since the ray is parallel to the axis)}$$

$$N'u' = Nu + y \frac{(N' - N)}{R} = 0 + y \frac{(-1 - 1)}{R} = \frac{-2y}{R}$$

thus

$$u' = \frac{N'u'}{N'} = \frac{N'u'}{-1} = \frac{2y}{R}$$

The final intercept length is

$$l' = \frac{y}{u'} = \frac{yR}{2y} = \frac{R}{2}$$

and we find that the focal point lies halfway between the mirror and its center of curvature.

The concave mirror is the equivalent of a positive converging lens and forms a real image of distant objects. The convex mirror forms a virtual image and is equivalent to a negative element. Because of the index sign reversal on reflection, the sign of the focal length is reversed also and the focal length of a mirror is given by

$$f = -\frac{1}{2} R$$

so that the sign conforms to the convention of positive for converging elements and negative for diverging elements.

Example F

Calculate the focal length of the Cassegrain mirror system shown in Fig. 2.14 if the radius of the primary mirror is 200 mm, the radius of the secondary mirror is 50 mm, and the mirrors are separated by 80 mm.

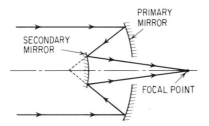

FIG. 2.14. Cassegrain mirror system. The image formed by the primary mirror is the virtual object for the secondary mirror.

Following our sign convention, the radii are both negative and the distance from primary to secondary mirror is also considered negative, since the light traverses this distance right to left. The index of the air is taken as +1.0 before the primary and after the secondary; between the two, the index is −1.0. Thus the optical data of the problem and the computation are set up and carried through as shown in Fig. 2.15.

Radius (R)		-200		-50	
Thickness (t)			-80		
Index (N)	$+1.0$		-1.0		$+1.0$
Ray Height (y)		1.0		$+0.2$	
Ray Slope $\times$ Index (Nu)	0		$+0.01$		$+0.002$

FIG. 2.15.

Careful attention to signs is necessary in this calculation to avoid mistakes.

The focal length of the system is given by $y_1/u_2' = 1.0/{+0.002} = 500$ mm. The final intercept distance (from R_2 to the focus) is equal to $y_2/u_2' = 0.2/{+0.002} = 100$ mm and the focal point lies 20 mm to the right of the primary mirror. Notice that the (second) principal plane is completely outside the system, 400 mm to the left of the secondary mirror.

2.10 Systems of Separated Components

It is often convenient to treat an optical system which is made up of separated elements or components (i.e., a group of elements treated as a unit), in terms of the component focal lengths and spacings, instead of handling the system by means of surface by surface calculation. To this end we can introduce the paraxial ray height y into the equations of Section 2.3, just as we did in Section 2.6.

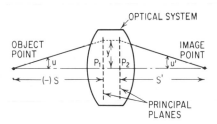

FIG. 2.16. The principal planes are planes of unit magnification, so that a ray appears to leave the second principal plane at the same height (y) that it appears to strike the first principal plane.

An optical component (which may be made up of a number of elements) is shown in Fig. 2.16 with its object a distance s from the first principal plane and its image a distance s' from the second principal plane. The principal planes are planes of unit magnification, in that the incident and emergent ray paths appear to strike (and emerge from) the same height on the first and second principal planes. Thus, in Fig. 2.16 a ray from the object point which would (if extended) strike the first principal plane at a distance y from the axis, emerges from the last surface of the system as if it were coming from the same height y on the second principal plane. For this reason we can write the following relationships:

$$u = \frac{y}{s} \quad \text{and} \quad u' = \frac{y}{s'}$$

and substitute $s = y/u$ and $s' = y/u'$ into Eq. 2.4:

$$\frac{1}{s'} = \frac{1}{s} + \frac{1}{f}$$

$$\frac{u'}{y} = \frac{u}{y} + \frac{1}{f} \tag{2.4}$$

$$u' = u + y/f$$

If we now replace the reciprocal focal length $(1/f)$ with ϕ, we get the first equation of the set:

$$u' = u + y\phi \tag{2.31}$$

The transfer equations to the next component in the system are the same as those used in the paraxial surface-by-surface raytrace of Section 2.6:

$$y_2 = y_1 - d u'_1 \tag{2.32}$$

$$u'_1 = u_2 \tag{2.33}$$

where y_1 and y_2 are the ray heights at the principal planes of components #1 and 2, u'_1 is the slope angle after passing through component #1, and d is the axial distance from the second principal plane of component #1 to the first principal plane of component #2.

These equations are equally applicable to systems composed of either thick or thin lenses. Obviously, when applied to thin lenses, d becomes the spacing between elements since the element and its principal planes are coincident.

The preceding equations may be used to derive compact expressions for the effective focal length and back focal length of two separated components. Let us assume that we have two lenses of powers ϕ_a and ϕ_b separated by a distance d (if the lenses are thin; if they are thick, d is the separation of their principal points). The system is sketched in Fig. 2.17.

Beginning with a ray parallel to the axis which strikes lens a at y_a, we have

$$u_a = 0$$

$$u'_a = 0 + y_a \phi_a \quad \text{by (Eqn 2.31)}$$

$$y_b = y_a - d y_a \phi_a = y_a (1 - d\phi_a) \quad \text{by (Eqn 2.32)}$$

$$u'_b = y_a \phi_a + y_a(1 - d\phi_a)\phi_b \text{ by (Eqn 2.31)}$$

$$= y_a(\phi_a + \phi_b - d\phi_a\phi_b)$$

The power (reciprocal focal length) of the system is given by

$$\phi_{ab} = \frac{1}{f_{ab}} = \frac{u'_b}{y_a} = \phi_a + \phi_b - d\phi_a\phi_b = \frac{1}{f_a} + \frac{1}{f_b} - \frac{d}{f_a f_b} \quad (2.34)$$

and thus

$$f_{ab} = \frac{f_a f_b}{f_a + f_b - d} \quad (2.35)$$

The back focal length is given by

$$\text{b.f.l.} = \frac{y_b}{u'_b} = \frac{y_a(1 - d\phi_a)}{y_a(\phi_a + \phi_b - d\phi_a\phi_b)}$$

$$= \frac{(1 - d/f_a)}{1/f_a + 1/f_b - d/f_a f_b} = \frac{f_b(f_a - d)}{f_a + f_b - d} \quad (2.36)$$

By substituting f_{ab}/f_a from Eq. 2.35, we get

$$\text{b.f.l.} = \frac{f_{ab}(f_a - d)}{f_a} \quad (2.36a)$$

The front focal length (ffl) for the system is found by reversing the raytrace (i.e., trace from right to left) or more simply by substituting f_b for f_a to get

$$(-)\text{f.f.l.} = \frac{f_{ab}(f_b - d)}{f_b} \quad (2.36b)$$

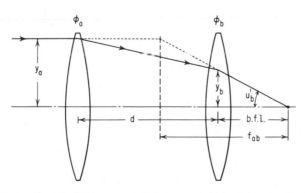

FIG. 2.17. Raytrace through the two separated components to determine focal length and back focus of the combination.

Frequently it is useful to be able to solve for the focal lengths of the components when the focal length, back focal length, and spacing are given for the system. Manipulation of Eqs. 2.35 and 2.36a will yield

$$f_a = \frac{df_{ab}}{f_{ab} - (\text{b.f.l.})} \tag{2.37}$$

$$f_b = \frac{-d(\text{b.f.l.})}{f_{ab} - (\text{b.f.l.}) - d} \tag{2.38}$$

2.11 The Optical Invariant

The Optical Invariant, or LaGrange Invariant, is a constant for a given optical system, and it is a very useful one. Its numerical value may be calculated in any of several ways, and the invariant may then be used to arrive at the value of other quantities without the necessity of certain intermediate operations which would otherwise be required.

Let us consider the application of Eq. 2.25a to the tracing of two rays through an optical system. One ray (the "axial" ray) is traced from the foot, or axial intercept, of the object; the other ray (the "oblique" ray) is traced from an off-axis point on the object. Figure 2.18 shows these two rays passing through a generalized system.

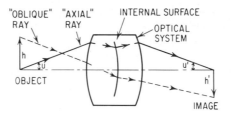

FIG. 2.18.

At any surface in the system, we write out Eq. 2.25a for each ray, using the subscript p to denote the data of the "oblique" ray.

For the "axial" ray $N'u' = Nu + y(N' - N)c$

For the "oblique" ray $N'u'_p = Nu_p + y_p(N' - N)c$

We now extract the common term $(N' - N)c$ from each equation and equate the two expressions:

$$(N' - N)c = \frac{N'u' - Nu}{y} = \frac{N'u'_p - Nu_p}{y_p}$$

Multiplying by yy_p and rearranging, we get

$$yNu_p - y_pNu = yN'u'_p - y_pN'u'$$

Note that on the left side of the equation the angles and indices are for the left side of the surface (that is, before refraction) and that on the right side of the equation the terms refer to the same quantities after refraction. Thus $y_pNu - yNu_p$ is a constant which is invariant across any surface.

By a similar series of operations based on Eq. 2.26, we can show that $(y_pNu - yNu_p)$ for a given surface is equal to $(y_pNu - yNu_p)$ for the next surface. Thus this term is not only invariant across the surface but also across the space between the surfaces; it is therefore invariant through the entire optical system.

$$\text{Invariant} \quad I = y_pNu - yNu_p \tag{2.39}$$

As an example of its application we now write the invariant for the object plane and image plane of Fig. 2.18. In the object plane $y_p = h, N = N, y = 0$, and we get

$$I = hNu - (0)Nu_p = hNu$$

In the image plane $y_p = h', N = N', y = 0$, and we get

$$I = h'N'u' - (0)N'u'_p = h'N'u'$$

Equating the two expressions gives

$$hNu = h'N'u' \tag{2.40}$$

which can be rearranged to give a very generalized expression for the magnification of an optical system

$$m = \frac{h'}{h} = \frac{Nu}{N'u'} \tag{2.41}$$

Equation 2.40 is, of course, valid only for the extended paraxial region; this relationship is sometimes applied to trigonometric calculations where it takes the form

$$hN \sin u = h'N' \sin u' \tag{2.42}$$

Example G

We can apply the invariant to the calculation made in Example D by assuming that only the axial ray has been traced. The axial ray slope at the object was $-0.0333...$ and the corresponding computed slope at the image was found to be $+0.047555...$ Since the object and image were both in air of index 1.0, we can find the image height from Eq. 2.41,

$$m = \frac{h'}{h} = \frac{h'}{20} = \frac{Nu}{N'u'} = \frac{1.0(-.0333...)}{1.0(+.047555...)}$$

$$h' = \frac{20(-0.0333...)}{(+0.047555)}$$

$$h' = -14.0187$$

This value agrees with the height found in Example D by tracing a ray from the tip of the object to the tip of the image. The saving of time by the elimination of the calculation of this extra ray indicates the usefulness of the invariant.

Another useful expression is derived when we consider the case of a lens with its object at infinity. At the first surface the invariant is

$$I = y_p N(0) - y_1 N u_p = -y_1 N u_p$$

since the "axial" ray from an infinitely distant object has a slope angle u of zero. At the image plane y_p is the image height h', and y for the "axial" ray is zero; thus

$$I = h'N'u' - (0)N'u_p' = h'N'u'$$

Equating the two expressions for I, we get

$$h'N'u' = -y_1 N u_p$$

$$h' = -u_p \frac{Ny_1}{N'u'} \tag{2.43}$$

which is useful for systems where the object and image are not in air. If both object and image are in air, we set $N = N' = 1.0$, and, recalling that $f = y_1/u'$, we find

$$h' = -u_p f \tag{2.44}$$

As one might expect from the preceding, a paraxial system is completely described by the ray data of any two unrelated rays. Thus, when we have traced two rays, we can determine the ray data of a third ray without further ray tracing, by using

$$\bar{y} = Ay_p + By \tag{2.45}$$

$$\bar{u} = Au_p + Bu \tag{2.46}$$

where $\bar{y}$ and $\bar{u}$ refer to the third ray, and y_p, u_p, y, and u are the ray data for the "oblique" and "axial" rays as before. The constants A and B are determined by solving Eqs. 2.45 and 2.46 to get

$$A = \frac{\bar{y}u - \bar{u}y}{uy_p - yu_p} = \frac{N}{I}(\bar{y}u - \bar{u}y) \tag{2.47}$$

$$B = \frac{\bar{u}y_p - \bar{y}u_p}{uy_p - yu_p} = \frac{N}{I}(\bar{u}y_p - \bar{y}u_p) \tag{2.48}$$

Equations 2.47 and 2.48 are evaluated for some surface in the optical system at which the height and slope data for all three rays are known (e.g., the first surface). The constants A and B are inserted into Eqs. 2.45 and 2.46; the values of $\bar{y}$ and $\bar{u}$ can then be determined for locations in the optical system at which the ray data for only the "axial" and "oblique" rays are known, by inserting this data in Eqs. 2.45 and 2.46.

As an example of the application of these equations, consider a system for which the "axial" and "oblique" rays have been traced for finite conjugates. The front, back, and effective focal lengths can be determined without additional ray tracing. We have values for the initial rays (y, u, y_p and u_p at the first surface) and for the final rays (y', u', y_p' and u_p' at the last surface); we wish to determine the final data ($\bar{y}'$ and $\bar{u}'$) for a third ray with starting data of $\bar{y} = 1$ and $\bar{u} = 0$. The application of Eqs. 2.45 through 2.48 will yield

$$efl = \frac{\bar{y}}{\bar{u}'} = \frac{uy_p - yu_p}{uu_p' - u_pu'} \tag{2.49}$$

$$bfl = \frac{\bar{y}'}{\bar{u}'} = \frac{uy_p' - u_py'}{uu_p' - u_pu'} \tag{2.50}$$

Reversing the process by setting $\bar{u}' = 0$ and $\bar{y}' = 1$, we get the (normally negative) value for the front focal length

$$ffl = \frac{\bar{y}}{\bar{u}} = \frac{u_p'y - u'y_p}{uu_p' - u_pu'} \tag{2.51}$$

2.12 Summary of Paraxial Equations and Conventions

Image position and size for an optical system in air:

$$f^2 = -xx'$$

$$\frac{1}{s'} = \frac{1}{f} + \frac{1}{s}$$

$$m = \frac{h'}{h}$$

$$m = \frac{f}{x} = \frac{-x'}{f} = \frac{s'}{s} = \frac{u}{u'}$$

Image position through a single surface:

$$\frac{N'}{l'} = \frac{(N' - N)}{R} + \frac{N}{l}$$

Paraxial raytrace through a series of surfaces:

$$u_1 = (y_1 - h)/l_1$$

$$N'u' = y(N' - N)c + Nu$$

$$y_2 = y_1 - tu'_1$$

$$u_2 = u'_1$$

$$h' = y_k - l'_k u'_k$$

Focal length of a thick lens in air:

$$\phi = \frac{1}{f} = (N - 1)\left[\frac{1}{R_1} - \frac{1}{R_2} + \frac{t(N - 1)}{R_1 R_2 N}\right]$$

$$\text{b.f.l.} = f - \frac{ft(N - 1)}{NR_1}$$

Focal length of a thin lens in air:

$$\phi = \frac{1}{f} = (N - 1)\left(\frac{1}{R_1} - \frac{1}{R_2}\right)$$

Paraxial raytrace through a series of separated components:

$$y = lu = l'u'$$

$$u' = u + y\phi$$

$$y_2 = y_1 - du'_1$$

$$u_2 = u'_1$$

Focal lengths of two separated components:

$$\phi_{ab} = \frac{1}{f_{ab}} = \phi_a + \phi_b - d\phi_a\phi_b$$

$$f_{ab} = \frac{f_a f_b}{f_a + f_b - d}$$

$$(\text{b.f.l.}) = \frac{f_b(f_a - d)}{f_a + f_b - d} = \frac{f_{ab}(f_a - d)}{f_a}$$

$$f_a = \frac{df_{ab}}{f_{ab} - (\text{b.f.l.})}$$

$$f_b = \frac{-d(\text{b.f.l.})}{f_{ab} - (\text{b.f.l.}) - d}$$

The optical invariant:

$$I = y_p Nu - yNu_p$$

$$hnu = h'N'u'$$

$$h' = -u_p \frac{Ny_1}{N'u'} \quad \text{object at infinity}$$

$$h' = -u_1 f \quad \text{object at infinity, system in air}$$

Sign Conventions

1. Light travels from left to right.
2. Focal length is positive for converging lenses.

3. Heights above the axis are positive.
4. Distances to the right are positive.
5. A radius or curvature is positive if the center of curvature is to the right of the surface.
6. Angles of incidence and refraction are positive if the ray is rotated clockwise to reach the normal.
7. Slope angles are positive if the ray is rotated counterclockwise to reach the axis.
8. After a reflection (when light direction is reversed), the signs of subsequent indices and spacings are reversed.

It may be noted that, although the discussions of this chapter have centered about spherical surfaces, and the equations derived have utilized the radii and curvatures of spherical surfaces, the paraxial expressions are equally valid for all surfaces of rotation centered on the optical axis when the osculating radius of the surface is used.

3

Aberrations

3.1

In Chapter 2 we discussed the image forming characteristics of optical systems, but limited our consideration to an infinitesimal thread-like region about the optical axis called the paraxial region. In this chapter we will consider, in general terms, the behavior of lenses with *finite* apertures and fields of view. It has been pointed out that well-corrected optical systems behave nearly according to the rules of paraxial imagery given in Chapter 2. This is another way of stating that a lens without aberrations forms an image of the size and in the location given by the equations for the paraxial or first order region. We shall measure the aberrations by the amount by which rays miss the paraxial image point.

It can be seen that aberrations may be determined by calculating the location of the paraxial image of an object point and then tracing a large number of rays (by the exact trigonometrical ray tracing equations of Chapter 10) to determine the amounts by which the rays depart from the paraxial image point. Stated this baldly, the mathematical determination of the aberrations of a lens which covered any reasonable field at a real aperture would seem a formidable task, involving an almost infinite amount of labor. However, by classifying the various types of image faults and by understanding the behavior of each type, the work of determining the aberrations of a lens system can be simplified greatly, since only a few rays need be traced to evaluate each aberration; thus the problem assumes more manageable proportions.

Seidel investigated and codified the primary aberrations and derived analytical expressions for their determination. For this reason, the primary image defects are usually referred to as the Seidel Aberrations.

3.2 The Seidel Aberrations   Third order aberrations

The Seidel Aberrations of a system in monochromatic light are called spherical aberration, coma, astigmatism, Petzval curvature and distortion. In this section we will define each aberration and discuss its characteristics, its representation, and its effect on the appearance of the image. Each aberration will be discussed as if it alone were present; obviously in practice one is far more likely to encounter aberrations in combination than singly.

3.2.1 Spherical Aberration

Spherical aberration can be defined as the variation of focus with aperture. Figure 3.1 is a somewhat exaggerated sketch of a simple lens forming an "image" of an axial object point a great distance away. Notice that the rays close to the optical axis come to a focus (intersect the axis) very near the paraxial focus position. As the ray height at the lens increases, the position of the ray intersection with the optical axis moves further and further from the paraxial focus. The distance from the paraxial focus to the axial intersection of the ray is called longitudinal spherical aberration. Transverse, or lateral, spherical aberration is the name given to the aberration when it is measured in the "vertical" direction. Thus, in Fig. 3.1 AB is the longitudinal, and AC the transverse spherical aberration of ray R.

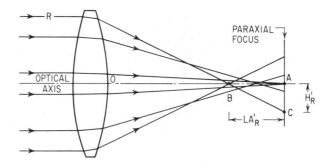

FIG. 3.1. A simple converging lens with undercorrected spherical aberration. The rays further from the axis are brought to a focus nearer the lens.

Since the magnitude of the aberration obviously depends on the height of the ray, it is convenient to specify the particular ray with which a certain amount of aberration is associated. For example, marginal spherical aberration refers to the aberration of the ray through the edge of the lens aperture.

Spherical aberration is determined by tracing a paraxial ray and a trigonometric ray from the same object point and determining their final intercept distances l' and L'. In Fig. 3.1, l' is distance OA and L' (for ray R) is distance OB. The longitudinal spherical aberration of the image point is abbreviated LA' and

$$LA' = l' - L' \qquad (3.1)$$

Transverse spherical aberration is related to LA' by the expression

$$TA' = LA' \tan U'_R = (l' - L') \tan U'_R \qquad (3.2)$$

where U'_R is the angle the ray makes with the axis. Using this sign convention, spherical aberration with a negative sign is called undercorrected spherical, since it is usually associated with simple uncorrected positive elements. Similarly, positive spherical is called overcorrected, and is generally associated with diverging elements.

The spherical aberration of a system is usually represented graphically. Longitudinal spherical is plotted against the ray height at the lens, as shown in Fig. 3.2a, and transverse spherical is plotted against the final slope of the ray, as shown in Fig. 3.2b. Figure 3.2b is called a ray intercept curve.

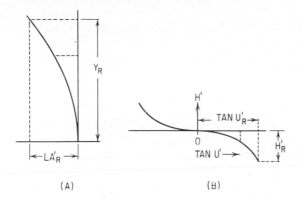

(A)   (B)

FIG. 3.2. Graphical representation of spherical aberration. (a) as a longitudinal aberration, in which the longitudinal spherical aberration (LA') is plotted against ray height (Y). (b) as a transverse aberration, in which the ray intercept height (H') at the paraxial reference plane is plotted against the final ray slope (TAN U').

For a given aperture and focal length, the amount of spherical aberration in a simple lens is a function of object position and the shape, or bending, of the lens. For example, a thin glass lens with its object at infinity has a minimum amount of spherical at a nearly plano convex shape, with the convex surface toward the object. A meniscus shape, either convex-concave or concave-convex has much more spherical aberration. If the object and image are of equal size (each being two focal lengths from the lens) then the shape which gives the minimum spherical is equiconvex.

The image of a point formed by a lens with spherical aberration is usually a bright dot surrounded by a halo of light; the effect of spherical on an extended image is to soften the contrast of the image and to blur its details.

In general a positive, converging lens will introduce undercorrected spherical aberration to a system and a negative lens the reverse, although there are certain exceptions to this.

3.2.2 Coma

Coma can be defined as the variation of magnification with aperture. Thus, when a bundle of oblique rays are incident on a lens with coma, the rays passing through the edge portions of the lens are imaged at a different height than those passing through the center portion. In Fig. 3.3, the upper and lower rim rays, A and B respectively, intersect the image plane below the ray P which passes through the center of the lens. The distance from P to the intersection of A and B is called the tangential coma of the lens, and is given by

$$\text{Coma}_T = H'_{AB} - H'_P \qquad (3.3)$$

where H'_{AB} is the height from the optical axis to the intersection of the upper and lower rim rays and H'_P is the height from the axis to the intersection of the ray P with the plane perpendicular to the axis and passing through the intersection of A and B.

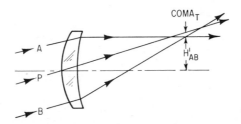

FIG. 3.3. In the presence of coma, the rays through the outer portions of the lens focus at a different height than the rays through the center of the lens.

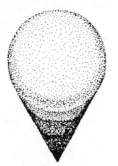

FIG. 3.4. The coma patch. The image of a point source is spread out into a comet shaped flare.

The appearance of a point image formed by a comatic lens is indicated in Fig. 3.4. Obviously the aberration is named after the comet shape of the figure.

Figure 3.5 indicates the relationship between the position at which the ray passes through the lens aperture and the location which it occupies in the coma patch. Figure 3.5a represents a head-on view of the lens aperture, with ray positions indicated by the letters A through H and A' through H', with the primed rays in the inner circle. The resultant coma patch is shown in Fig. 3.5b with the ray locations marked with corresponding letters. Notice that the rays which formed a circle on the aperture also form a circle in the coma patch, but as the rays go around the aperture circle once, they go around the image circle twice. The primed rays of the smaller circle in the aperture also form a correspondingly smaller circle in the image, and the central ray, P, is at the point of the figure. Thus the comatic image can be viewed as being made up of a series of different sized circles arranged tangent to a 60° angle.

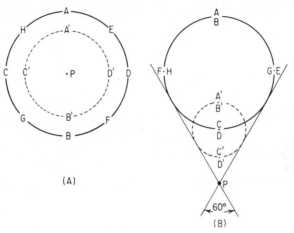

(A)

(B)

FIG. 3.5. The relationship between the position of a ray in the lens aperture and its position in the coma patch. (a) view of the lens aperture with rays indicated by letters. (b) the letters indicate the positions of the corresponding rays in the image figure. Note that the diameters of the circles in the image are proportional to the *square* of the diameters in the aperture.

In Fig. 3.5b the distance from P to AB is the tangential coma of Eq. 3.3. The distance from P to CD is called the sagittal coma and is one third as large as the tangential coma. About 55% of all the energy in the coma patch is concentrated in the small triangular area between P and CD; thus the sagittal coma is a somewhat better measure of the effective size of the image than is the tangential coma.

Coma is a particularly disturbing aberration since its flare is non-symmetrical. Its presence is very detrimental to accurate determination of the image position since it is much more difficult to locate the "center of gravity" of a coma patch than for a circular blur such as that produced by spherical aberration.

Coma varies with the shape of the lens element and also with the position of any apertures or diaphragms which limit the bundle of rays forming the image.

3.2.3 Astigmatism and Field Curvature

In the preceding section on coma, we introduced the terms tangential and sagittal; a fuller discussion of these terms is appropriate at this point. If a lens system is represented by a drawing of its axial section, rays which lie in the plane of the drawing are called meridional or tangential rays. Thus rays A, P and B of Fig. 3.5 are tangential rays. Similarly, the plane through the axis is referred to as the meridional or tangential plane, as may *any* plane through the axis.

Rays which do not lie in a meridional plane are called skew rays. The oblique meridional ray through the center of the aperture of a lens system is called the principal, or chief, ray. If we imagine a plane passing through the chief ray and perpendicular to the meridional plane, then all the (skew) rays from the object which lie in this plane are sagittal rays. Thus in Fig. 3.5 all the rays except A, A', P, B' and B are skew rays and the sagittal rays are C, C', D' and D.

Now the image of a point source formed by a fan of rays in the tangential plane will be a line image; this line, called the tangential image, is perpendicular to the tangential plane, that is, it lies in the sagittal plane. Conversely, the image formed by the rays of the sagittal fan is a line which lies in the tangential plane.

Astigmatism occurs when the tangential and sagittal (sometimes called radial) images do not coincide. In the presence of astigmatism, the image of a point source is not a point, but takes the form of two separate lines as shown in Fig. 3.6. Between the astigmatic focii the image is an elliptical or circular blur.

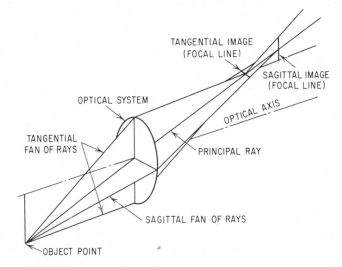

FIG. 3.6. Astigmatism.

Unless a lens is poorly made, there is no astigmatism when an axial point is imaged. As the imaged point moves further from the axis, the amount of astigmatism gradually increases. Off-axis images seldom lie exactly in a true plane; when there is primary astigmatism in a lens system, the images lie on curved surfaces which are paraboloidal in shape. The shape of these image surfaces is indicated for a simple lens in Fig. 3.7.

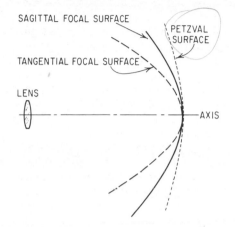

FIG. 3.7. The primary astigmatism of a simple lens. The tangential image is three times as far from the Petzval surface as the sagittal image.

The amount of astigmatism in a lens is a function of the shape of the lens and its distance from the aperture or diaphragm which limits the size of the bundle of rays passing through the lens. In the case of a simple lens whose own diameter limits the size of the ray bundle, the astigmatism is equal to the square of the distance from the axis to the image (i.e., the image height) divided by the focal length of the element.

Every optical system has associated with it a sort of basic field curvature, called the Petzval curvature, which is a function of the index of refraction of the lens elements and their surface curvatures. When there is no astigmatism, the sagittal and tangential image surfaces coincide with each other and lie on the Petzval surface. When there is primary astigmatism present, the tangential image surface lies three times as far from the Petzval surface as the sagittal image; note that both image surfaces are on the same side of the Petzval surface, as indicated in Fig. 3.7.

When the tangential image is to the left of the sagittal image (and both are to the left of the Petzval surface) the astigmatism is called undercorrected or inward (toward the lens) curving. When the order is reversed, the astigmatism is overcorrected, or backward curving. In Fig. 3.7, the astigmatism is undercorrected and all three surfaces are inward curving.

Positive lenses introduce inward curvature of the Petzval surface to a system and negative lenses introduce backward curvature. The Petzval curvature of a thin simple element is equal to one half the square of the image height divided by the focal length and index of the element.

3.2.4 Distortion

When the image of an off-axis point is formed further from the axis or closer to the axis than the image height given by the paraxial expressions of Chapter 2, the image of an extended object is said to be distorted. The amount of distortion is the displacement of the image from the paraxial position, and can be expressed either directly or as a percentage of the paraxial image height.

The amount of distortion ordinarily increases as the image size increases; the distortion itself usually increases as the cube of the image height (percentage distortion increases as the square). Thus, if a centered rectillinear object is imaged by a system afflicted with distortion, it can be seen that the images of the corners will be displaced more (in proportion) than the images of the points making up the sides. Figure 3.8 shows the appearance of a square figure imaged by a lens system with distortion. In Fig. 3.8a the distrotion is such that the images are displaced outward from the correct position, resulting in a flaring or pointing of the corners. This is overcorrected or pin-cushion distortion. In Fig. 3.8b the distortion is of the opposite type and the corners of the square are pulled inward more than the sides; this is negative or barrel distortion.

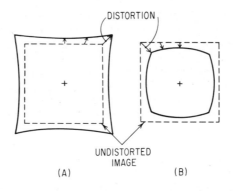

FIG. 3.8. Distortion. (a) positive, or pincushion distortion. (b) negative, or barrel distortion. The sides of the image are curved because the amount of distortion varies as the cube of the distance from the axis. Thus, in the case of a square, the corners are distorted $2\sqrt{2}$ as much as the center of the sides.

A little study of the matter will show that a system which produces distortion of one sign will produce distortion of the opposite sign when object and image are interchanged. Thus a camera lens with barrel distortion will have pin-cushion distortion if used as a projection lens (i.e. when the film is replaced by a slide). Obviously if the same lens is used both to photograph and to project the slide, the projected image will be rectilinear (free of distortion) since the distortion in the slide will be exactly cancelled out upon projection.

3.3 Chromatic Aberrations

Because of the fact that the index of refraction varies as a function of the wavelength of light, the properties of optical elements also vary with wavelength. Longitudinal chromatic aberration is the variation of focus (or image position) with wavelength. In general the index of refraction of optical materials is higher for short wavelengths than for long wavelengths; this causes the short wavelengths to be more strongly refracted at each surface of a lens so that, in a simple positive lens for example, the blue light rays are brought to a focus closer to the lens than the red rays. The distance along the axis between the two focus points is the longitudinal or axial chromatic aberration. Figure 3.9 shows the chromatic aberration of a simple positive element. When the short wavelength rays are brought to a focus to the left of the long wavelength rays, the chromatic is termed undercorrected.

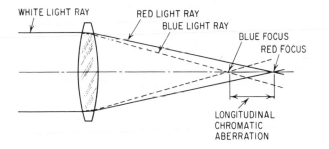

FIG. 3.9 The undercorrected longitudinal chromatic aberration of a simple lens is due to the blue rays undergoing a greater refraction than the red rays.

The image of an axial point in the presence of chromatic aberration is a central bright dot surrounded by a halo. The rays of light which are in focus, and those which are nearly in focus, form the bright dot. The out of focus rays form the halo. Thus, in an undercorrected visual instrument, the image would have a yellowish dot (formed by the orange, yellow and green rays) and a purplish halo (due to the red and blue rays). If the screen on which the image is formed is moved toward the lens, the central dot will become blue; if it is moved away the central dot will become red.

When a lens system forms images of different sizes for different wavelengths, or spreads the image of an off axis point into a rainbow, the difference between the image heights for different colors is called Lateral Color or Chromatic Difference of Magnification. In Fig. 3.10 a simple lens with a displaced diaphragm is shown forming an image of an off axis point. Since the diaphragm limits the rays which reach the lens, the ray bundle from the off axis point strikes the lens above the axis and is bent downward as well as being brought to a focus. The blue rays are bent more than the red and thus form their image nearer the axis.

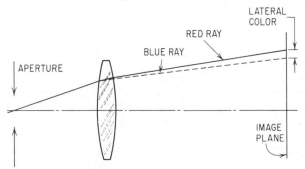

FIG. 3.10. Lateral color, or chromatic difference of magnification results in different sized images for different wavelengths.

The chromatic variation of index also produces a variation of the monochromatic aberrations discussed in Section 3.2. Since each aberration results from the manner in which the rays are bent at the surfaces of the optical system, it is to be expected that, since rays of different color are bent differently, the aberrations will be somewhat different for each color. In general this proves to be the case, and the effects are of practical importance when the basic aberrations are well corrected.

3.4 The Effect of Lens Shape and Stop Position on the Aberrations

A consideration of either the thick-lens focal length equation

$$\frac{1}{f} = (N - 1)\left[\frac{1}{R_1} - \frac{1}{R_2} + \frac{(N - 1)}{N} \frac{t}{R_1 R_2}\right]$$

or the thin-lens focal length equation

$$\frac{1}{f} = (N - 1)\left[\frac{1}{R_1} - \frac{1}{R_2}\right] = (N - 1)(C_1 - C_2)$$

reveals that for a given index and thickness, there is an infinite number of combinations of R_1 and R_2 which will produce a given focal length. Thus a lens of some desired power may take on a number of different shapes or "bendings". The aberrations of the lens are changed markedly as the shape is changed; this effect is the basic tool of optical design.

As an illustrative example, we will consider the aberrations of a thin positive lens made of borosilicate crown glass with a focal length of 100 mm and a clear aperture of 10 mm (a speed of $f/10$) which is to image an infinitely distant object over a field of view of $\pm 17°$. A typical borosilicate crown is 517:645 which has an index of 1.517 for the sodium D line ($\lambda = 5893 A°$), an index of 1.51461 for C light ($\lambda = 6563 A°$) and 1.52262 for F light ($\lambda = 4861 A°$).

(The aberration data which are presented in the following paragraphs were calculated by means of the thin lens third order aberration equations of Chapter 10.)

If we first assume that the stop or limiting aperture is in coincidence with the lens, we find that several aberrations do *not* vary as the lens shape is varied. Axial chromatic aberration is constant at a value of -1.55 mm (undercorrected); thus the blue focus (F light) is 1.55 mm nearer the lens than the red focus (C light). The astigmatism and field curvature are also constant. At the edge of the field (30 mm from the axis) the sagittal focus is 7.5 mm closer to the lens than the paraxial focus and the tangential focus is 16.5 mm inside the paraxial focus. Two aberrations, distortion and lateral color, are zero when the stop is at the lens.

Spherical aberration and coma, however, vary greatly as the lens shape is changed. Figure 3.11 shows the amount of these two aberrations plotted against the curvature of the first surface of the lens. Notice tha coma varies linearly with lens shape, taking a large positive value when the lens is a meniscus with both surfaces concave toward the object. As the lens is bent through plano-convex, convex-plano and convex meniscus shapes, the amount of coma becomes more negative, assuming a zero value near the convex-plano form.

The spherical aberration of this lens is always undercorrected; its plot has the shape of a parabola with a vertical axis. Notice that the spherical aberration reaches a minimum (or more accurately, a maximum) value at approximately the same shape for which the coma is zero. This, then, is the shape that one would select if the lens were to be used as a telescope objective to cover a rather small field of view.

Let us now select a particular shape for the lens, say $C_1 = -.0193$ and investigate the effect of placing the stop away from the lens, as shown in Fig. 3.12. The spherical and axial chromatic aberrations are completely unchanged by shifting the stop, since the axial rays strike the lens in exactly the same manner regardless of where the stop is located. The lateral color and distortion,

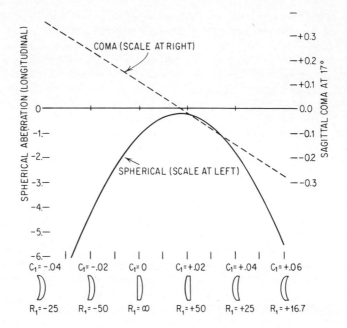

FIG. 3.11. Spherical aberration and coma as a function of lens shape. Data plotted are for a 100 mm focal length lens (with the stop at the lens) at $f/10$ covering $\pm 17°$ field.

however, take on positive values when the stop is behind the lens and negative when it is before the lens. Figure 3.13 shows a plot of lateral color, distortion, coma, and tangential field curvature as a function of the stop position. The most pronounced effects of moving

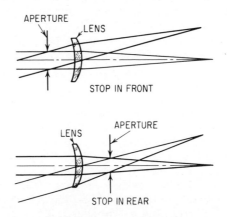

FIG. 3.12. The aperture stop away from the lens. Notice that the oblique ray bundle passes through an entirely different part of the lens when the stop is in front of the lens than when it is behind the lens.

the stop are found in the variations of coma and astigmatism. As the stop is moved toward the object, the coma decreases linearly with stop position, and has a zero value when the stop is about 18.5 mm in front of the lens. The astigmatism becomes less negative so that the position of the tangential image approaches the paraxial focal plane. Since astigmatism is a quadratic function of stop position, the tangential field curvature (x_t) plots as a parabola. Notice that the parabola has a maximum at the same stop position for which the coma is zero. This is called the "natural" position of the stop, and for all lenses with undercorrected primary spherical aberration, the natural, or coma-free, stop position produces a more backward curving (or less inward curving) field than any other stop position.

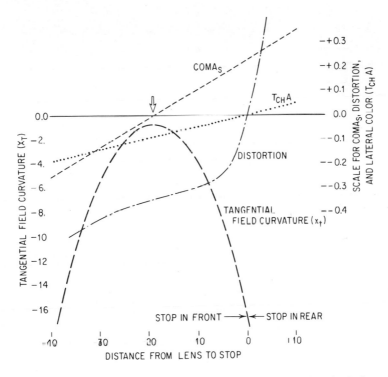

FIG. 3.13. Effect of shifting the stop position on the aberrations of a simple lens. The arrow indicates the "natural" stop position where coma is zero. (efl = 100. C_1 = -.0193 Speed = $f/10$ Field = ± 17°.)

Figure 3.11 showed the effect of lens shape with the stop fixed in contact with the lens, and Fig. 3.13 showed the effect of stop position with the lens shape held constant. There is a "natural" stop position for each shape of the simple lens we are considering. In Fig. 3.14, the aberrations of the lens have again been plotted against the lens shape; however, in this figure, the aberration values are those which occur when the stop is in the "natural" position. Thus for each bending, the coma has been removed by choosing this stop position, and the field is as far backward curving as possible.

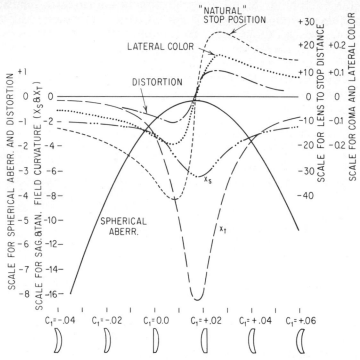

FIG. 3.14. The variation of the aberrations with lens shape when the stop is located in the "natural" (coma free) position for each shape. Data is for 100 mm $f/10$ lens covering $\pm 17°$ field, made from BSC-2 glass (517 : 645).

Notice that the shape which produces minimum spherical aberration also produces the maximum field curvature, so that this shape, which gives the best image near the axis, is not suitable for wide field coverage. The meniscus shapes at either side of the figure represent a much better choice for this purpose, for although the spherical aberration is much larger at these bendings, the field is much more nearly flat. This is the type of lens used in inexpensive box cameras at speeds of $f/11$ or $f/16$.

3.5 Aberration Variation with Aperture and Field

In the preceding section, we considered the effect of lens shape and aperture position on the aberrations of a simple lens, and in that discussion we assumed that the lens operated at a fixed aperture of $f/10$ (stop diameter of 10 mm) and covered a fixed field of ±17° (field diameter of 60 mm). It is often useful to know how the aberrations of such a lens vary when the size of the aperture or field is changed.

Figure 3.15 lists the relationships between the primary aberrations and the semi-aperture y (in column one) and the image height h (in column two). To illustrate the use of this table, let us assume that we have a lens whose aberrations are known; we wish to determine the size of the aberrations if the aperture diameter is increased by 50% and the field coverage reduced by 50%. The new y will be 1.5 times the original and the new h will be 0.5 times the original.

Aberration	vs Aperture	vs Field Size
Spherical (longitudinal)	y^2	—
Spherical (transverse)	y^3	—
Coma	y^2	h
Petzval curvature	—	h^2
Astigmatism	—	h^2
Length of astigmatic lines	y	h^2
Distortion (linear)	—	h^3
Distortion (percentage)	—	h^2
Axial Chromatic (longitudinal)	—	—
Lateral Chromatic	—	h

FIG. 3.15

Since longitudinal spherical aberration is shown to vary with y^2, the 1.5 times increase in aperture will cause the spherical to be $(1.5)^2$ or 2.25 times as large. Similarly transverse spherical, which varies as y^3, will be $(1.5)^3$ or 3.375 times larger.

Coma varies as y^2 and h; thus, the coma will be $(1.5)^2 \times 0.5$ or 1.125 times as large. The Petzval curvature and astigmatism, which vary with h^2 will be reduced to $(0.5)^2$ or 0.25 of their previous value, while the length of the astigmatic lines will be $1.5(0.5)^2$ or 0.375 of their original length.

The aberrations of a lens also depend on the position of the object and image. A lens which is well corrected for an infinitely distant object, for example, may be very poorly corrected if used to image a nearby object. This is because the ray paths and incidence angles change as the object position changes.

It should be obvious that if *all* the dimensions of an optical system are scaled up or down, the *linear* aberrations are also scaled in exactly the same proportion. Thus if the simple lens used as the example in Section 3.4 were increased in focal length to 200 mm, its aperture increased to 20 mm and the field coverage increased to 120 mm, then the aberrations would all be doubled. Note, however, that the speed, or f/number, would remain at $f/10$

and the angular coverage would remain at ±17°. The percentage distortion would not be changed.

Aberrations are occasionally expressed as angular aberrations. For example, the transverse spherical aberration of a system subtends an angle from the second principal point of the system; this angle is the angular spherical aberration. Note that the angular aberrations are not changed by scaling the size of the optical system.

3.6 Optical Path Difference

Aberrations can also be described in terms of the wave nature of light. In Chapter 1, it was pointed out that the light waves converging to form a "perfect" image were spherical in shape. Thus when aberrations are present in a lens system, the waves converging on an image point are deformed from the ideal shape (which is a sphere centered on the image point). For example, in the presence of undercorrected spherical aberration the wave front is curled inward at the edges, as shown in Fig. 3.16. This can be understood if we remember that a ray is the path of a point on the wave front and is also normal to the wave. Thus, if the ray is to intersect the axis in front of the paraxial focus, the section of the wave front associated with the ray must be curled inward. The wavefront shown is "ahead" of the reference sphere; the distance by which it is ahead is called the Optical Path Difference or O.P.D., and is customarily expressed in units of wavelengths. The wave fronts associated with axial aberrations are symmetrical figures of rotation, in contrast to the off-axis aberrations such as coma and astigmatism. For example, the wave front for astigmatism would be a section of a torus (the outer surface of a doughnut) with different radii in the prime meridians.

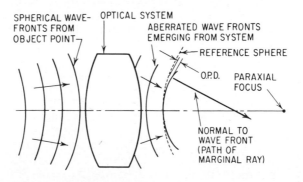

FIG. 3.16. The optical path difference (O.P.D.) is the distance between the emerging wave front and a reference sphere (centered in the image plane) which coincides with the wave front at the axis. The O.P.D. is thus the difference between the marginal and axial paths through the system for an axial point.

3.7 Aberration Correction and Residuals

Section 3.4 indicated two methods which are used to control aberrations in simple optical systems, namely lens shape and stop position. For many applications a higher level of correction is needed, and it is then necessary to combine optical elements with aberrations of opposite signs so that the aberrations contributed to the system by one element are cancelled out, or corrected, by the others. A typical example is the achromatic doublet used for telescope objectives, shown in Fig. 3.17. A single positive element would be afflicted with both undercorrected spherical aberration and undercorrected chromatic aberration. In a negative element, on the other hand, both aberrations are overcorrected. In the doublet a positive element is combined with a less powerful negative element in such a way that the aberrations of each balance out. The positive lens is made of a glass with a low chromatic dispersion (crown) and the negative element of a glass with a high dispersion (flint). Thus, the negative element has a greater amount of chromatic aberration per unit of power, by virtue of its greater dispersion, than the crown element. The relative powers of the elements are chosen so that the chromatic exactly cancels while the focusing power of the crown element dominates.

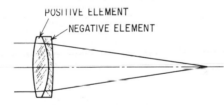

FIG. 3.17. Achromatic doublet telescope objective. The powers and shapes of the two elements are so arranged that each cancels the aberrations of the other.

The situation with regard to spherical aberration is quite analogous except that element power, shape, and index of refraction are involved instead of power and dispersion as in chromatic.

Aberration correction usually is exact only for one zone of the aperture of a lens or for one angle of obliquity, because the aberrations of the individual elements do not balance out exactly for all zones and angles. Thus, while the spherical aberration of a lens may be corrected to zero for the rays through the edge of the aperture, the rays through the other zones of the aperture usually do not come to a focus at the paraxial image point. A typical spherical aberration plot for a "corrected" lens is shown in Fig. 3.18. Notice that the rays through only one zone of the lens intersect the paraxial focus. Rays through the smaller zones focus nearer the lens system and have undercorrected spherical;

rays above the corrected zone show overcorrected spherical. The undercorrected aberration is called residual or zonal aberration; Fig. 3.18 would be said to show an undercorrected zonal aberration. This is the usual state of affairs for most optical systems. Occasionally a system is designed with an overcorrected spherical zone, but this is rare.

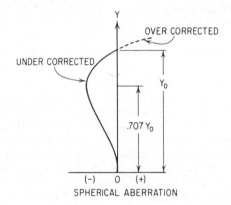

FIG. 3.18. Plot of longitudinal spherical aberration vs. ray height for a "corrected" lens. For most lenses, the maximum undercorrection occurs for the ray whose height is .707 that of the ray with zero spherical.

Chromatic aberration has residuals which take two different forms. The correction of chromatic aberration is accomplished by making the focii of two different wavelengths coincide.

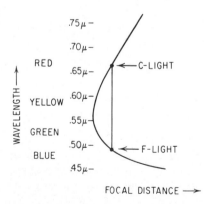

FIG. 3.19. The secondary spectrum of a typical doublet achromat, corrected so that C and F light are joined at a common focus. The distance from the common focus of C and F to the minimum of the curve (in the yellow green at about 0.55μ) is called the secondary spectrum.

However, due to the nature of the great majority of optical materials, the dispersion characteristics of the positive and negative elements used in an achromat do not "match up", so that the focal points of other wavelengths do not coincide with the common focal point of the two selected colors. This difference in focal distance is called secondary spectrum. Figure 3.19 shows a plot of back focal distance vs. wavelength for a typical achromatic lens, in which the rays for C light (red) and F light (blue) are brought to a common focus. The yellow rays come to a focus about 1/2400th of the focal length ahead of the C-F focal point.

The second major chromatic residual may be regarded as a variation of chromatic aberration with ray height, or as a variation of spherical aberration with wavelength, and is called spherochromatism. In ordinary spherochromatism, the spherical aberration in blue light is overcorrected and the spherical in red light is undercorrected (when the spherical aberration for the yellow light is corrected). Figure 3.20 is a spherical aberration plot in three wavelengths for a typical achromatic doublet of large aperture. The correction has been adjusted so that the red and blue rays striking the lens at a height of .707 of the marginal ray height are brought to a common focus. The distance between the yellow focus and the red-blue focus at this height is, of course, the secondary spectrum discussed above. Notice that above this .707 zone the chromatic is overcorrected and below it is undercorrected, so that one half of the area of the lens aperture is overcorrected and one half undercorrected.

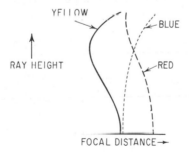

FIG. 3.20. Spherochromatism. The longitudinal aberration of a "corrected" lens is shown for three wavelengths. The marginal spherical for yellow light is corrected, but is over-corrected for blue light and undercorrected for red. The chromatic aberration is corrected at the zone, but is overcorrected above it and undercorrected below. A transverse plot of these aberrations is shown in Fig. 3.23 K.

The other aberrations have similar residuals. Coma may be completely corrected for a certain field angle, but will often be overcorrected above this obliquity and undercorrected below it. Coma may also undergo a change of sign with aperture, with the central part of the aperture overcorrected and the outer zone undercorrected.

Astigmatism usually varies markedly with field angle. Figure 3.21 shows a plot of the sagittal and tangential field curvatures for a typical photographic anastigmat, in which the astigmatism is zero for one zone of the field. This point is called the node, and typically the two focal surfaces separate quite rapidly outside the node.

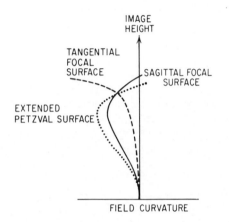

FIG. 3.21. Field curvature of a photographic anastigmat. The astigmatism has been corrected for one zone of the field, but is overcorrected inside this zone and undercorrected beyond it.

3.8 Ray Intercept Curves and the "Orders" of Aberrations

When the image plane intersection heights of a fan of meridional rays are plotted against the slope of the rays as they emerge from the lens, the resultant curve is called a ray intercept curve, an $H' - \tan U'$ curve, or sometimes (erroneously) a rim ray curve. The shape of the intercept curve not only indicates the amount of spreading or blurring of the image directly, but also can serve to indicate which aberrations are present. Figure 3.2b for example, shows simple undercorrected spherical aberration.

In Fig. 3.22, an oblique fan of rays from a distant object point is brought to a perfect focus at point P. If the reference plane passes through P, it is apparent that the $H' - \tan U'$ curve will be a straight horizontal line. However, if the reference plane is behind P (as shown) then the ray intercept curve becomes a tilted straight line since the height, H', decreases as $\tan U'$ increases. Thus it is apparent that shifting the reference plane (or focusing the system) is equivalent to a rotation of the $H' - \tan U'$ curve. A valuable feature of this type of aberration representation is that one can immediately assess the effects of refocusing the optical system by a simple rotation of the figure. Notice that the slope of the line $(\Delta H'/\Delta \tan U')$ is equal to the distance (δ) from the reference plane to the point of focus, so that for an oblique ray fan the tangential field curvature is equal to the slope of the ray intercept curve.

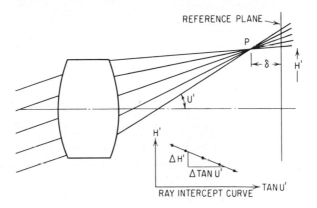

FIG. 3.22. The ray intercept curve (H' vs. TAN U') of a point which does not lie in the in the reference plane is a tilted straight line. The slope of the curve ($\Delta H'/\Delta$ TAN U') is equal to the distance from the reference plane to the point of focus (δ). Note that δ is equal to X_T, the tangential field curvature, if the paraxial focal plane is chosen as the reference plane.

Figure 3.23 shows a number of intercept curves, each labeled with the aberration represented. The generation of these curves can be readily understood by sketching the ray paths for each aberration and then plotting the intersection height and slope angle for each ray as a point of the curve. Distortion is not shown in Fig. 3.23; it would be represented as a vertical displacement of the curve from the paraxial image height h'. Lateral color would be represented by curves for two colors which were vertically displaced from each other.

It is apparent that the ray intercept curves which are "odd" functions, that is, the curves which have a rotational symmetry about the origin, can be represented mathematically by an equation of the form

$$y = a + bx + cx^3 + dx^5 + \cdots$$

or

$$H' = a + b \tan U' + c \tan^3 U' + d \tan^5 U' + \cdots \qquad (3.4)$$

All the ray intercept curves for axial image points are of this type. Since the curve for an axial image must have $H' = 0$ when

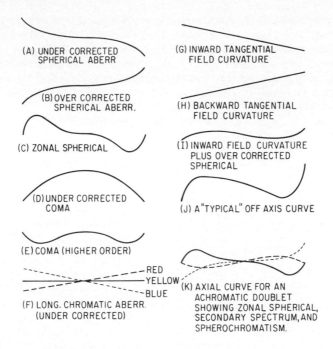

(A) UNDER CORRECTED SPHERICAL ABERR

(B) OVER CORRECTED SPHERICAL ABERR.

(C) ZONAL SPHERICAL

(D) UNDER CORRECTED COMA

(E) COMA (HIGHER ORDER)

(F) LONG. CHROMATIC ABERR. (UNDER CORRECTED)

RED
YELLOW
BLUE

(G) INWARD TANGENTIAL FIELD CURVATURE

(H) BACKWARD TANGENTIAL FIELD CURVATURE

(I) INWARD FIELD CURVATURE PLUS OVER CORRECTED SPHERICAL

(J) A "TYPICAL" OFF AXIS CURVE

(K) AXIAL CURVE FOR AN ACHROMATIC DOUBLET SHOWING ZONAL SPHERICAL, SECONDARY SPECTRUM, AND SPHEROCHROMATISM.

FIG. 3.23. Ray intercept curves for various aberrations. The ordinate for each curve is H', the height at which the ray intersects the (paraxial) image plane; the abcissa is TAN U', the final slope of the ray with respect to the optical axis.

$\tan U' = 0$, it is apparent that the constant a must be a zero. It is also apparent that the constant b for this case represents the amount the reference plane is displaced from the paraxial image plane. Thus the curve for lateral spherical aberration plotted with respect to the paraxial focus can be expressed by the equation

$$TA' = c \tan^3 U' + d \tan^5 U' + e \tan^7 U' + \cdots \qquad (3.5)$$

It is, of course, possible to represent the curve by an odd power expansion in terms of the final angle U', or $\sin U'$, or the ray height at the lens (Y), or even the initial slope of the ray at the object (U_0) instead of $\tan U'$ (the constants will, of course, be different for each).

For simple uncorrected lenses the first term of Eq. 3.5 is usually adequate to describe the aberration. For the great majority of "corrected" lenses the first two terms are dominant; in a few cases three terms (and rarely four) are necessary to satisfactorily represent the aberration. As examples, Figs. 3.2b, 3.23a, and 3.23b can be represented by $TA' = c \tan^3 U'$, and this type of aberration is called third-order spherical. Figure 3.23c, however, would require two terms of the expansion to represent it adequately; thus $TA' = c \tan^3 U' + d \tan^5 U'$. The amount of aberration represented by the second term is called the fifth-order aberration. Similarly, the aberration represented by the third term of Eq. 3.5 is called the seventh-order aberration. The fifth-, seventh-, ninth-, etc. order aberrations are collectively referred to as high-order aberrations.

As will be shown in a subsequent chapter, it is possible to calculate the amount of the primary, or third-order aberrations, without trigonometric raytracing, that is, by means of data from a paraxial raytrace. This type of aberration analysis is called third-order theory. The name "first-order optics" given to that part of geometrical optics devoted to locating the paraxial image is also dereived from this power series expansion, since the first-order term of the expansion results purely from a longitudinal displacement of the reference plane from the paraxial focus.

4

Prisms and Mirrors

4.1

In most optical systems, prisms serve one of two major functions. In spectral instruments (spectroscopes, spectrographs, spectrophotometers, etc.) their function is to disperse the light or radiation; that is, to separate the different wavelengths. In other applications, prisms are used to displace, deviate or to reorient a beam of light or an image. In this type of use, the prism is carefully arranged so that it will *not* separate the different colors.

4.2 Dispersing Prisms

In a typical dispersing prism, as shown in Fig. 4.1, a light ray strikes the first surface at an angle of incidence I_1 and is refracted downward, making an angle of refraction I'_1 with the normal to the surface. The ray is thus deviated through an angle of $(I_1 - I'_1)$ at this surface. At the second surface the ray is deviated through an angle $(I'_2 - I_2)$, so that the total deviation of the ray is given by

$$D = (I_1 - I'_1) + (I'_2 - I_2) \qquad (4.1)$$

From the geometry of the figure it can be seen that angle I_2 is equal to $(A - I'_1)$ where A is the vertex angle of the prism; making this substitution in Eq. 4.1 we get

$$D = I_1 + I'_2 - A \qquad (4.2)$$

To compute the deviation produced by the prism we can readily determine the angles in Eq. 4.2 by Snell's Law (Eq. 1.3) as follows (where N is the prism index):

$$\sin I'_1 = \frac{1}{N} \sin I_1 \qquad (4.3)$$

$$I_2 = A - I'_1 \qquad (4.4)$$

$$\sin I'_2 = N \sin I_2 \qquad (4.5)$$

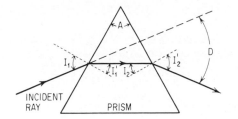

FIG. 4.1. The deviation of a light ray by a refracting prism.

While it is ordinarily much more convenient to calculate the deviation step by step, using the equations above, it is possible to combine them into a single expression for D, in terms of I_1, A and N as follows:

$$D = I_1 \quad A + \arcsin\left[(N^2 - \sin^2 I_1)^{\frac{1}{2}} \sin A - \cos A \sin I_1\right] \quad (4.6)$$

It is apparent that the deviation is a function of the prism index and that the deviation will be increased as the index as the index is raised. For most materials, the index of refraction is higher for short wavelengths (blue light) than for long wavelengths (red light). Therefore, the deviation angle will be greater for blue light than red, as indicated in Fig. 4.2. This variation of the deviation angle with wavelength is called the dispersion of the prism. An expression for the dispersion can be found by differentiating the preceding equations with respect to the index N, assuming that I_1 is constant, yielding,

$$dD = \frac{\cos I_2 \tan I'_1 + \sin I_2}{\cos I'_2} \, dN \quad (4.7)$$

The dispersion with respect to wavelength is simply $dD/d\lambda$ and is obtained by dividing both sides of Eq. 4.7 by $d\lambda$. The resulting $dN/d\lambda$ term on the right is the dispersion of the prism material.

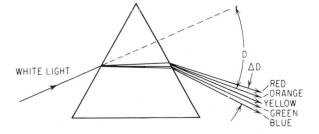

FIG. 4.2. The dispersion of white light into its component wavelengths by a refracting prism. (Highly exaggerated).

4.3 The Thin Prism

If all the angles involved in the prism are very small, we can, as in the paraxial case for lenses, substitute the angle itself for its sine. This case occurs when the prism angle A is small and when the ray is almost at normal incidence to the prism faces. Under these conditions, we can write

$$i'_1 = i_1/N$$

$$i_2 = A - i'_1 = A - i_1/N$$

$$i'_2 = Ni_2 = NA - i_1$$

$$D = i_1 + i'_2 - A = i_1 + NA - i_1 - A$$

and finally

$$D = A(N - 1) \tag{4.8}$$

This expression is of great utility in evaluating the effects of a small prismatic error in the construction of an optical system since it allows the resultant deviation of the light beam to be determined quite readily.

The dispersion of a thin prism is obtained by differentiating Eq. 4.8 with respect to N, which gives $dD = A\,dN$. If we substitute A from Eq. 4.8 we get

$$dD = D \frac{dN}{(N - 1)} \tag{4.9}$$

Now the fraction $(N - 1)/\Delta N$ is one of the basic numbers used to characterize optical materials. It is called the reciprocal relative dispersion, Abbe V number, or V-value. Ordinarily N is taken as the index for the sodium D line $(.5893\mu)$ and ΔN is the index difference between the hydrogen $F(.4861\mu)$ and $C(.6563\mu)$ lines.

$$V = \frac{N_D - 1}{N_F - N_C} \tag{4.10}$$

Making the substitution of $1/V$ for $dN/(N - 1)$ in Eq. 4.9 we get

$$dD = \frac{D}{V} \tag{4.11}$$

which allows us to immediately evaluate the chromatic dispersion produced by a thin prism.

4.4 Minimum Deviation

The deviation of a prism is a function of the initial angle of incidence I_1. It can be shown that the deviation is at a minimum when the ray passes symmetrically through the prism. In this case $I_1 = I'_2 = \frac{1}{2}(A + D)$ and $I'_1 = I_2 = A/2$, so that if we know the prism angle A and the minimum deviation D_0 it is a simple matter to compute the index of the prism from

$$ N = \frac{\sin I_1}{\sin I'_1} = \frac{\sin \frac{1}{2}(A + D_0)}{\sin \frac{1}{2} A} \qquad (4.12) $$

This is a widely used method for measurement of index, since the minimum deviation position is readily determined on a spectrometer. This position for the prism is also approximated in most spectral instruments because it allows the largest diameter beam to pass through a given prism and also produces the smallest amount of loss due to surface reflections.

4.5 The Achromatic Prism and the Direct Vision Prism

It is occasionally useful to produce an angular deviation of a light beam without introducing any chromatic dispersion. This can be done by combining two prisms, one of high dispersion glass and the other of low dispersion glass. We desire the sum of their deviations to equal $D_{1,2}$ and the sum of their dispersions to equal zero. Using the equations for thin prisms (Eqs. 4.8 and 4.11) we can express these requirements as follows:

Deviation $D_{1,2} = D_1 + D_2 = A_1(N_1 - 1) + A_2(N_2 - 1)$

Dispersion $dD_{1,2} = dD_1 + dD_2 = 0 = \dfrac{D_1}{V_1} + \dfrac{D_2}{V_2}$

$$ = \frac{A_1(N_1 - 1)}{V_1} + \frac{A_2(N_2 - 1)}{V_2} $$

A simultaneous solution for the angles of the two prisms gives.

$$A_1 = \frac{D_{1,2}V_1}{(N_1 - 1)(V_1 - V_2)}$$

$$A_2 = \frac{D_{1,2}V_2}{(N_2 - 1)(V_2 - V_1)}$$

(4.13)

It is apparent that the prism angles will have opposite signs and that the prism with the larger V-value (smaller relative dispersion) will have the larger angle. A sketch of an achromatic prism is shown in Fig. 4.3.

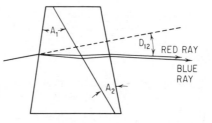

FIG. 4.3. An achromatic prism. The red and blue rays emerge parallel to each other; no chromatic dispersion is introduced by the deviation.

In the direct vision prism it is desired to produce a dispersion without deviating the ray. By setting the deviation, $D_{1,2}$, equal to zero and preserving the dispersion term, $dD_{1,2}$, in the preceding equations we can solve for the angles of two prisms which will produce the desired result. The solution is

$$A_1 = \frac{dD_{1,2}V_1V_2}{(N_1 - 1)(V_2 - V_1)}$$

$$A_2 = \frac{dD_{1,2}V_1V_2}{(N_2 - 1)(V_1 - V_2)}$$

(4.14)

A two-element *direct vision prism* is shown in Fig. 4.4a. In order to obtain a large enough dispersion for practical purposes it is often necessary to use more than two prisms. Figure 4.4b shows the application of such a prism to a hand spectroscope.

Since Eqs. 4.13 and 4.14 were derived using the equations for thin prisms it is obvious that the values of the component prism angles which they give will be approximations to the exact values when the prisms are other than thin. For exact work, these approximate values must be adjusted by exact ray tracing based on Snell's Law.

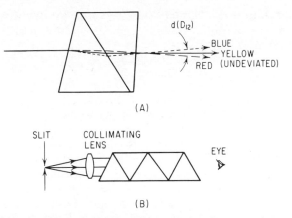

(A)

(B)

FIG. 4.4. a) A direct vision prism disperses the light into its spectral components without deviation of the beam. b) Hand spectroscope. The collimating lens produces a magnified image of the slit at infinity for easy viewing. The prism then disperses the light into a spectrum without deviation of the yellow ray.

4.6 Total Internal Reflection

When a light ray passes from a higher index to a lower index, the ray is refracted away from the normal to the surface as shown in Fig. 4.5a. As the angle of incidence is increased, the angle of refraction increases at a greater rate, in accordance with Snell's Law $(N > N')$:

$$\sin I' = \frac{N}{N'} \sin I$$

When the angle of incidence reaches a value such that $\sin I = N'/N$, then $\sin I' = 1.0$ and $I' = 90°$. At this point none of the light is

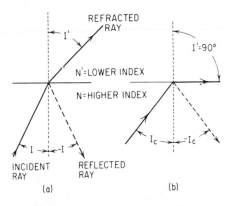

FIG. 4.5. Total internal reflection occurs when a ray, passing from a higher to a lower index of refraction, has an angle of incidence whose sine equals or exceeds N'/N.

transmitted through the surface; the ray is totally reflected back into the denser medium, as is any ray which makes a greater angle to the normal. The angle

$$I_c = \text{arc sin } \frac{N'}{N} \tag{4.15}$$

is called the critical angle and for an ordinary air glass surface has a value of about 42° if the index of the glass is 1.5; for an index of 1.7, the critical angle is near 36°.

For practical purposes, if the boundary surface is smooth and clean, 100% of the energy is redirected along the totally reflected ray. However, it should be noted that the electromagnetic field associated with the light actually does penetrate the surface for a relatively short distance. If there is anything near the other side of the boundary surface, the total internal reflection can be "frustrated" to some extent and a portion of the energy will be transmitted. Since the distance of effective penetration is only to the order of the wavelength of the light involved, this phenomenon has been used as the basis of a light valve or modulator. In the German "Licht-Sprecher", an external piece of glass was placed in contact with the reflecting face of a prism to frustrate the reflection, and then moved an extremely short distance away (e.g. a few microns) to reinstate the reflection.

It should also be noted that the reflection of a totally reflecting surface is decreased by aluminizing or silvering the surface. When this is done, the reflectance drops from 100% to the reflectance of the coating applied to the surface.

4.7 Reflection from a Plane Surface

Since the prism systems which are discussed in the balance of this chapter are primarily reflecting prisms (the majority of which can be replaced by a system of plane mirrors), we shall first discuss the imaging properties of a plane reflecting surface. Rays originating at an object are reflected according to the law of reflection, which states that both the incident and reflected rays lie in the plane of incidence and that both rays make equal angles with the normal to the surface. The normal to the surface is the perpendicular at the point where the ray strikes the surface, and the plane of incidence is that plane containing the incident ray and the normal.

In Fig. 4.6, the plane of the page is the plane of incidence. Two rays from point P are shown reflected from the surface MM'. By extending the rays backward, it can be seen that after reflection they appear to be coming from point P', which is a virtual image of point P. Both P and P' lie on the same normal to the surface (POP') and the distance OP is exactly the same as the distance OP'.

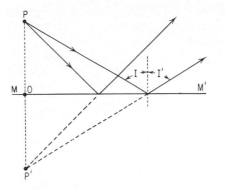

FIG. 4.6. A plane reflecting surface forms a virtual image of an object point. Object and image are equidistant from the reflecting surface and both lie on the same normal to the surface.

If we now consider an extended object such as the arrow *AB* in Fig. 4.7, we can readily locate the position of its image by using the principles of the preceding paragraph to locate the images of points *A* and *B*. An observer at *E* looking directly at the arrow would see the arrowhead *A* at the top of the arrow. However, in the reflected image, the arrowhead (*A′*) is at the bottom of the arrow. The image of the arrow has been reoriented (or inverted) by the reflection.

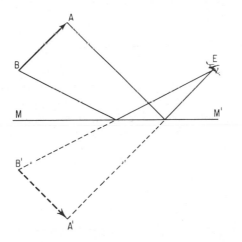

FIG. 4.7. The reflected image *A′B′* of the arrow *AB* appears inverted to an observer at *E*.

If we add a cross piece *CD* to the arrow, the image is formed as shown in Fig. 4.8, and although the image of the arrow has been inverted, the image of the cross piece has the same left to right orientation as the object.

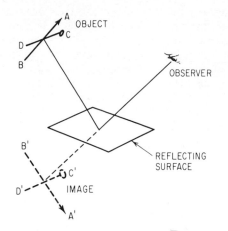

FIG. 4.8. The reflected image is inverted top to bottom, but not left to right.

The preceding discussion has treated reflection from the standpoint of an observer viewing a reflected image. Since the path of light rays is completely reversible, we can equally well consider point P' in Fig. 4.6 to be an image formed by a lens at the right. Then P would be the reflected image of P'. Similarly in Figs. 4.7 and 4.8, we may replace the eye with a lens whose image is the primed figure ($A'B'$ or $A'B'C'D'$) and view the unprimed figures as their reflected images.

A point worth noting is that reflection constitutes a sort of "folding" of the ray paths. In Fig. 4.9, the lens images the arrow at AB. If we now insert reflecting surface MM', the reflected image is at $A'B'$. Notice that if the page were folded along MM', the arrow AB and the solid line rays would exactly coincide with the arrow $A'B'$ and the reflected (dashed) rays. It is frequently convenient to "unfold" a complex reflecting system; one advantage of this device is that an accurate drawing of the ray paths becomes a simple matter of straight lines.

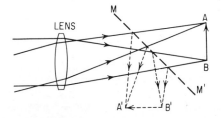

FIG. 4.9. The reflecting surface MM' folds the optical system. Note that if the page is folded along MM', the rays and images coincide.

A useful technique to determine the image orientation after passage through a system of reflectors is to imagine that the image is a transverse arrow, or pencil, which is bounced off the reflecting surface, much as a stick would be bounced off a wall. Figure 4.10 illustrates the technique. The first illustration shows the pencil approaching and striking the reflecting surface, the second shows the point bouncing off the reflector and the blunt end continuing in the original direction, and the third shows the pencil in the new orientation after the reflection. If the process is repeated with the pencil perpendicular to the plane of the paper, the orientation of the other meridian of the image can be determined. The procedure can then be repeated through each reflection in the system.

FIG. 4.10. A useful technique in determining the orientation of a reflected image is to visualize the image as a pencil "bouncing" off a solid wall as it moves along the system axis.

A card marked with the arrow and cross bar of Fig. 4.11 is also useful for this purpose. The reader's attention is directed to the fact that the initial orientation of the pencil, or pattern, is chosen so that one meridian of the pattern coincides with the plane of incidence. In the majority of reflecting systems, one or the other of the meridians will be in the plane of incidence throughout the system, and the application of this technique is straightforward. Where this is not the case, the card can be marked with a second set of meridians so that the second set is aligned with the plane of incidence. This second set can then be carried through the reflection as before; the orientation of the final image is of course given by the original set of markings. Fig. 4.20b exemplifies this method.

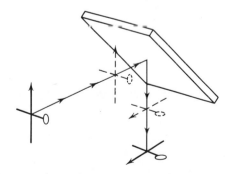

FIG. 4.11. Image orientation after reflection.

4.8 Plane Parallel Plates

Figure 4.12 shows a lens which, in air, would form an image at P. The insertion of the plane parallel plate between the lens and P displaces the image to P'. If we trace the path of the light rays through the plate, we first notice that the ray emerging from the plate has exactly the same slope angle that it had before passing through the plate, since by Snell's Law, $\sin I'_1 = (1/N) \sin I_1$, and $I_2 = I'_1$ (since the surfaces are parallel). Thus, $\sin I_2 = \sin I'_1 = (1/N) \sin I_1 = (1/N) \sin I'_2$, and $I_1 = I'_2$. Therefore, the effective focal length of the lens system, and the size of the image, are unchanged by the insertion of the plate.

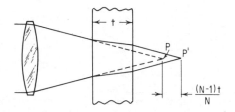

FIG. 4.12. The longitudinal displacement of an image by a plane parallel glass plate.

The amount of longitudinal displacement of the image is readily determined by application of the paraxial ray tracing formulae of Chapter 2, and is equal to $(N - 1)t/N$. The effective thickness of the plate compared to air (the equivalent air thickness) is less than the actual thickness t by the amount of this shift. The equivalent air thickness is thus found by subtracting the displacement from the thickness and is equal to t/N. The concept of equivalent thickness is useful when one wishes to determine whether a certain size prism can be fitted into the available air space of an optical system, and also in prism system design.

If the plate is rotated through an angle I as shown in Fig. 4.13, it can be seen that the "axis ray" is laterally displaced by an amount D, which is given by

$$D = t \cos I (\tan I - \tan I')$$

or

$$D = t \sin I \left[1 - \sqrt{\frac{1 - \sin^2 I}{N^2 - \sin^2 I}} \right]$$

For small angles, we can make the usual substitution of the angle for its sine or tangent to get

$$d = \frac{tI(N - 1)}{N}$$

This lateral displacement of a tilted plate is made use of in high speed cameras (where the rotating plate displaces the image an amount approximately equal to the travel of the continuously moving film) and in optical micrometers. The optical micrometer is usually placed in front of a telescope and used to displace the line of sight. The amount of displacement is read off a calibrated drum connected to the mechanism which tilts the plate.

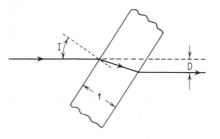

FIG. 4.13. The lateral displacement of a ray by a tilted plane parallel plate.

When used in parallel light, a plane parallel plate is free of aberrations (since the rays enter and leave at the same angles). However, if the plate is inserted in a convergent or divergent beam, it does introduce aberrations. The longitudinal image displacement $(N - 1)t/N$ is greater for short wavelength light (higher index) than for long, so that overcorrected chromatic aberration is introduced. The amount of displacement is also greater for rays making large angles with the axis; this is, of course, overcorrected spherical aberration. When the plate is tilted, the image formed by the meridional rays is shifted backward while the image formed by the sagittal rays (in a plane perpendicular to the page in the figures) is not, so that astigmatism is introduced.

The amount of aberration introduced by a plane parallel plate can be computed by the formulae below. Reference to Fig. 4.14 will indicate the meanings of the symbols

U and u — slope angle of the ray to the axis

U_p and u_p — the tilt of the plate

t — thickness of the plate

N — index of the plate

V — Abbe V number $(N_D - 1)/(N_F - N_C)$

Chromatic Aberration $= l'_F - l'_C = \dfrac{t(N - 1)}{N^2 V}$

$$\text{Spherical Aberration} = L' - l' = \frac{t}{N}\left[1 - \frac{N\cos U}{\sqrt{N^2 - \sin^2 U}}\right] \text{ (exact)}$$

$$= \frac{tu^2(N^2 - 1)}{2N^3} \text{ (third order)}$$

$$\text{Astigmatism} = l'_s - l'_t = \frac{t}{\sqrt{N^2 - \sin^2 U_p}}\left[\frac{N^2\cos^2 U_p}{(N^2 - \sin^2 U_p)} - 1\right] \text{(exact)}$$

$$= \frac{tu_p^2(N^2 - 1)}{N^3} \text{ (third order)}$$

$$\text{Sagittal Coma} = \frac{tu^2 u_p(N^2 - 1)}{2N^3} \text{ (third order)}$$

$$\text{Lateral chromatic} = \frac{tu_p(N - 1)}{NV} \text{ (third order)}$$

These expressions are extremely useful in estimating the effect that the introduction of a plate or a prism system will have on the state of correction of an optical system.

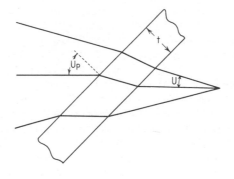

FIG. 4.14.

4.9 The Right Angle Prism

The right angle prism, with angles of 45°-90°-45°, is the building block of most non-dispersing prism systems. Figure 4.15 shows a parallel bundle of rays passing through such a prism, entering through one face, reflecting from the hypotenuse face and leaving through the second face. If the rays are normally incident

on the face of the prism, they are deviated through an angle of 90°. At the
hypotenuse face, the rays have an angle
of incidence of 45° so that they are subject to total internal reflection. If the
entrance and exit faces are low reflection coated, this makes the prism a
highly efficient reflector for visual
usage since the only losses are the absorption of the material and the reflection losses at the faces which total
only a few per cent. (In the ultraviolet
and infrared portions of the specrum,
the absorption of a prism may be quite

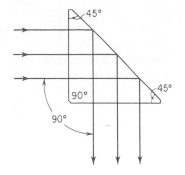

FIG. 4.15. Right angle prism.

objectionable.) It can be seen that the total internal reflection is
limited to rays which have angles of incidence greater than the
critical angle, and many prism systems are made of high index
glass to permit total reflection
over larger angles.

By unfolding the prism, as
indicated by the dashed lines in
Fig. 4.16, it is apparent that the
prism is the equivalent of a glass
block with parallel faces, with a
thickness equal to the length of
the entrance or exit faces. The
equivalent air thickness of the
block is, of course, this thickness
divided by the index of the prism.

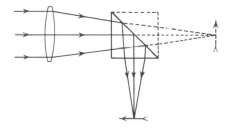

FIG. 4.16.

If the 45°–90°–45° prism is used with the light beam incident on
the hypotenuse face as shown in Fig. 4.17, the light is totally reflected twice and the rays emerge in the opposite direction, having
been deviated through 180°. Figure 4.17 also indicates the unfolded
prism path and the image orientation of this prism. Notice that the
image has been inverted, top to bottom, but not left to right.

This prism is a constant deviation prism. Regardless of the
angle at which a ray enters the prism, the emergent ray will be parallel, as shown in Fig. 4.18a. This characteristic is a property of the

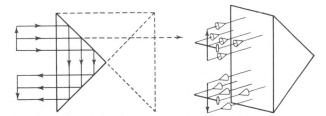

FIG. 4.17. Right angle prism used with hypotenuse as entrance and exit face.

two reflecting surfaces of the prism. A system which directs the light ray back on itself is called a retro-director; this prism is a retrodirector in one meridian only. (Another of the many constant deviation systems possible with two reflectors is the 90° deviation arrangement shown in Fig. 4.18b, where the reflecting surfaces are at 45° to each other.) A prism made by cutting off one corner of a cube, so that there are three mutually perpendicular reflecting surfaces, is retrodirective in both meridians. The corner cube reflector will return all the light rays striking it back toward their source.

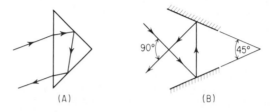

(A)   (B)

FIG. 4.18. a) The right angle prism used in the manner shown is a constant deviation prism, in that each ray is reflected through exactly 180°. The entering and emergent paths are parallel, regardless of the initial angle the ray makes with the prism. b) A pair of constant deviation mirrors. In this case, the deviation produced by the two reflections is always exactly 90°.

A third orientation of the 45°-90°-45° prism is shown in Fig. 4.19, in which the bundle of rays arrives parallel to the hypotenuse face of the prism. After being refracted downward at the entrance face, the rays are reflected upward from the hypotenuse and emerge after a second refraction at the exit face. The unfolded path of the rays (shown in dashed lines) indicates that this prism is the equivalent of a plane parallel plate which is tilted with respect to the axis of the bundle, whereas in the preceding examples the prism faces have been normal to the axis. If this prism is used in a convergent light beam, it will introduce a substantial amount of

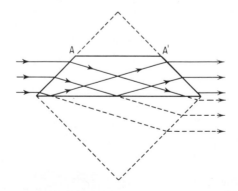

FIG. 4.19. The Dove prism. The dashed lines show that the dove prism is equivalent to a tilted plate and will introduce astigmatism when used in convergent or divergent beams.

astigmatism (roughly equal to one quarter of its thickness). For this reason, this prism, which is known as a Dove Prism, is used almost exclusively in parallel light. Since the apex of the prism is not used by the light beam, the prism is usually truncated at AA'.

The Dove prism has a very interesting effect on the orientation of the image. In Fig. 4.20a, the arrow and cross bar pattern is shown to be inverted from top to bottom but not left to right. If the prism is rotated 45°, as in Fig. 4.20b, the image is rotated through 90°; if the prism is rotated 90° as in Fig. 4.20c, the pattern is rotated 180°. Thus, the image is rotated twice as fast as the prism. (The analysis of the image orientation in Fig. 4.20b is an example of the use of an auxiliary pattern as described in Section 4.7. The auxiliary pattern is shown in dotted lines in Fig. 4.20b).)

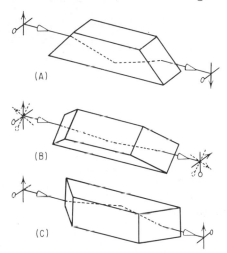

FIG. 4.20. The orientation of an image by a Dove prism. a) original position, b) prism rotated 45°; image is rotated 90° c) prism rotated 90°; image is rotated 180°. Note that the dotted arrow and crossbar in b) is oriented so that the arrow is in the plane of incidence to simplify the analysis of the image orientation.

The length of the Dove prism is 4 to 5 times the diameter of the bundle of rays which it will transmit. If two Dove prisms are cemented hypotenuse to hypotenuse (after silvering or aluminizing these faces), the aperture is thereby doubled with no increase in length. The Double Dove prism is used in parallel light as is the Dove. It must be precisely fabricated to avoid producing two slightly separated images. When the Double Dove is rotated, or tipped, about its center it can be used as a scanner to change the direction of sight of a telescope or periscope.

4.10 The Roof Prism

If the hypotenuse face of a right angle prism is replaced by a "roof", that is, two surfaces at 90° whose intersection lies in the hypotenuse, the prism is called a roof or Amici prism. Face and side views of a roof prism are shown in Fig. 4.21. The addition of the roof to the prism serves to introduce an extra inversion to the image, as can be seen by comparing the final orientation of the cross bar in Fig. 4.11 with that in Fig. 4.22a. This can be understood by tracing the path of the dashed ray in Fig. 4.22a which connects the circles in the arrow and cross bar figures before and after passing through the prism.

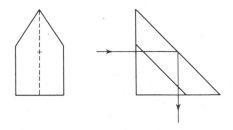

FIG. 4.21. Roof prism, or Amici prism.

In practice, the Amici prism is usually fabricated with the corners cut off, as shown in Fig. 4.22b, in order to reduce the size and weight of the prism. The 90° roof angle must be made to a high order of accuracy. If there is an error in the roof angle, the beam is split into two beams which diverge at an angle which is six times the error. Thus, to avoid any apparent doubling of the image, the roof angle is usually made accurate to one or two seconds of arc.

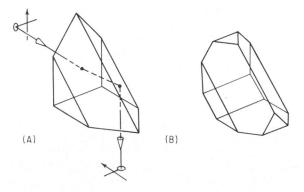

(A)   (B)

FIG. 4.22. Amici prism a) showing a single ray path through the prism and indicating the image orientation, b) with truncated corners to reduce weight without sacrifice of useful aperture.

4.11 Erecting Prism Systems

In an ordinary telescope, the objective lens forms an inverted image of the object, which is then viewed through the eyepiece. The image seen by the eye is upside down and reversed from left to right, as indicated in Fig. 4.23. To eliminate the inconvenience of viewing an inverted image, an erecting system is often provided to reinvert the image to its proper orientation. This may be a lens system or a prism system.

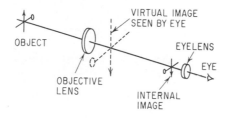

FIG. 4.23. In a simple telescope, the objective lens forms a real, inverted internal image of the object, which is re-imaged by the eyelens. The image seen by the eye is a virtual inverted image of the object.

4.11.1

The most commonly used prism erecting system is the Porro prism of the first type, illustrated in Fig. 4.24. The Porro system consists of two right angle prisms oriented at 90° to each other. The first prism inverts the image from top to bottom and the second prism reverses it from left to right. The optical axis is displaced laterally, but is not deviated. One can see that if this system is inserted into the telescope of Fig. 4.23, the final image will have the same orientation as the object. Although the prism system is ordinarily inserted between the objective and eyepiece (to minimize its size), it will erect the image reagrdloss of where it is placed in the system.

The Porro prism (first type) owes its popularity to the fact that the 45°-90°-45° prisms are relatively easy and inexpensive to manufacture, with no critical tolerances. However, if the prisms are not mounted so that their roof edges are exactly at 90° to each other, the final image will be rotated through twice the angular mounting error. This is of special importance in binocular systems where the image presented to one eye must be identical to that presented to the other.

A shallow ground slot is often cut across the center of the hypotenuse face of each prism to prevent unwanted reflections from this face.

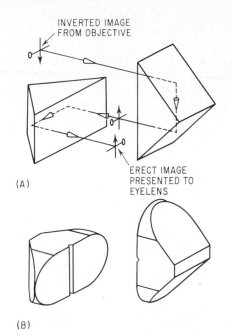

INVERTED IMAGE
FROM OBJECTIVE

ERECT IMAGE
PRESENTED TO
EYELENS

(A)

(B)

FIG. 4.24. Porro prism system (first type) a) indicating the way the Porro system erects an inverted image. b) Porro prisms are usually fabricated with rounded ends to save space and weight. Note that the spacing between the prisms has been shown increased for clarity.

4.11.2

The Porro prism of the second type is shown in Fig. 4.25, and serves the same purpose as the Porro #1 system. Both Porro systems function by total internal reflection so that no silvering is required. It is common to round off the ends of the prisms to conserve space and weight.

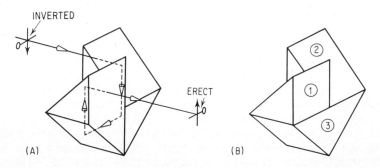

INVERTED

ERECT

(A)   (B)

FIG. 4.25. Porro prism system (second type) a) indicating the erection of an inverted image. This system is shown made from two prisms in a) and from three prisms in b).

The second Porro is somewhat more difficult to fabricate than the first type, but in some applications its compactness, and the fact that the two prisms can be readily cemented together, offer compensating advantages. The Porro #2 may also be made in three pieces, by cementing two small right angle prisms on the hypotenuse of a large right angle prism as indicated in Fig. 4.25b.

4.11.3

The Abbe (or Konig, or Brashear-Hastings) prism (Fig. 4.26) is an erecting prism which can be used when it is desired to erect the image without displacing the axis as the Porro prisms do. The roof is necessary to provide the left to right reversal of the image; the roof angle must be made accurately to avoid image doubling.

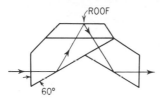

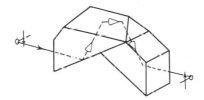

FIG. 4.26. Abbe prism. Used as an in-line erecting system, it does not displace the axis as the Porro systems do, nor does it materially displace the image longitudinally.

If this prism is made without the roof, it will invert the image in one meridian only, just as the Dove prism. However, since its entrance and exit faces are normal to the system axis, it may be used in a converging beam without introducing astigmatism.

4.11.4 Other Erecting Prisms

Among the many prisms designed to erect an image are those sketched in Fig. 4.27. The fact that the image is inverted and reversed left to right after passing through these prisms may be verified by the methods outlined in Section 4.7. Notice that each prism has been arranged so that the axial ray enters and leaves the prism normal to the prism faces and that all reflections are total internal reflections. In the Leman and Goerz prisms, the axis is displaced but not deviated. In the Schmidt and Modified Amici prisms, the axis is deviated through a definite angle, which can be selected by the designer (within the limits allowed by total internal reflection).

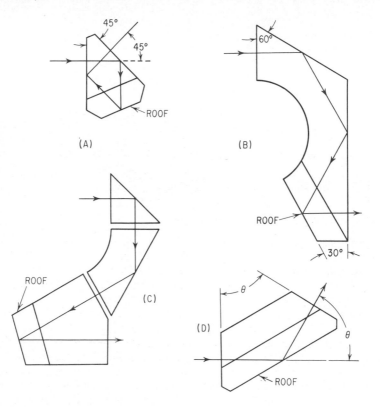

FIG. 4.27. Erecting prisms. a) Schmidt Prism b) Leman (or Springer) Prism c) Goerz Prism d) Modified Amici prism.

4.12 Inversion Prisms

The Dove prism (Figs. 4.19 and 4.20) and the roofless Abbe prism mentioned in Section 4.11.3 are examples of prisms which invert the image in one meridian but not the other. The plane mirror and the right angle prism (Figs. 4.11 and 4.16) are also simple inversion systems. Figure 4.28 shows the above prisms plus the Pechan prism, which is a relatively compact prism for this purpose. Notice that the addition of a ''roof'' to any of these prisms will convert it to an erecting system.

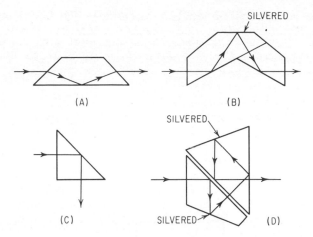

FIG. 4.28. Inversion Prisms. a) Dove Prism b) Reversion Prism c) Right Angle Prism d) Pechan Prism.

4.13 The Penta Prism

The Penta Prism, Fig. 4.29a, will neither invert nor reverse the image. Its function is to deviate the line of sight by 90°. It has the valuable property of being a constant deviation prism, in that it deviates the line of sight through the same angle regardless of its orientation to the line of sight.

Most of the prism systems described in this chapter could be replaced by a series of plane mirrors and this is sometimes done for reasons of weight and/or economy. However, a prism, as a monolithic glass block, is a very stable system and is not as subject to environmental variation of angles as is an assemblage of mirrors on a metal support block.

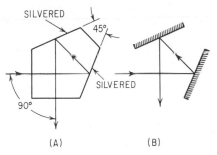

FIG. 4.29. The Penta Prism (a) and its equivalent mirror system (b).

The Penta Prism is used where it is desirable to produce an exact 90° deviation, without having to orient the prism precisely.

The end reflectors of small rangefinders are often of this type, and in optical tooling and precise alignment work, the Penta prism is useful to establish an exact 90° angle. In large range finders, however, the prism is replaced by two mirrors (Fig. 4.29b), securely cemented to a block in order to avoid the weight, absorption and cost of a large block of solid glass.

Occasionally a roof is substituted for one of the reflecting faces of the Penta prism to invert the image in one meridian.

4.14 Rhomboids and Beam Splitters

The Rhomboid Prism is a simple means of displacing the line of sight without affecting the orientation of the image or deviating the line of sight. The Rhomboid prism and its mirror system equivalent are shown in Fig. 4.30.

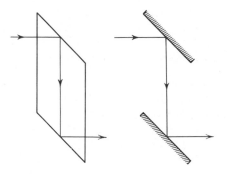

FIG. 4.30. a) Rhomboid Prism b) An equivalent mirror system. Both systems displace the optical axis without deviation or reorientation of the image.

A beamsplitter is frequently useful for the purpose of combining two beams (or images) into one, or for separating one beam into two. A thin plate of glass with one surface coated with a semi-reflecting coating, as shown in Fig. 4.31a, can be used for this purpose, but it suffers from two drawbacks. First, if the angle of convergence of the beam were large, it would introduce astigmatism, and second, the reflection from the second surface, although faint, would produce a ghost image displaced from the primary image. (Note that in parallel light neither of these objections is valid, provided the surfaces of the plate are accurately parallel). The beamsplitter cube (Fig. 4.31b) avoids these difficulties. It is composed of two right angle prisms cemented together. The hypotenuse of one prism is coated with a semi-reflecting coating before cementing. Where the weight or absorption of the cube cannot be tolerated, a pellicle is often used as a semi-reflector. A pellicle is a thin membrane (usually a plastic) stretched over a frame; by virtue of its extreme thinness, the astigmatism and ghost displacement are reduced to acceptable values.

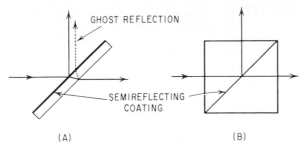

(A)   (B)

FIG. 4.31. Beam Splitters. a) A thin parallel plate is convenient but may be objectionable because of ghosting and astigmatism, unless used in parallel light. b) Beam splitting cube has a semi-reflecting coating applied to one of the diagonal faces before cementing.

In Fig. 4.32, two binocular eyepiece prism systems are sketched. Both serve the same function, namely splitting the light beam from an objective lens into two parts. The two beams are displaced sufficiently so that they can be presented to two eyepieces and both eyes may simultaneously view the same subject. Notice that in both systems, extra glass has been added to the left hand path so that the amount of glass in each path is identical; in this way the aberrations introduced by the glass are the same for each path. Most of the glass in these systems could be dispensed with if desired, since each of them is equivalent to a beam splitting cube plus three reflectors. In System 4.32b, the two halves can be rotated about the objective axis to vary the spacing between the eyepieces as shown in Fig. 4.32c. Notice that the image is not rotated by this procedure, but retains its original orientation.

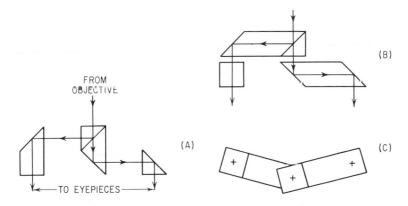

FIG. 4.32. Prism systems for binocular eyepiece instruments. System a) can be adjusted to match the user's eye separation by sliding both outer prisms in or out; this defocuses the instrument. Sketch c) shows how the halves of b) can be rotated about the objective axis to make this adjustment.

4.15 Plane Mirrors

In the preceding discussions we have indicated several times that reflecting prisms may be replaced by mirrors. For most applications, it is necessary that the mirrors be first surface mirrors, as opposed to ordinary second surface mirrors. The two types are sketched in Fig. 4.33. The first surface mirror is usually preferable because it does not produce a ghost image as does the second surface mirror. In addition, the second surface mirror requires the processing of an extra surface in its fabrication. It also requires the light to pass through a thickness of glass which may introduce aberrations and which will absorb energy in ultraviolet and infrared applications. The second surface mirror can be made more durable, however, since its reflecting coating can be protected from the elements by electro-deposited and painted coverings. First surface mirrors are usually made with vacuum deposited aluminum films protected by a thin transparent overcoating of silicon monoxide or magnesium fluoride.

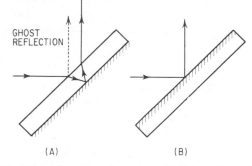

FIG. 4.33. a) Second surface mirror b) First surface mirror

4.16 The Design of Prism and Reflector Systems

Ordinarily it is required of a prism (or reflector) system that it produce an image with a certain orientation and with the emergent beam of light redirected in a given manner. The design effort is usually best begun by establishing the minimum number of reflectors which will produce the desired result. This is most simply (and perhaps best) accomplished by straightforward trial and error. A rough perspective sketch is made to indicate the reflections necessary to locate the image in its desired position. The orientation of the image is then checked by the technique of Section 4.7; reflectors are added in various orientations until the image orientation is correct. Usually several roughly equivalent schemes are possible, and a selection can be made based on the requirements of the application.

When the reflection system is completed, the optical system is unfolded, that is, sketched with the optical axis as a straight line. The object, image, and lens apertures are added to the sketch and the necesssary sizes for the reflectors are determined in both meridians. If the system is to be composed of prisms, the unfolded layout is repeated with the axial distances adjusted to the "equivalent air thickness" (t/N) for that portion of the system which is glass.

As an example of reflector system design, let us consider the problem presented by Fig. 4.34. The object at A is to be projected by an ordinary lens B onto a screen at S. The plane of S is parallel to the original projection axis and its center is above the axis by some amount Y. The required orientations of object and image are shown in the sketch.

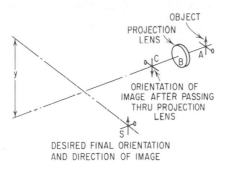

FIG. 4.34.

We begin by noting that the image formed by the projection lens will be inverted in both meridians with respect to the object, as shown at C in Fig. 4.34. Now, passing to Fig. 4.35, let us consider the effect of a mirror placed at D. Of the four directions shown as possible reflections at D, the upward reflection labeled D_1 seems the most promising since it sends the light in a direction that it must eventually take, so we shall pursue this line. Using similar reasoning at E, we should be inclined to select E_2; however, the image at E_2 is rotated 90° from our desired orientation. (At this point, we note that a Pechan prism with reflecting face at 45° to our meridians could produce the necessary 90° rotation). Selecting E_1 on the basis that its image orientation is closest to the desideratum, we consider a reflection at F. Again F_3 is in the proper direction, but the image is reversed from left to right. Case F_1 has the proper orientation, but the light is traveling away from the screen. If we add a mirror to reverse the direction of propagation, we will have both orientation and direction as required. To accomplish this without directing the light back through F, we must resort to a figure 4 arrangement as shown in Fig. 4.36, which diagrams the entire system.

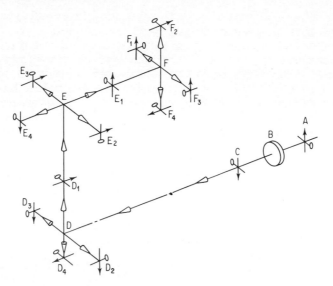

FIG. 4.35.

It is quite apparent that Fig. 4.36 represents only one of the many possible arrangements of mirrors which could be utilized to accomplish this same end result. The reader may also have noticed that the discussion has been limited to reflections for which the plane of incidence lay in one of the Cartesian reference planes, and also that first consideration was given to reflections which deviated the axis by 90°. For the novice, these restrictions have much to recommend them; one is well advised to keep first trials of this type as simple and uncomplicated as possible. Further, the reduction

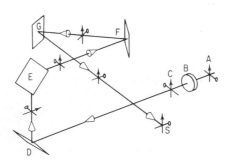

FIG. 4.36.

of the system to practice is much simplified if compound angles are avoided. If our problem had required that the final image be rotated 45°, then we would necessarily have had to depart from the Cartesian planes to achieve the desired result.

The Porro erecting prism (Fig. 4.37a) will serve as an illustrative example of the "unfolding" technique used in the design of prism systems. The prisms have been unfolded in Fig. 4.37b (for clarity, the second prism is shown rotated 90° about the axis). Each prism can be seen to be the equivalent of a glass block whose thickness is twice the size of its end face. Notice that the rays from the lens are refracted at each air-glass surface of the system and that the image has been displaced to the right by the prisms.

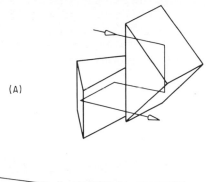

(A)

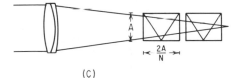

(B)

(C)

FIG. 4.37. a) Porro Prism System (first type) b) Unfolded prisms. Dashed lines indicate path rays would take without prisms. Solid line shows the displacement of the focal point by the prisms. c) The prisms are drawn to their equivalent air thickness so that the rays can be drawn as straight lines.

In Fig. 4.37c, the prisms are drawn with their "equivalent air thickness" as discussed in Section 4.8. This allows us to draw the light rays through the prism as straight lines, simplifying the construction considerably.

Now let us suppose that we are to design the minimum size Porro system for a binocular. The objective lens has a focal length of 7 in., an aperture of 2 in. and is to cover a 5/8 in.-diameter field, as sketched in Fig. 4.38a. We first note that the proportions of face width to "equivalent air thickness" for each prism (Fig. 4.36a) are $A:2A/N = 1:2/N$, or, if we assume an index of 1.50, 3:4. We begin the design from the image and work toward the objective. Placing the exit face of the prism one-half inch from the image (to allow for clearance and to keep the glass surface well out of the focal plane), we construct the dashed line shown in

Fig. 4.38a with a slope of 3:8 (one-half the face-to-equivalent-thickness ratio) starting from the axial intercept of the exit face. This line is of course the locus of the corners of a family of prisms of various sizes, and the point where it intersects the extreme clearance ray defines the minimum size prism which will transmit the entire cone of light from the objective. For practical purposes, the prism should be made slightly larger than this to allow for bevels and mounting shoulders.

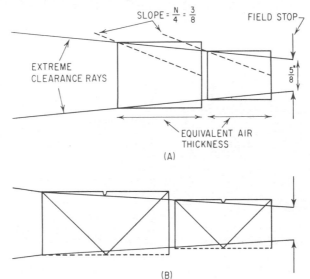

FIG. 4.38. The layout of a minimum size prism system is shown in a) The extreme clearance rays connect the rim of the objective with the edge of the field of view. The intersection of the dashed lines (see text) with these rays locates the corner of the smallest prism which will pass the full image cone. In b) the prisms are drawn to scale, showing their true thickness.

The procedure is now repeated for the other prism; an air space is left between the two to allow for the mounting plate to which both prisms are to be fastened. In Fig. 4.38b, the system is drawn to scale, with the prism blocks expanded to their true length. The reason for the ground slot usually cut into the hypotenuse faces of Porro prisms can be understood from an examination of the unfolded drawings. Light rays from outside the desired field of view can be reflected (by total internal reflection) from these faces back into the field where are quite annoying; the slot intercepts these rays as they graze along the hypotenuse.

4.17 Analysis of Fabrication Errors

The effects produced by errors in prism angles (due to manufacturing tolerances) are readily analyzed. Such angular errors can be treated as equivalent to the rotation of a reflecting surface from its nominal position, and/or the addition of a thin dispersing prism to the system.

As an example, consider the right angle prism shown in Fig. 4.39 and assume that the upper 45° angle is too large by ϵ and that the lower 45° angle is too small by ϵ.

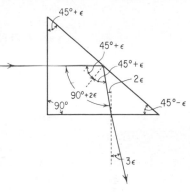

A ray normal to the entrance face will make an angle of incidence of 45°+ϵ at the hypotenuse; the angle of reflection will then be 45° + ϵ and the ray will be reflected through an angle of 90° + 2ϵ. Thus rotating the reflecting face through ϵ has introduced an error of 2ϵ in direction of the ray.

At the exit face, the ray has an angle of incidence of 2ϵ and, if the prism index is 1.5, an angle of refraction of 3ϵ. Thus the total deviation of the ray from its nominal direction is 3ϵ. Also, since the ray has been deviated by refraction at this surface through an angle ϵ, the ray will be dispersed and spread out into a spectrum subtending an angle of ϵ/V according to Eq. 4.11.

FIG. 4.39. The passage of a ray through a right angle prism whose hypotenuse face is tilted from its proper position by a small angle ϵ. After reflection, the ray is deviated by 2ϵ; this is increased to 3ϵ (or 2$N\epsilon$) by refraction at the exit face.

5

The Eye and Color

5.1

A knowledge of the characteristics of the human eye is important to the practice of optical engineering because the majority of optical systems utilize the eye as the final element of the system in one way or another. Thus, it is vital that the designer of an optical system understand what the eye can and cannot accomplish. For example, if a visual optical system is required to recognize a certain size target, or to measure to a certain degree of accuracy, the magnification of the image presented to the eye must be sufficient to allow the eye to detect the necessary details. On the other hand, it would be wasteful to design a system with a perfection of image rendition which the eye could not utilize.

The human eye is a living optical system and its characteristics vary widely from individual to individual. For a given individual, the characteristics may vary from day to day, indeed from hour to hour. Therefore, the data presented in this chapter must be considered as central values in a range of values; in fact, some data are useful only as an indication of the order of magnitude of a certain characteristic. The conditions under which the eye is used play a large role in determining the behavior of the eye and must *always* be taken into account.

In physiological optics, the unit of measure for the power of a lens or optical system is the Diopter, the abbreviation for which is D. The diopter power of a lens is simply the reciprocal of its effective focal length, when the focal length is expressed in meters. For example, a lens with a one meter focal length has a power of one diopter; a one-half meter focal length, two diopters; and a lens of one inch focal length has a power of 40 diopters (or more exactly, $39.37\ D$). For a single surface, the dioptric power is given by $(N' - N)/R$, with R the radius in meters. A one diopter prism produces a deviation of one centimeter in a one meter distance, that is, a deviation of 0.01 radians or about 0.57 degrees.

5.2 The Structure of the Eye

The eyeball is a tough plastic-like shell filled largely with a jelly-like substance under sufficient pressure to maintain its shape. It rides in a bony socket of the skull on pads of flesh and fat. It is held in place and rotated by six muscles.

Figure 5.1 is a horizontal section of the right eye; the nose is to the left of the figure. The outer shell (sclera) is white and opaque except for the cornea, which is clear. The cornea supplies most of the refractive power of the eye. Behind the cornea is the aqueous humor, which (as its name implies) is a watery fluid. The iris, which gives the eye its color, is capable of expanding or contracting to control the amount of light admitted to the eye. The pupil formed by the iris can range in diameter from 8 mm in very dim light to less than 2 mm under very bright conditions. The lens of the eye is a flexible capsule suspended by a multitude of fibers, or ligaments, around its periphery. The eye is focused by changing the shape of the lens. When the muscles to which the suspensory ligaments are connected are relaxed, the lens has its flattest shape and the normal eye is focused at infinity. When these muscles contract, the lens bulges, so that its radii are stronger and the eye is focused for nearby objects. This process is called accomodation.

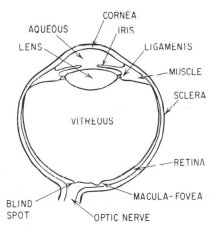

FIG. 5.1. Schematic Horizontal Section of Right Eyeball (from above).

Behind the lens is the vitreous humor, a material with the consistency of jelly. All of the optical elements of the eye are largely water; in fact, a reasonable simulation of the optics of the eye can be made by considering the eye as a single refracting surface of water ($N_D = 1.333$, $V = 55$).

The following table lists typical values for the radii, thicknesses and indices of the optical surfaces of the eye. These, of course, vary from individual to individual.

R_1 (air to cornea) + 7.8 mm	t_1 (cornea) 0.6	N_1 1.376
R_2 (cornea to aqueous) + 6.4 mm	t_2 (aqueous) 3.0	N_2 1.336
R_3 (aqueous to lens) + 10.1 mm	t_3 (lens) 4.0	N_3 1.386 - 1.406
R_4 (lens to vitreous) − 6.1 mm	t_4 (vitreous) 16.9	N_4 1.337

The principal points are located 1.5 and 1.8 mm behind the cornea and the nodal points are 7.1 and 7.4 mm behind the cornea. The first focal point is 15.6 mm outside the eye, the second is of course at the retina. The distance from the second nodal point to the retina is 17.1 mm; thus the retinal size of an image can be found by multiplying the angular subtense of the object in radians (from the first nodal point) by this distance. When the eye accommodates (focuses), the lens becomes nearly equiconvex with radii of about 5.3 mm, and the nodal points move a few millimeters toward the retina. The center of rotation of the eyeball is 13 to 16 mm behind the cornea.

The retina contains blood vessels, nerve fibers, the light sensitive rod and cone cells and a pigment layer, in that order in the direction that the light travels. The optic nerve and the associated blind spot are located where the nerve fibers leave the eyeball and proceed to the brain. Slightly to the temporal (outer) side of the optical axis of the eye is the macula; the center of the macula is the fovea. At the fovea, the structure of the retina thins out and, in the central 0.3 mm diameter, only cones are present. The fovea is the center of sharp vision. Outside this area rods begin to appear; further away only rods are present.

There are about six million cones in the retina, about 120 million rods and only about one million nerve fibers. The cones of the fovea are 1 to 1.5 microns in diameter and are about 2 to 2.5 microns apart. The rods are about 2 microns in diameter. In the outer portions of the retina, the sensitive cells are more widely spaced and are multiply connected to nerve fibers (several hundred to a fiber), accounting for the less distinct vision in this area of the retina.

The field of vision of an eye approximates an ellipse about 130° high by about 200° wide. The binocular field of vision, seen by both eyes simultaneously, is approximately circular and about 130° in diameter.

5.3 Characteristics of the Eye

Visual Acuity

The characteristic of the eye which is probably of greatest interest to the optical engineer is its ability to recognize small, fine details. Visual Acuity (V.A.) is defined and measured in terms of the angular size of the smallest character that can be recognized. The characters most frequently used to test V.A. are upper case letters or a heavy ring with a break in the outline. Many upper case letters can be considered as made up of five elements; for example, the letter E has three bars and two spaces. Visual Acuity is the reciprocal of the angular size (in minutes of arc) of one of the elements of the letter. Normal V.A. is considered to be 1.0, that is, when the smallest recognizable letter subtends an angular height of five minutes from the eye and each element of the letter subtends one minute. Acuity is frequently expressed as the ratio between the distance to the target (usually 20 feet) and the distance at which the target element would subtend one minute. Thus, a V.A. of one-half, or 20/40, indicates that the minimum recognizable letter subtends 10 minutes and its elements two minutes. In the Landolt broken ring test, the width of the ring and the width of the break correspond to the letter element size and recognition consists of determining the orientation of the break. Visual Acuity occasionally reaches 2 or 3 in unusual individuals under ideal conditions.

As indicated above, the normal Visual Acuity is one minute and this is the value for the resolution of the eye which is conventionally assumed in connection with the design of optical instruments. However, this is the value of V.A. under what might be termed "normal conditions," and it is the value only for that part of the field of view which corresponds to the fovea of the retina. Outside the fovea, the acuity drops rapidly as indicated in Fig. 5.2, which is a plot of Visual Acuity (relative to that at the fovea, which is arbitrarily set at unity) vs the angular position of the test target in the field of view.

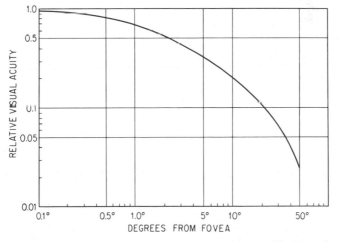

FIG. 5.2. Variation of visual acuity with retinal position of image. Visual acuity is given relative to that at Fovea.

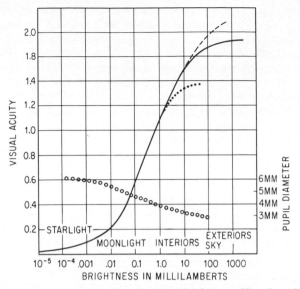

FIG. 5.3. Visual acuity as a function of object brightness. Visual acuity in reciprocal minutes. The dashed and dotted lines show the effect of increased and decreased (respectively) surround brightness (one millilambert is approx. the brightness of a perfect diffuser illuminated by one foot candle). The open circle curve indicates the diameter of the pupil; pupil diameters are larger in the young and smaller in the old, especially at lower brightnesses.

As the brightness of a scene is diminished, the iris opens wider and the rods take over from the cones. At low illuminations, the eye is color blind and the fovea becomes a blind spot, since the cones lack the necessary sensitivity to respond to low levels of illumination. One result of this process is that the Visual Acuity drops as the illumination drops. This relationship is plotted in Fig. 5.3, which also indicates the normal pupil size. Note that the brightness of the area surrounding the test target affects the acuity. A uniform illumination seems to maximize the acuity. Fig. 5.4 shows that, as might be expected, reducing the contrast of the target will also reduce the acuity.

Because the eye has about 1 to 1-1/2 diopters of chromatic aberration, V. A. is affected by the wavelength of light illuminating the target. Normally, V. A. is given for white light. In monochromatic light, the acuity is very slightly higher for the yellow and yellow-green wavelengths and slightly

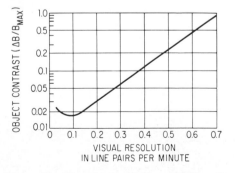

FIG. 5.4. The object contrast $(\Delta B/B_{max})$ necessary for the eye to resolve a pattern of alternating bright and dark bars of equal width.

lower for red wavelengths. In blue (or far red) light, V. A. may be 10-20% lower, and in violet light the reduction in V. A. is 20-30%. The chromatic of the eye can be corrected or doubled (by external lenses) without detection; a quadrupling is noticeable.

Other Types of Acuity

Vernier Acuity is the ability of the eye to align two objects, such as two straight lines, a line and a cross hair, a line between two parallel lines, etc. In making settings of this type, the eye is extremely capable. In instrument design, it can be safely assumed that the average person can repeat vernier settings to better than five seconds of arc and that he will be accurate to about ten seconds of arc. Exceptional individuals may do as well as one or two seconds. Thus, the vernier acuity is five or ten times the visual acuity. Vernier acuity is best when setting one line between two, next best setting a line on cross hairs or aligning two butting lines, and less effective in superimposing two lines.

The narrowest black line on a bright field that the eye can detect subtends an angle of from 1/2 to one second of arc. In conditions of reversed contrast, i.e. a bright line or bright spot, the size of the line is not as important as its brightness. The governing factor is the amount of energy which reaches and triggers the retinal cell into responding. The minimum level seems to be 50 to 100 quanta incident on the cornea (only a few percent of the energy incident on the cornea actually reaches the cell).

The eye is capable of detecting motion to the order of 10 seconds of arc. The slowest motion that the eye will detect is one or two minutes of arc per second of time. At the other extreme, a point moving faster than 200° per second will blur into a streak.

The eyes judge distance from a number of clues. Accommodation, convergence (the turning in of the eyes to view a near object), haze, perspective, experience, etc., each play a part. Three dimensional or stereo vision results from the separation of the two eyes which causes each eye to see a slightly different picture of an object. The amount of stereo parallax which can be detected is as small as 2 to 4 seconds. In a clue-less surround two rods at a 20 foot distance can be adjusted so that they are equally distant from the observer to within about one inch.

Sensitivity

The lowest level of brightness which can be seen or detected is determined by the light level to which the eye has become accustomed. When the illumination level is reduced, the pupil of the eye expands, admitting more light, and the retina becomes more sensitive (by switching from cone vision to rod vision and also by an electrochemical mechanism involving rhodopsin, the visual purple pigment). This process is called dark adaptation. Figure 5.5 illustrates the adaptation process as a function of the length of time that the eye is in darkness. The "fovea only" curve indicates that after a few minutes, the level of brightness detectable by the portion of the retina used for distinct vision is as low as it will ever get; at lower levels of illumination, only the outer portions of the retina are useful. Figure 5.5 is for a target which subtends about 2°; the threshold brightness is lower for larger targets and higher for smaller targets. As indicated by the dashed lines, the conditions of the test have a great bearing on the threshold of vision, and the data of Fig. 5.5 should be regarded as indicating only an order of magnitude for the threshold.

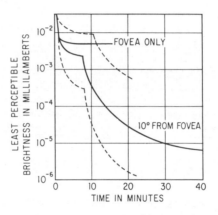

FIG. 5.5. The threshold of vision. The minimum brightness perceptible drops sharply with time as the eye adapts itself to darkness. The upper and lower dashed curves show the effect of high and low illumination levels (respectively) before adaptation begins. For areas subtending more than 5° the threshold is almost constant, but rises rapidly as target size is reduced. Curves shown above are for a target subtending about 2°.

The eye is a poor photometer; it is very inaccurate at judging the absolute level of brightness. However, it is an excellent instrument for comparison purposes, and can be used to match the brightness or color of two adjacent areas with a high degree of precision. Figure 5.6 indicates the brightness difference that the eye can detect as a function of the absolute brightness of the test areas. At ordinary brightness levels, a brightness difference of about 2% is detectable. (Note that in comparison photometry, in which the eye is called upon to match two areas, the precision of setting is increased by making a series of readings. In half the readings, the brightness of the variable area is raised until an apparent match is obtained; in the other half of the readings, the brightness is lowered to obtain the apparent match. The average is then much more accurate than either set.) Contrast sensitivity is best when there is no visible dividing line between the two areas under comparison. When the areas are separated, or if the demarcation between areas is not distinct, contrast sensitivity drops markedly.

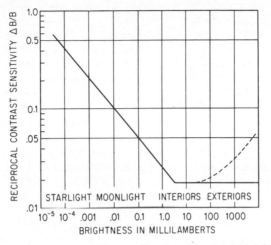

FIG. 5.6. The contrast sensitivity of the eye as a function of field brightness. The smallest perceptible difference in brightness between two adjacent fields (ΔB) as a fraction of the field brightness (B) remains quite constant for brightnesses above 1 milliambert if the field is large. The dashed line indicates the contrast sensitivity for a dark surrounded field. (One milliambert is approximately the brightness of a perfect diffuser illuminated by one foot candle).

Figure 5.7 indicates the capability of the normal eye as a comparison colorimeter. Again the eye is poor at determining the absolute wavelength of a color, but quite good at determining a color match; wavelength differences of a few millimicrons are detectable under suitable conditions. The comments of the preceding paragraph regarding dividing lines between test areas apply to color sensitivity as well.

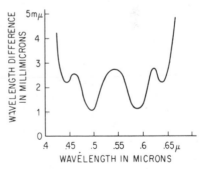

The sensitivity of the eye to light is a function of the wavelength of the light. Under normal conditions of illumination, the eye is most sensitive to yellow-green light at a wavelength of 0.55 microns, and its sensitivity drops off on either side of this peak. For most purposes the sensitivity of the eye may be considered to extend from 0.4 microns to 0.7 microns. Thus, in designing an optical

FIG. 5.7. Sensitivity of the eye to color differences. The amount by which two colors must differ for the difference to be detectable in a side by side comparison is plotted as a function of wavelength.

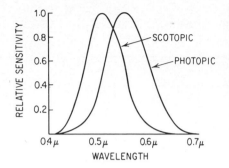

FIG. 5.8. The relative sensitivity of the eye to different wavelengths for normal levels of illumination (photopic vision) and under conditions of dark adaption (scotopic vision).

instrument for visual use, the monochromatic aberrations are corrected for a wavelength of 0.55 or 0.59 microns and chromatic aberration is corrected by bringing the red and blue wavelengths to a common focus. The wavelengths usually chosen are either $e(0.5461\mu)$ or $D(0.5893\mu)$ for the yellow, $C(0.6563\mu)$ for the red and $F(0.4861\mu)$ for the blue.

Figure 5.8 shows the sensitivity of the eye as a function of wavelength for normal levels of illumination and also for the dark adapted eye. Notice that the peak sensitivity for the dark adapted eye shifts toward the blue end of the spectrum, to a value near 0.51 microns. This "Purkinje Shift" is due to the differing chromatic sensitivities of the rods and cones of the retina, as shown in Fig. 5.9, where the two curves are plotted in a way to emphasize the greater sensitivity of the rods. Figure 5.10A is a standardized plot of ocular sensitivity which is used in colorimetry determinations. The long wavelength portion of this curve (Fig. 5.10B) is useful in estimating the visibility of near infrared search lights (as used on sniper-scopes, etc.) under conditions where security is desired.

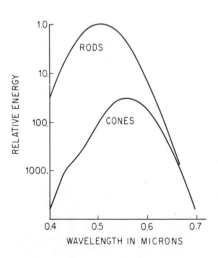

FIG. 5.9. The relative sensitivity of rods and cones as a function of wavelength.

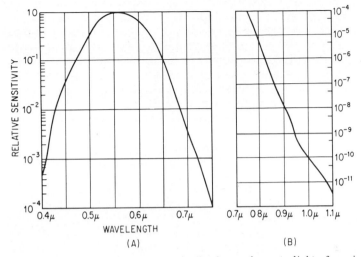

FIG. 5.10. a) Relative sensitivity of a standardized normal eye to light of varying wave lengths. b) Sensitivity in the near infrared.

5.4 Defects of the Eye

Nearsightedness (myopia) is a defect of focus resulting from too much power in the lens and cornea and/or too long an eyeball. The result is that the image of a distant object falls ahead of the retina and cannot be focused sharply. Since myopia results from an excessive amount of positive power, it is corrected by placing a negative lens before the eye. The power of the negative lens is chosen so that its image is formed at the most distant point on which the myopic eye can focus. A person with 2 diopters of myopia cannot see clearly beyond one-half meter (20 inches) and a minus two diopter lens (focal length equals minus one-half meter or minus 20 inches) is used to correct for this amount of myopia. The onset of myopia frequently coincides with adolescence, when growth is most rapid.

Farsightedness (hyperopia) is the reverse of myopia and results from too short an eye and/or too little power in the refracting elements of the eye. The image of a distant object is formed (when the eye is relaxed) behind the retina. Hyperopia is corrected by the use of a positive lens. Obviously the farsighted individual can, to the extent that his power of accommodation will allow, refocus his eye to bring the image onto the retina.

Astigmatism is a difference in the power of the eye from meridian to meridian and usually results from an imperfectly formed cornea, which has a stronger radius in one direction than in the other. Astigmatism of the eye is corrected by the use of toroidal surfaces on the spectacle lenses.

The chromatic aberration of the eye was discussed in Section 5.3; most eyes have some undercorrected spherical aberration as well. The lens of the eye has aspheric surfaces and a higher index of refraction in the central core of the lens than in the outer portions; both of these factors reduce the power of the system at the margin of the lens and tend to correct the heavy undercorrected spherical from the cornea. A few persons have overcorrected spherical. In most people, the spherical tends toward overcorrection with accommodation, since the lens bulges more at the center than at the edge when the eye focuses on a near point. As much as ±2 diopters of spherical have been measured; however, like chromatic aberration, spherical seems to have little effect on the resolution of the eye.

Presbyopia is the inability to accommodate (focus) and results from the hardening of the material of the lens which comes with age. Figure 5.11 indicates the (typical) relationship between age and the power of accommodation. When the eye can no longer accommodate to reading distance (two or three diopters), it is necessary to wear positive lenses to read comfortably.

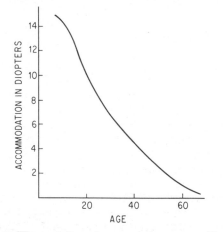

FIG. 5.11. The variation of accomodation power with age.

Keratoconus is a conical shaped cornea and can be corrected by contact lenses which effectively overlay a new spherical surface on the cornea. An opaque or cloudy lens (cataract) is frequently removed surgically to restore vision. The resultant loss of power is made up by an extremely strong positive spectacle lens, but such an aphakic eye, lacking a lens, cannot accommodate. Also, the change in retinal image size due to the shift in refractive power from inside to outside the eye (due to the strong spectacle lens) will preclude binocular vision if only one eye is lensless.

Aniseikonia is the name given to a disparity in retinal image size from one eye to the other, occurring in otherwise normal eyes, and results in lack of binocular vision if larger than a few percent. Aniseikonia can be corrected by special thick lenses which are effectively low power telescopes whose magnifications balance out the difference in magnification.

In instrument design, a number of additional factors should be taken into consideration, especially for binocular instruments. An adjustment must be provided for the variation in interpupillary distance, so that both sides of the instrument can be aligned with the pupils of the eyes. This distance ranges from two to three inches. Both halves of a binocular instrument must have the same magnification (within 1/2 to 2 percent, depending on the individual's tolerance) and both halves must have their axes parallel (to within 1/4 to 1/2 prism diopter). Each side must be independently focusable to allow for variations in focus between the two eyes. A focus adjustment of ±4 diopters will take care of the requirements of all but a few percent of the population; ±2 diopters will satisfy about 85%. The depth of field of the eye (the distance on either side of the point of best focus through which vision is distinct) is about ±1/4 diopter.

5.5 Color and Colorimetry

Color can be regarded as an intrinsic physical property of an object (or radiation) or as a visual sensation. As a sensation, it results from three different types of receptor cells in the retina, each of which responds to a different portion of the visual spectrum. As a physical property, color is determined by the wavelength distribution of the transmitted or reflected light. A spectrophotometer is an instrument which can determine the transmission or reflection of a material as a function of the wavelength of the light incident upon it. Briefly, a spectrophotometer consists of a device for producing monochromatic light (or nearly so) of known wavelengths and a detector cell which is calibrated to measure the relative intensity of the light falling on it. In use, the monochromator is set at some wavelength and the intensity of the light on the cell is measured. The sample is then inserted between the monochromator and the cell and the intensity is read again. The quotient of the two intensity readings is the transmission of the sample. The process is repeated throughout the wavelength region of interest and the results are plotted as shown in Fig. 5.12, which gives the transmission characteristics of several typical glass filters.

Spectrophotometric curves provide the information necessary to determine the intrinsic color of a material. Because the response of the eye varies with wavelength, as indicated in Figs. 5.8, 5.9 and 5.10, the apparent color (i.e., the sensation) will depend on the chromatic distribution of the light illuminating the sample. Thus, two color samples which "match" under incandescent light (which is yellowish) may not match when compared under the light from a blue sky. The only way that one can be certain of a color match under all conditions of illumination is if the spectrophotometric curves are identical throughout the visible spectrum.

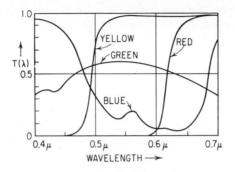

FIG. 5.12. Spectral transmission curves for several optical glass filters. Each curve is labeled with the apparent visual color. Note that the blue glass transmits in the far red region as well as in the blue; if it transmitted more red and less yellow, it would appear purple.

This leads to difficulties in color specification, since, even if a set of master standard color samples (whose color characteristics were permanent) were available, it would still be necessary to specify the type of illumination under which a comparison was to be made. While the spectrophotometric curve is a complete description of an object's color, it is difficult to interpret such curves readily, and they can be somewhat deceptive if one is called on to say just what the apparent color will be under a given set of conditions.

Color can be specified by three factors; dominant wavelength, purity, and brightness. Dominant wavelength is the wavelength of pure spectral light which has the same hue as the sample. Purity is the proportion of pure spectral light of the dominant wavelength which is mixed with "white" light of a specified nature to match the tint or saturation of the sample. Brightness is the integrated spectral response of the eye to the light from the sample and can be expressed as:

$$B = \int B(\lambda) \cdot V(\lambda) \, d\lambda$$

where $V(\lambda)$ is the relative visual response function of Fig. 5.10 and $B(\lambda)$ is the spectral brightness function of the sample in suitable units of energy per solid angle per unit of area per unit of wavelength. Alternatively, brightness can be expressed as a percentage transmission or reflection by substituting $T(\lambda)$ or $R(\lambda)$ in the above expression.

Given any three different ("Primary") colored sources of light, it is possible to mix them in proportions that will match any sample. (Depending on the three colors chosen, one or more of the colors may have to be mixed negatively, that is, added to the sample, to make a match between the sample and the adjacent test patch.) Thus, a color can be specified in terms of X, Y and Z, the amounts of the three primaries required to match it. For convenience, the coordinates:

$$x = \frac{X}{X + Y + Z} \tag{5.1}$$

$$y = \frac{Y}{X + Y + Z} \tag{5.2}$$

$$z = \frac{Z}{X + Y + Z} \tag{5.3}$$

where z is redundant since $x + y + z = 1$, can be used to determine the dominant wavelength and purity of a color. This is done by first determining the x and y coordinates for the pure spectral colors and plotting them as shown in Fig. 5.13. If a "white" point is plotted at C, a chromaticity diagram of this type has a number of useful characteristics. If the x and y coordinates of a sample color are plotted on the diagram (point G in Fig. 5.13), a line through the "white" point C and the color point G, when extended to the line which is the locus of the pure spectral colors, will indicate the dominant wavelength of G. The exitation purity of the color represented by G is given by the distance CG divided by the distance DC (expressed as a percentage). If two colors are plotted, the line drawn between the two points is the locus of all the colors that can be obtained by mixing the two colors.

The nonspectral colors (purples) are represented in Fig. 5.13 by the straight dashed line connecting the extreme ends of the spectrum (since purple is a mixture of red and blue light) and

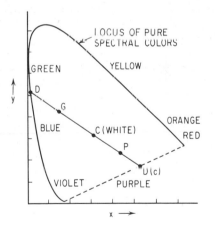

FIG. 5.13. Chromaticity diagram. The dominant wavelength of the color represented by G is determined by extending the line through G and C to the boundary line which is locus of pure spectral colors.

their dominant wavelengths are specified in terms of their complimentary colors, found by extending the line between points P and C in Fig. 5.13 to $D(c)$ and to D.

This system of colorimetry has been internationally standardized. Three primary colors (imaginary to avoid negative mixtures as would be necessary with real colors) have been specified and a standard visual response (Fig. 5.10a) has been determined by extensive testing. The three primary colors are indicated in Fig. 5.14, which shows, for each spectral wavelength, the amounts of each of the primary colors required to match the spectral color. Thus, the coordinates of the pure spectral colors in Fig. 5.13 are determined by substituting the values of $\bar{x}, \bar{y}$, and $\bar{z}$ from Fig. 5.14 into Eqns. 5.1 and 5.2 (setting $X = \bar{x}$, $Y = \bar{y}$ and $Z = \bar{z}$).

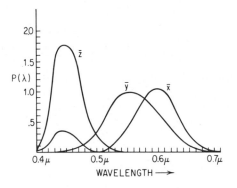

FIG. 5.14. The amounts of the standardized I.C.I. primaries ($\bar{x}$: "red-purple", $\bar{y}$: "green", $\bar{z}$: "blue") which, when combined, will produce a pure spectral color of the wavelength indicated.

Three standard illuminants were also established and called A (a gas filled tungsten lamp at $2848°K$), B (to simulate noon sunlight) and C (to simulate average daylight). The spectral characteristics of these sources are plotted in Fig. 5.15. Illuminant C is most commonly utilized as the reference illuminant for color specification.

From this information, one can determine the color specification in terms of dominant wavelength, purity and brightness for a color sample for which the spectral characteristics are available.

As an example, assume that we desire to determine the color specification for a filter illuminated by Illuminant A. The spectral transmission curve $T(\lambda)$ is determined by spectrophotometry (e.g. Fig. 5.12). The energy distribution of the light (from illuminant A) transmitted by the filter is found by multiplying, at each wavelength, the transmission $T(\lambda)$ by the value of $E(\lambda)$ for illuminant A (Fig. 5.15).

The amounts of the primary colors necessary to match this distribution are then found from:

$$X = \int P(\lambda)_x \, E(\lambda)_A \, T(\lambda) \, d\lambda \qquad (5.4)$$

$$Y = \int P(\lambda)_y \, E(\lambda)_A \, T(\lambda) \, d\lambda \qquad (5.5)$$

$$Z = \int P(\lambda)_z \, E(\lambda)_A \, T(\lambda) \, d\lambda \qquad (5.6)$$

where $P(\lambda)_x$, $P(\lambda)_y$ and $P(\lambda)_z$ are shown in Fig. 5.14. The integration is, of course, handled numerically since P, E, and T are not ordinary functions. Then X, Y, and Z are substituted into Eqs. 5.1 and 5.2, and the resulting x and y coordinates are plotted on the chromaticity chart. The dominant wavelength and purity are determined as described previously. Brightness, in terms of effective visual transmission (or reflection), can be determined by evaluating Eq. 5.5 with $T(\lambda) = 1.0$ for all wavelengths. The ratio of Y calculated for the filter to Y for $T(\lambda) = 1.0$ is then the effective visual transmission. This is true because $P(\lambda)_y$ is the same as the visual response curve of Fig. 5.10.

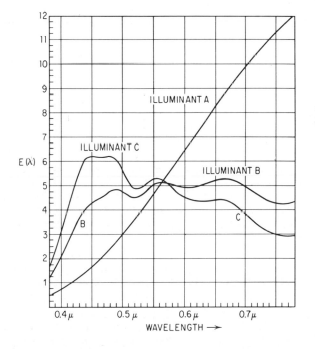

FIG. 5.15. The spectral energy distribution of the standard I.C.I. illuminants. Illuminant A is a Tungsten lamp, B simulates noon sunlight and C simulates average daylight.

Table of Selected Ordinates
Wavelengths in millimicrons (10^{-6} millimeters)

Ordinate Number	For Illuminant A			For Illuminant B			For Illuminant C		
	λx	λy	λz	λx	λy	λz	λx	λy	λz
1	444	488	416	428	472	415	424	466	414
2	517	508	425	442	494	423	436	489	422
3	544	517	429	454	506	427	444	500	426
4	554	524	433	468	513	430	452	509	429
5	561	530	436	528	520	433	461	515	432
6	567	535	439	543	525	435	474	521	434
7	572	539	441	552	529	438	531	525	437
8	576	544	444	558	534	440	544	530	439
9	580	548	446	564	538	442	552	534	441
10	584	552	448	569	541	444	559	538	443
11	587	555	450	573	545	446	564	541	444
12	590	559	453	577	549	448	569	545	446
13	594	563	455	581	552	450	573	548	448
14	597	566	457	585	556	452	577	552	450
15	599	570	459	588	559	454	581	555	452
16	602	573	461	591	562	456	585	558	454
17	605	577	463	595	566	458	589	562	456
18	608	580	465	598	569	460	592	565	458
19	611	584	467	601	573	462	596	569	460
20	614	588	469	605	577	464	600	573	462
21	617	592	472	608	581	466	603	576	464
22	620	596	474	612	585	468	607	580	466
23	623	600	477	615	589	471	611	585	469
24	627	605	480	619	594	474	615	590	471
25	631	610	483	623	599	477	619	595	474
26	635	615	488	628	605	480	624	601	478
27	640	622	493	633	612	484	630	608	482
28	647	629	499	640	620	490	637	616	487
29	656	640	508	649	631	499	646	627	495
30	673	659	527	666	651	515	663	647	511
Factors	.03661	.03333	.01185	.03303	.03333	.02842	.03269	.03333	.03938

FIG. 5.16.

The labor of wavelength by wavelength numerical integration can be alleviated by the use of the *selected ordinate method.* The table of selected ordinates (Fig. 5.16) lists wavelength values for which the integral of the product $P(\lambda)_i E(\lambda)_I$ has been divided into equal increments. To use this method, the value of the transmission $T(\lambda)$ (or reflection, brightness, etc., as appropriate) corresponding to each wavelength (ordinate) in the table is determined from the spectrophotometric data for the sample. These values for the X, Y, and Z ordinates are summed independently and each sum is multiplied by the factor given at the foot of the table column, to get $X = \text{factor} \cdot \Sigma T(\lambda)$ etc. Substitution of X, Y, and Z into Eqs. 5.1 and 5.2 then yields x and y, which are plotted as before. Note that the value of Y yields the effective visual transmission directly.

EXAMPLE   119

Figure 5.17 is a chromaticity chart of modest size which will permit wavelength and purity determinations to a reasonable degree of precision. More accurate charts and tables are published (e.g., Reference 6) in colorimetry handbooks.

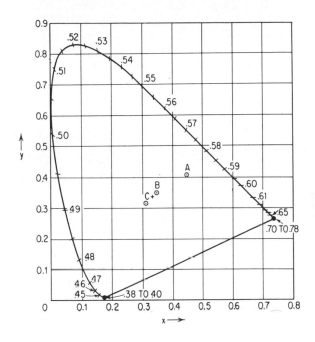

FIG. 5.17. Chromaticity Chart. Wavelengths along the spectral line are in microns. Points A, B and C indicate the coordinates of I.C.I. Illuminants A, (.448, .408), B (.348, .352) and C (.310, .316).

For materials whose spectral characteristics are smooth, the labor may be further reduced by using every second or third ordinate and multiplying the factors by 2 or 3, respectively.

Example

Let us determine the color specifications for an imaginary filter whose transmission at any wavelength is given by its wavelength in microns, so that $T = \lambda$. The spectrophotometric curve will then be a straight line running from $T = 0.4$ at a wavelength of 400 millimicrons to $T = 0.7$ at 700 millimicrons. Since the curve is smooth, we will use only 10 ordinates, using numbers 2, 5, 8, etc. Referring to the table of Fig. 5.16, the second ordinate under the X column for illuminant C lists $\lambda = 436$ mμ; thus, the transmission at this wavelength for our imaginary filter will be

T = 0.436; similarly for the fifth ordinate T = .461 and so on. The calculation is indicated in the following table:

Ordinate No.	$T(\lambda x)$	$T(\lambda y)$	$T(\lambda z)$
2	.436	.489	.422
5	.461	.515	.432
8	.544	.530	.439
11	.564	.541	.444
14	.577	.552	.450
17	.589	.562	.456
20	.600	.573	.462
23	.611	.585	.469
26	.624	.601	.478
29	.646	.627	.495
Totals	5.652	5.575	4.547

The values of X, Y and Z are found by multiplying the column totals by three times the factors at the foot of each column of the table (Fig. 5.16), giving:

$$X = 3 \times .03269 \times 5.652 = 0.5543$$
$$Y = 3 \times .03333 \times 5.575 = 0.5575$$
$$Z = 3 \times .03938 \times 4.547 = 0.5372$$

Substituting into Eqs. 5.1 and 5.2 to get the coordinates, we find:

$$x = \frac{0.5543}{0.5543 + 0.5575 + 0.5372} = 0.336$$

$$y = \frac{0.5575}{1.649} = 0.338$$

We now plot x and y on the chromaticity chart of Fig. 5.17, where the point is shown as the small cross between the points representing illuminants B and C. Extending a line from C through the cross to the boundary, we pick off the dominant wavelength as 0.58 microns. Measuring the distance from C to the cross and dividing by the distance from C to the boundary, we get a purity of 13%. The effective visual transmission is given by Y above as 55.8%. Evidently our hypothetical filter is nearly white with a slight yellow-orange cast.

This system of color specification is extremely valuable in that it not only puts matters on a convenient numerical basis, but allows tolerances to be readily specified and maintained, on the basis of physical measurement rather than visual comparison.

References

1. Hartridge, "Recent Advances in the Physiology of Vision," Blakiston, 1950.
2. Davson, "The Physiology of the Eye," Blakiston, 1950.
3. Adler, "Physiology of the Eye - Clinical Application," Mosby, 1959.
4. Zoethout, "Physiological Optics," Professional Press, 1939.
5. M.I.T. Handbook of Colorimetry, Technology Press, 1936.
6. American Institute of Physics Handbook, McGraw-Hill, 1963.
7. Hardy and Perrin, "Principles of Optics," McGraw-Hill, 1932.
8. Judd and Wyszecki, "Color in Business, Science and Industry," Wiley, 1963.
9. Judd, "Colorimetry," N.B.S. Circular 478, U.S. Government Printing Office, 1950.

Exercises

1. Determine the color (dominant wavelength) of the blue sky. Use illuminant B (noon sun) and assume a filter whose transmission is given by $T(\lambda) = \lambda^{-4}$ (The blue color of the sky is due to the scattering of sunlight by the atmosphere. The amount of scattering is inversely proportional to the fourth power of the wavelength of the light being scattered). Use 10 ordinates. Ans. 0.48μ

2. What power telescope is necessary to enable a person with "normal" visual acuity to read letters 1 mm high at a distance of 300 feet? (tangent of one minute of arc is .0003) Ans. 135×

3. What power corrective lens would be prescribed for a nearsighted person who could not focus clearly on an object more than 5 inches away? Ans. −8 diopters

4. Assuming a depth of focus of ±1/4 diopter, over what range of distance is vision perfectly clear when the eye is focused at 10 inches? Ans. 1-1/4 inches

5. It is desired to set an optical vernier to a precision of 0.0001 in. Assuming that the vernier projects the image of a ruled scale onto a screen which is viewed from a distance of 10 inches and that the setting is made by aligning a scale line with a cross hair on the screen, what magnification must the projection lens of the optical vernier have? Use 10 seconds of arc for the vernier acuity. (Tangent of one second is .000005) Ans. 5 power

6. A convex reflector of radius of curvature =10 in. is mounted on a spindle and rotated. a) What is the largest amount that its center of curvature can be displaced from the axis of rotation without the motion of the reflected image of a distant object being detected by the naked eye? Assume the reflected image is viewed from 10 inches. b) What are the fastest and slowest speeds of rotation at which the motion caused by a decentration of 0.02 in. can be detected?
    Ans. a) 0.00025 in., b) 3 to 5 rpm, 300 to 500 rps.

7. Find the effective visual transmission of illuminant C through a filter whose transmission is given by:
    from $.4\mu$ to $.55\mu$ $T = 1.0 - (\lambda - .4)/.15$
    from $.55\mu$ to $.7\mu$ $T = (\lambda - .55)/.15$
    Ans. 22%

8. a) If a plane parallel plate is specified to have zero, ±10 milli-diopters, power, what is the shortest tolerable focal length it may have? b) Assuming one surface is truly flat, what is the strongest (shortest) acceptable radius for the other surface if the index of refraction is 1.6? c) If the piece has a diameter of 20 mm., how many Newton's rings will be visible when this surface is tested against a true flat? (Use $\lambda = 0.55$ microns. One fringe occurs for each $\lambda/2$ change in thickness of the air space)
    Ans. a) ±100 meters, b) ±60 meters, c) 3 rings

6

Stops and Apertures

6.1

In every optical system, there are apertures (or stops) which limit the passage of energy through the system. These apertures are the clear diameters of the lenses and diaphragms in the system. One of these apertures will determine the diameter of the cone of energy which the system will accept from an axial point on the object. This stop is termed the "aperture stop" and its size determines the illumination (irradiance) at the image. Another stop will usually limit the size or angular extent of the object which the system will image. This stop is called the "field stop." The importance of these stops to the photometry (radiometry) and performance of the system cannot be overemphasized.

The elements of an ordinary box camera system are sketched in Fig. 6.1 and illustrate both aperture and field stops in their most basic forms. The diaphragm in front of the lens limits the diameter of the bundle of rays that the system can accept and is thus the aperture stop. The mask adjacent to the film determines the angular field coverage of the system and is quite apparently the field stop of the camera.

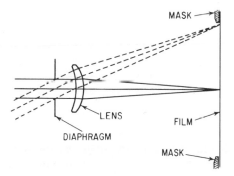

FIG. 6.1. The elements of a simple box camera illustrate the functions of elementary aperture and field stops (the diaphragm and mask respectively.)

Not all systems are as obvious as the box camera, however, and we will now consider more complex arrangements. Because the theory of stops is readily explained by the use of a concrete example, the following discussions will be with reference to Fig. 6.2, which is a highly exaggerated sketch of a telescopic system focused on an object at a finite distance. The system shown consists of an objective lens, erector lens, eyelens, and two internal diaphragms. The objective forms an inverted image of the object. This image is then re-imaged by the erector lens at the first focal point of the eyelens, so that the eyelens forms the final image of the object at infinity.

6.2 The Aperture Stop

By following the path of the axial rays (designated by solid lines) in Fig. 6.2, it can be seen that diaphragm #1 is the aperture of the system which limits the size of the axial cone of energy from the object. All of the other elements of the system are large enough to accept a bigger cone. Thus, diaphragm #1 is the aperture stop of the system.

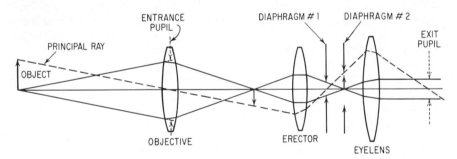
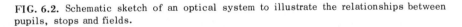

FIG. 6.2. Schematic sketch of an optical system to illustrate the relationships between pupils, stops and fields.

The ray through the center of the aperture stop is called the principal or chief ray, and is shown in the figure as a dashed line. The entrance and exit pupils of the system are the images of the aperture stop in object and image space respectively. That is, the entrance pupil is the image of the aperture stop as it would be seen if viewed from the axial point on the object; the exit pupil is the aperture stop image as it would be seen if viewed from the final image plane (in this case, at an infinite distance). In the system of Fig. 6.2, the entrance pupil lies within the objective lens and the exit pupil lies to the right of the eyelens. Notice that the initial and final intersections of the dashed principal ray with the axis locate the pupils, and that the diameter of the axial cone of rays at the pupils indicates their diameters. It can be seen that, for any point on the object, the amount of radiation accepted by, and emitted from, the system is determined by the size and location of the pupils.

6.3 The Field Stop

By following the path of the principal ray in Fig. 6.2, it can be seen that another principal ray starting from a point in the object which is further from the axis would be prevented from passing through the system by diaphragm #2. Thus, diaphragm #2 is the field stop of this system. The images of the field stop in object and image space are called the entrance and exit windows respectively. In the system of Fig. 6.2, the entrance window is coincident with the object and the exit window is at infinity (which is coincident with the image). Note that the windows of a system do not coincide with the object and image unless the field stop lies in the plane of a real image formed by the system.

The angular field of view is determined by the size of the field stop, and is the angle which the entrance or exit window subtends from the entrance or exit pupil respectively. The angular field in object space is frequently different from that in image space. (Alternate definition: the angular field of view is the angle subtended by the object or image from the first or second nodal point of the system respectively. Thus, for non-telescopic systems in air, object and image field angles are equal.)

6.4 Vignetting

The optical system of Fig. 6.2 was deliberately chosen as an ideal case in which the roles played by the various elements of the system are definite and clear cut. This is not usually the situation in real optical systems, since the diaphragms and lens apertures often play dual roles.

Consider the system shown in Fig. 6.3, consisting of two positive lenses, A and B. For the axial bundle of rays, the situation is clear; the aperture stop is the clear aperture of lens A, the entrance pupil is at A, and the exit pupil is the image, formed by lens B, of the diameter of lens A.

Some distance off the axis, however, the situation is markedly different. The cone of energy accepted from point D is limited on its lower edge by the lower rim of lens A and on its upper edge by the upper rim of lens B. The size of the accepted cone of energy from point D is less than it would be if the diameter of lens A were the only limiting agency. This effect is called *vignetting*, and it causes a reduction in the illumination at the image D'. It is apparent that for some object point still further from the axis than point D, no energy at all would pass through the system.

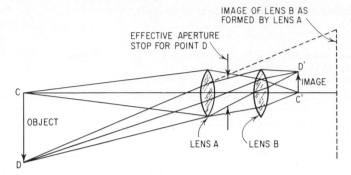

FIG. 6.3. Vignetting in a system of separated components. The cone of rays from point D is limited by the lower rim of lens A and the upper rim of B, and is smaller than the cone accepted from point C. Note that the upper ray from D just passes through the image of Lens B which is formed by Lens A.

The appearance of the system when viewed from point D is shown in Fig. 6.4. The entrance pupil has become the common area of two circles, one the clear diameter of lens A, and the other the diameter of lens B as imaged by lens A. The dashed lines in Fig. 6.3 indicate the location and size of this image of B, and the arrows indicate the effective aperture stop which has a size, shape, and position completely different than that for the axial case.

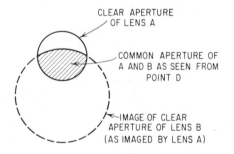

FIG. 6.4. The apertures of the optical system of Fig. 6.3 as they are seen from point D.

Example A

Let us determine the pupils, windows and fields of an optical system of the type shown in Fig. 6.2, assuming the lenses to be "thin lenses." The elements of the system are as follows:

objective: clear aperture = $2.3''$;
                effective focal length = $10''$
erector: clear aperture = $1.7''$;
                effective focal length = $2''$
eyelens: clear aperture = $1.3''$;
                effective focal length = $1''$

diaphragm #1: clear aperture = $0.25''$
diaphragm #2: clear aperture = $0.7''$
distance, object to objective: $50.''$
distance, objective to erector: $16.5''$
distance, erector to eyelens: $5''$
distance, erector to diaphragm #1: $2.38''$
distance, erector to diaphragm #2: $4''$

We begin the analysis by tracing a paraxial ray from the object point on the axis, using the thin lens ray tracing equations (2.31 and 2.32) of Chapter 2. We insert two zero power elements in the system to represent the diaphragms, so that we can determine the ray heights at the diaphragms. We assume a nominal ray height of $+1.0$ at the objective lens, giving $u_1 = +1.0/(-50.) = -.02$. The calculation is shown in the table of Fig. 6.5, lines 3 and 4.

	Object Plane	Objective Lens	Erector Lens	Diaphragm #1	Diaphragm #2	Eyelens
1. $\phi = 1/f$		$+0.1$	$+0.5$	0.0	0.0	$+1.0$
2. d		50.	16.5	2.38	1.62	1.0
3. y	0.0	$+1.0$	$-.32$	$-.1296$	0.0	$+0.8$
4. u		$-.02$	$+.08$	$-.08$	$-.08$	-0.8	0.0
5. CA		2.3	1.7	0.25	0.7	1.3
6. CA/y		2.3	5.31	1.929	Infinity	16.25
7. $y_0 = .9645\,y$	0.0	$+.9654$	$-.3086$	$-.125$	0.0	$+.07716$
8. $u_0 = .9645\,u$		$-.01929$	$+.07716$	$-.07716$	$-.07716$	$-.07716$	0.0
9. y_p	$+1.4$	$+.1631$	$-.5142$	0.0	$+.35$	$+.5660$
10. u_p		$+.02474$	$+.04105$	$-.21605$	$-.21605$	$-.21605$	$+.35$
11. $y_p + y_0$		$+1.1276$	$-.8228$	$-.125$	$+.35$	$+.6432$
12. $y_p - y_0$		$-.8013$	$-.2056$	$+.125$	$+.35$	$+.4889$

FIG. 6.5. Tabulation of the raytrace data for Example A.

To determine which element of the system limits the diameter of the cone of rays, we add to our tabulation lines 5 and 6, showing the clear aperture of each element (CA) and the ratio of the clear aperture to the height that the axial ray strikes the element (CA/y). The element for which this ratio is the smallest, in this case diaphragm #1, is the aperture stop. Because of the linear nature of the paraxial equations, we can get the y and u values for any other axial ray by multiplying each entry in lines 3 and 4 by the same constant. If we use for the constant the value of $\frac{1}{2}CA/y$ for diaphragm #1, (.9645), we will get the data for a ray which just passes through the rim of diaphragm #1. This ray data is shown in lines 7 and 8 of the table. A comparison of the new y values of line 7 with the clear apertures of line 5 indicates that the ray will pass through all the other elements with room to spare.

To determine the locations of the pupils, we trace a ray through the center of the aperture stop, (diaphragm #1) in each direction. The data of such a ray is shown in lines 9 and 10 of the table. We then determine the axial intersections of this ray in object and image space and find that the (apparent) entrance pupil is located $+.1631/+.02474 = +6.594$ inches to the right of the objective lens (note that this differs from Fig. 6.2) and that the exit pupil is $.566/+.35 = +1.617$ inches to the right of the eyelens.

The diameter of the pupils is found from the ray data of lines 7 and 8 by determining the ray height in the plane of the pupils. Thus, the diameter of the entrance pupil is $2(.9645 + .01929 \times 6.594)$ or $2.183''$ and the diameter of the exit pupil is $2(.07716 - 0.0 \times 1.617)$ or $0.154''$.

A comparison of the values of CA/y_p would indicate that diaphragm #2 is the field stop. (The ray data in lines 9 and 10 of Fig. 6.5 have already been adjusted so that y_p at diaphragm #2 is equal to half of its clear aperture, in a manner analogous to that by which lines 7 and 8 were derived from lines 3 and 4.) The field of view is given by the slope of the principal ray which just skims through the field stop. This is the ray of lines 9 and 10; the object field is $\pm.02474$ radians and the image field is $\pm.35$ radians. The linear size of the object field is twice the height at which this ray strikes the image plane, or 2.8 inches.

A check for vignetting could be made by tracing rays from an object point at the edge of the field through the upper and power rims of the entrance pupil. Again, because of the linearity of the paraxial equations, we can avoid this labor, since the height of the upper rim ray at an element is given by $y_p + y_0$ and that of the lower rim ray by $y_p - y_0$. (The values of y_0 and y_p are taken from the ray trace data which has been adjusted, i.e., lines 7 and 9.) This data is tabulated in lines 11 and 12 and a comparison with the clear apertures of the elements indicates that these rays pass through the system without vignetting.

An alternate technique for determining the aperture stop is to calculate the size and position of the image of *each* diameter of the system as seen from the object, that is, as imaged by all the elements ahead of (or to the left of) the diameter. Then the diameter whose image subtends the smallest angle from the object is the aperture stop. A scale drawing of the images is handy when this technique is used.

6.5 Glare Stops and Baffles

A glare stop is essentially an auxiliary diaphragm located at an image of the aperture stop for the purpose of blocking out stray radiation. Figure 6.6 shows an erecting telescope in which the primary aperture stop is at the objective lens. Energy from sources outside the desired field of view, passing through the objective and reflecting from an internal wall, shield, or supporting member, can create a glare which reduces the contrast of the image formed by the system. This radiation can be blocked out by an internal diaphragm which is an accurate image of the objective aperture. Since the stray radiation will appear to be coming from the wall, and thus from outside the objective aperture, it will be imaged on the opaque portion of the diaphragm. Another glare stop could conceivably be located at the exit pupil of this particular system, since it is real and accessible; however, it would make visual use of the instrument quite inconvenient.

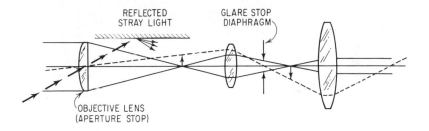

FIG. 6.6. Stray light, reflected from an inside wall of the telescope, is intercepted by the glare stop, which is located at the internal image of the objective lens.

In an analogous manner, field stops could be placed at both internal images to further reduce stray radiation. The principle here is straightforward. Once the primary field and aperture stops of a system are determined, auxiliary stops may be located at images of the primary stops to cut out glare. If the glare stops are accurately located and are the same size as the images of the primary stops, they do not reduce the field or illumination, nor do they introduce vignetting.

Baffles are often used to reduce the amount of radiation that is reflected from walls, etc. in a system. Figure 6.7 shows a simple radiometer consisting of a collector lens and a detector in a housing. Assume that radiation from a powerful source (such as the sun) outside the field of view reflects from the inner walls of the mount onto the detector and obscures the measurement of radiation from the desired target, as shown in the upper half of the sketch. Under these conditions, there is no possibility of using an internal glare stop (since there is no internal image of the entrance pupil) and the internal walls of the mount must be baffled as shown in the lower half of the sketch (although an external hood or sunshade could also be used if circumstances permit).

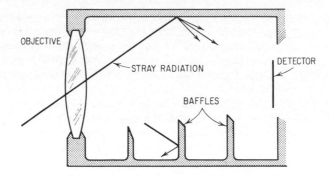

FIG. 6.7. Stray (undesired) radiation from outside the useful field of this simple radiometer can be reflected from the inner walls of the housing and degrade the function of the system. Sharp edged baffles, shown in the lower portion, trap this radiation and prevent the detector from "seeing" a directly illuminated surface.

The key to the efficient use of baffles is to arrange them so that no part of the detector can "see" a surface which is directly illuminated. The method of laying out a set of baffles is illustrated in Fig. 6.8. The dotted lines from the rim of the lens to the edge of the detector indicate the necessary clearance space, into which the baffles cannot intrude without obstructing part of the radiation from the desired field of view. The dashed line AA' is a "line of sight" from the detector to the point on the wall where the extraneous radiation begins. The first baffle is erected to the intersection of AA' with the dotted clearance line. Solid line BB' indicates the path of stray light from the top of the lens to the wall. The area from Baffle #1 to B' is thus shadowed and "safe" for the detector to "see". The dashed line from B' to A is thus the safe line of sight, and baffle #2 at the intersection of AB' and the clearance line will prevent the detector from "seeing" the illuminated wall beyond B'. This procedure is repeated until the entire side wall is protected. Note that the inside edges of the baffles should be sharp and their surfaces rough and blackened.

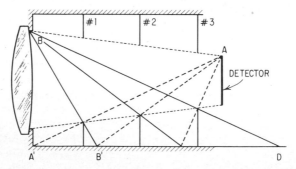

FIG. 6.8. Construction for the systematic layout of baffles. See text. Note that Baffle No. 3 shields the wall back to Point D; thus, all three baffles could be shifted foreward somewhat, so that their coverages overlap.

This type of baffling can be quite expensive and is not necessary in many cases. Frequently internal scattering can be sufficiently reduced by scoring or threading the offending internal surfaces of the mount. In this way, the reflections are broken up and scattered, reducing the amount of reflection and destroying any glare images. The use of a flat black paint is also advisable, although care must be taken to be sure that the paint remains both matte and black at near-grazing angles of incidence. Minnesota Mining and Manufacturing (M.M.M.) "Velvet" is excellent for this purpose.

6.6 The Telecentric Stop

A telecentric stop is an aperture stop which is located at the focal point of an optical system. It is widely utilized in optical systems designed for metrology (e.g. comparators and contour projectors) because it tends to reduce the measurement error caused by a slight defocusing of the system. Figure 6.9A shows a schematic telecentric system. Note that the dashed principal ray is parallel to the axis to the left of the lens. If this sytem is used to project an image of a scale (or some other object), it can be seen that a small defocusing displacement of the scale does not change the height of the scale at which the principal ray strikes. Contrast this with Fig. 6.9b where the stop is at the lens, and the defocusing causes a proportional error in the ray height. The telecentric stop is also used where it is desired to project the image of an object with depth (along the axis), since it yields less confusing images of the edges of such an object.

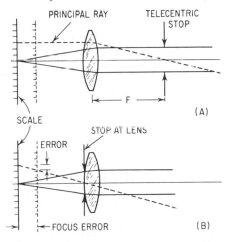

FIG. 6.9. The telecentric stop is located at the focal point of the projection system shown, so that the principal ray is parallel to the axis at the object. When the object is slightly out of focus (dotted) there is no error in the size of the projected image as there is in the system with the stop at the lens, shown in the lower sketch.

6.7 Apertures and Image Illumination

When a lens forms an image of an extended object, the amount of energy collected from a small area of the object is directly proportional to the area of the clear aperture, or entrance pupil, of the lens. At the image, the illumination (energy per unit area) is inversely proportional to the image area over which this object is spread. Now the aperture area is proportional to the square of the pupil diameter and the image area is proportional to the square of the image distance, or focal length. Thus, the square of the ratio of these two dimensions is a measure of the relative illumination produced in the image.

The ratio of the focal length to the clear aperture of a lens system is called the relative aperture, f-number, or speed, and (other factors being equal), the illumination in an image is inversely proportional to the square of this ratio. The relative aperture is given by:

$$f\text{-number} = \text{E.F.L.}/\text{clear aperture} \qquad (6.1)$$

As an example, an $8''$ focal length lens with a $1''$ clear aperture has an f-number of 8; this is customarily written $f/8$.

Another way of expressing this relationship is by the numerical aperture, which is the index of refraction (of the medium in which the image lies) times the sine of the half angle of the cone of illumination.

$$\text{Numerical aperture} = \text{N.A.} = N'\sin U' \qquad (6.2)$$

Numerical aperture and f-number are obviously two methods of defining the same characteristic of a system. Numerical aperture is more conveniently used for systems that work at finite conjugates (such as microscope objectives) and the f-number is applied to systems for use with distant objects (such as camera lenses and telescope objectives). For aplanatic systems (i.e., systems corrected for coma and spherical aberration) with infinite object distances, the two quantities are related by:

$$f\text{-number} = \frac{1}{2\,\text{N.A.}} \qquad (6.3)$$

For off-axis image points, even when there is no vignetting, the illumination is usually lower than for the point on the axis. Figure 6.10 is a schematic drawing showing the relationship between exit pupil and image plane for point A on axis and point H off axis. The illumination at an image point is proportional to the solid angle which the exit pupil subtends from the point. From

Fig. 6.10, it is apparent that for small values of ϕ, $\phi' = \phi\cos^2\theta$, and that $OA = OH\cos\theta$. Thus, the solid angle subtended by the pupil from H is reduced by a factor of $\cos^3\theta$ from that subtended at A. Now the illumination so far has been considered in a plane normal to the direction of propagation; it is apparent that at H the energy is spread over an area which is proportionately larger than at A because the cone strikes the surface at an angle (θ) from the normal; thus, a fourth $\cos\theta$ factor must be added, and we find that

$$\text{(illumination at } H) = \cos^4\theta \text{ (illumination at } A) \tag{6.4}$$

The importance of this effect on wide angle lenses can be judged from the fact that $\cos^4 30° = 0.56$, $\cos^4 45° = 0.25$ and $\cos^4 60° = .06$. It can be seen that the illumination on the film in a wide angle camera will fall off quite rapidly.

Note that the preceding has been based on the assumption that the pupil diameter is constant (with respect to θ) and that θ is the angle formed in image space (although many people mistakenly apply it to the field angle in object space). The "cosine fourth" law can be modified if the construction of the lens is such that the apparent size of the pupil increases for off axis points, or if a sufficiently large amount of barrel distortion is introduced to hold θ to smaller values than one would expect from the cor-

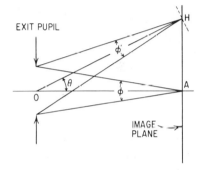

FIG. 6.10. Relationship between exit pupil and image points, used to demonstrate that the illumination at H is $\cos^4\theta$ times that at A.

responding field angle in object space. Certain extreme wide angle camera lenses make use of these principles to increase off axis illumination. The $\cos^4$ effect is in addition to any illumination reduction caused by vignetting.

6.8 Depth of Focus

The concept of depth of focus rests on the assumption that for a given optical system, there exists a blur (due to defocusing) of small enough size such that it will not adversely affect the performance of the system. The depth of focus is the amount by which the image may be shifted longitudinally with respect to some reference plane (e.g. film, reticle) and introduce no more than the acceptable blur. The depth of field is the amount by which the object may be shifted before the acceptable blur is produced. The size of the acceptable blur may be specified as the linear diameter of the blur spot (as is common in photographic applications) (Fig. 6.11)

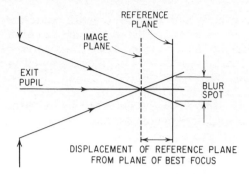

FIG. 6.11. When an optical system is defocused, the image of a point becomes a blurred spot. The size of the blur is determined by the relative aperture of the system and the focus shift.

or as an angular blur, i.e. the angular subtense of the blur spot from the lens. Thus, the linear and angular blurs (B and β respectively) and the distance D are related by

$$\beta = \frac{B}{D} = \frac{B'}{D'} \tag{6.5}$$

for a system in air, where the primed symbols refer to the image-side quantities.

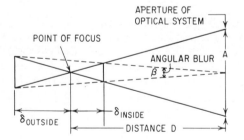

FIG. 6.12. Relationships used to determine the longitudinal depth of field in terms of a tolerable angular blur.

From Fig. 6.12, it can be seen that the depth of field δ for a system with a clear aperture A can be obtained from the relationship

$$\frac{\delta}{\beta(D \pm \delta)} = \frac{D}{A}$$

This expression can be solved for the depth of field, giving

$$\delta = \frac{D^2\beta}{(A \pm D\beta)} \tag{6.6}$$

Note that the depth of field toward the optical system is smaller than that away from the system. When δ is small in comparison with the distance D, this reduces to

$$\delta = \frac{D^2\beta}{A} \tag{6.7}$$

For the image side, the relationship is

$$\delta' = \frac{D'^2\beta}{A} = \frac{F^2\beta}{A} = F\beta\,(f/\#) \tag{6.8}$$

where the second and third forms of the right hand side apply when the image is at the focal point of the system, and F is the system focal length.

The depth of focus in terms of linear blur-spot size B can be obtained by substituting Eq. 6.5 into the above. Also, note that the depth of field δ and the depth of focus δ' are related by the longitudinal magnification of the system, so that

$$\delta' = \overline{m}\,\delta = m^2\delta \tag{6.9}$$

The hyperfocal distance of a system is the distance at which the system must be focused so that the depth of field extends to infinity. If $(D + \delta)$ equals infinity, then β is equal to A/D, so that

$$D\,(\text{hyperfocal}) = \frac{A}{\beta} \tag{6.10}$$

6.9 Diffraction Effects of Apertures

Even if we assume that an infinitely small point source of light is possible, no lens system can form a true point image, even though the lens be perfectly made and absolutely free of aberrations. This results from the fact that light does not really travel in straight-line rays, but behaves as a wave motion, bending around corners and obstructions to a small but finite degree.

According to Huygen's principle of light wave propagation, each point on a wave front may be considered as a source of spherical wavelets; these wavelets reinforce or interfere with each other to form the new wave front. When the original wave front is infinite in extent, the new wave front is simply the envelope of the wavelets in the direction of propagation. At the other extreme, when the wave front is limited by an aperture to a very small size (say to the order of a half wavelength), the new wave front becomes spherical about the aperture. Figure 6.13 shows a plane wave front incident on a slit *AC*, which is in front of a perfect lens. The lens is focused on a screen, *EF*. We wish to determine the nature of the illumination on the screen. Since the lens of Fig. 6.13 is assumed perfect, the optical path lengths *AE, BE* and *CE* are all equal and the waves will arrive in phase at *E*, reinforcing each other to produce a bright area. For wavelets starting from the plane wave front in a direction indicated by angle *a*, the paths are different; path *AF* differs from path *CF* by the distance *CD*. If *CD* is an integral number of wavelengths, the wavelets from *A* and *C* will reinforce at point *F*. If *CD* is an odd number of half wavelengths, a cancellation will occur. The illumination at *F* will be the summation of the contributions from each incremental segment of the slit, taking the phase relationships into account. It can be readily demonstrated that when *CD* is an integral number of wavelengths, the illumination at *F* is zero, as follows: if *CD* is one wavelength, then *BG* is one-half wavelength and the wavelets from *A* and *B* cancel. Similarly, the wavelets from the points just below *A* and *B* cancel and so on down the width of the slit. If *CD* is *N* wavelengths, we divide the slit into 2*N* parts (instead of two parts) and apply the same reasoning. Thus, there is a dark zone at *F* when

$$\text{Sin } a = \frac{N\lambda}{w}$$

where *N* is any integer, λ is the wavelength of the light and *w* is the width of the slit.

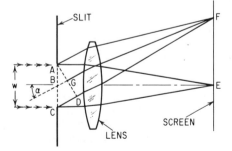

FIG. 6.13.

Thus, the illumination in the plane *EF* is a series of light and dark bands. The central bright band is the most intense, and the bands on either side are successively less intense. One can realize that the intensity should diminish by considering the situation when CD is 1.5λ, 2.5λ, etc. When CD is 1.5λ, the wavelets from two thirds of the slit can be shown (as in the preceding paragraph) to interfere and cancel out, leaving the wavelets from one third of the aperture; when CD is 2.5λ, only one-fifth of the slit is uncancelled. Since the "uncancelled" wavelets are neither exactly in nor exactly out of phase, the illumination at the corresponding points on the screen will be less than one-third or one-fifth of that in the central band.

For a more rigorous mathematical development of the subject, the reader is referred to the references following this chapter. The mathematical approach is one of integration over the aperture, combined with a suitable technique for the addition of the wavelets which are neither in nor out of phase. This approach can be applied to rectangular and circular apertures as well as to slits.

For a rectangular aperture, the illumination on the screen is given by

$$I = I_0 \frac{\mathrm{Sin}^2 m_1}{m_1^2} \cdot \frac{\mathrm{Sin}^2 m_2}{m_2^2} \tag{6.11}$$

$$m = \frac{\pi w \,\mathrm{Sin}\, a}{\lambda} \tag{6.12}$$

In these expressions λ is the wavelength, w the width of the exit aperture, a the angle subtended by the point on the screen, m_1 and m_2 correspond to the two principal dimensions, w_1 and w_2, of the rectangular aperture and I_0 is the illumination at the center of the pattern.

When the aperture is circular, the illumination is given by

$$I = I_0 \left[1 - \frac{1}{2}\left(\frac{m}{2}\right)^2 + \frac{1}{3}\left(\frac{m^2}{2^2\,2!}\right)^2 - \frac{1}{4}\left(\frac{m^3}{2^3\,3!}\right)^2 + \frac{1}{5}\left(\frac{m^4}{2^4\,4!}\right)^2 - \cdots \right]^2 \tag{6.13}$$

where m is given by Eq. 6.12 with the obvious substitution of the diameter of the circular exit aperture for the width, w. The illumination pattern consists of a bright central spot of light surrounded by rings of rapidly decreasing intensity. This pattern is called the Airy Disc.

We can convert from angle a to Z, the radial distance from the center of the pattern, by reference to Fig. 6.14. If the optical system is reasonably aberration free, then

$$\ell' = \frac{w}{2\,\mathrm{Sin}\,U'}$$

and to a close approximation, when a is small

$$Z = \frac{\ell' a}{N'} = \frac{aw}{2N' \sin U'}$$ (6.14)

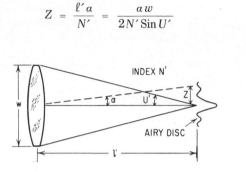

FIG. 6.14.

The table of Fig. 6.15 lists the characteristics of the diffraction patterns for circular and slit apertures. The table is derived from Eqs. 6.11 and 6.13, but the data is given in terms of Z and $\sin U'$ rather than a and w.

Ring (or band)	Circular Aperture			Slit Aperture	
	Z	Peak Illumination	Energy in Ring	Z	Peak Illumination
Central Maximum	0	1.0	83.9 %	0	1.0
1st Dark Ring	.61 $\lambda/N' \sin U'$	0.0		0.5 $\lambda/N' \sin U'$	0.0
1st Bright Ring	.82 $\lambda/N' \sin U'$	0.017	7.1 %	0.72 $\lambda/N' \sin U'$	0.047
2nd Dark Ring	1.12 $\lambda/N' \sin U'$	0.0		1.0 $\lambda/N' \sin U'$	0.0
2nd Bright Ring	1.33 $\lambda/N' \sin U'$	0.0041	2.8 %	1.23 $\lambda/N' \sin U'$	0.017
3rd Dark Ring	1.62 $\lambda/N' \sin U'$	0.0		1.5 $\lambda/N' \sin U'$	0.0
3rd Bright Ring	1.85 $\lambda/N' \sin U'$	0.0016	1.5 %	1.74 $\lambda/N' \sin U'$	0.0083
4th Dark Ring	2.12 $\lambda/N' \sin U'$	0.0		2.0 $\lambda/N' \sin U'$	0.0
4th Bright Ring	2.36 $\lambda/N' \sin U'$	0.00078	1.0 %	2.24 $\lambda/N' \sin U'$	0.0050
5th Dark Ring	2.62 $\lambda/N' \sin U'$			2.5 $\lambda/N' \sin U'$	0.0

FIG. 6.15. Tabulation of the size of and distribution of energy in the diffraction pattern at the focus of a perfect lens.

Notice that 84% of the energy in the Airy Disc is contained in the central spot, and that the illumination in the central spot is almost 60 times that in the first bright ring. Ordinarily the central spot and the first two bright rings dominate the appearance of the pattern, the other rings being too faint to notice. The illumination in an Airy Disc is plotted in Fig. 6.16. One should bear in mind the fact that these energy distributions apply to perfect, aberration-free systems. The presence of aberrations will, of course, modify the distribution.

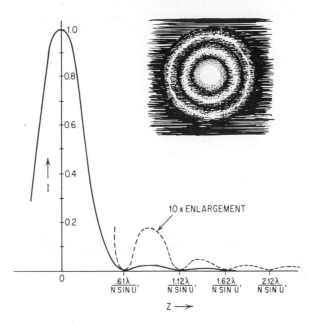

FIG. 6.16. The distribution of illumination in the Airy Disc. The appearance of the Airy Disc is shown in the upper right.

6.10 Resolution of Optical Systems

The diffraction pattern resulting from the finite aperture of an optical system establishes a limit to the performance which we can expect from even the best optical device. Consider an optical system which images two equally bright point sources of light. Each point is imaged as an Airy Disc, and if the points are close, the diffraction patterns will overlap. When the separation is such that it is just possible to determine that there are two points and not one, the points are said to be resolved. Figure 6.17 indicates the summation of the two diffraction patterns for various amounts of separation. When the image points are closer than $0.5\lambda/\text{N.A.}$ (N.A. is the Numerical Aperture of the system and equals $N'\sin U'$), the central maxima of both patterns blend into one and the combined patterns may appear to be due to a single source. At a separation of $0.5\lambda/\text{N.A.}$ the duplicity of the image points is detectable, although there is no minimum between the maxima from the two patterns. This is Sparrow's criterion for resolution. When the image separation reaches $0.61\lambda/\text{N.A.}$ the maximum of one pattern coincides with the first dark ring of the other and there is a clear indication of two separate maxima in the combined pattern. This is Lord

Rayleigh's criterion for resolution and is the most widely used value for the limiting resolution of an optical system.*

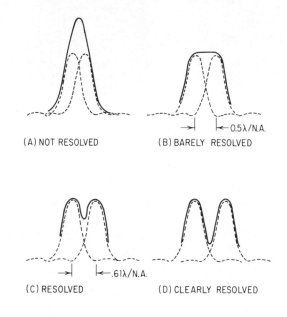

FIG. 6.17. The dashed lines represent the diffraction patterns of two point images at various separations. The solid line indicates the combined diffraction pattern. Case b) is the Sparrow criterion for resolution. Case c) is the Rayleigh criterion.

From the tabulation of Fig. 6.15, we find that the distance from the center of the Airy Disc to the first dark ring is given by

$$Z = \frac{0.61\lambda}{N' \sin U'} = \frac{0.61\lambda}{N.A.} \tag{6.15}$$

This is the separation of two image points corresponding to the Rayleigh criterion for resolution. This expression is useful in determining the limiting resolution for microscopes and the like. For resolution at the image, the N.A. of the image cone is used; for resolution at the object, the N.A. of the object cone is used.

To evaluate the performance limits of telescopes and other systems working at long object distances, an expression for the angular separation of the object points is more useful. Rearranging

---

*The diffraction pattern of two point images will always differ somewhat from the diffraction pattern of a single point. It is thus possible to detect the presence of two points (as opposed to one) in cases where the two points cannot be resolved or separated. This is the source of the occasional claims that a system "exceeds the theoretical limit of resolution."

Eq. 6.14 and substituting the limiting value of Z from Eq. 6.15, we get, in radian measure,

$$a = \frac{1.22\lambda}{w} \tag{6.16}$$

For ordinary visual instruments, λ may be taken as 0.55 microns, and using $4.85 \cdot 10^{-6}$ radians for one second of arc, we find that

$$a = \frac{5.5}{w} \text{ seconds of arc} \tag{6.17}$$

when w is expressed in inches. By a series of careful observations, the astronomer Dawes found that two stars of equal brightness could be resolved when their separation was $4.6/w$ seconds. Notice that if the Sparrow criterion is used instead of the Rayleigh criterion in Eq. 6.17, the limiting resolution angle is $4.5/w$ seconds, which is in close agreement with Dawes' findings.

It is worth emphasizing here that the resolution limit is a direct function of wavelength and an inverse function of the aperture of the system. Thus, the limiting resolution is improved by reducing the wavelength or by increasing the aperture. Note that focal length or working distance do not directly affect the resolution.

In an instrument such as a spectroscope, where it is desired to separate one wavelength from another, the measure of resolution is the smallest wavelength difference, $d\lambda$, which can be resolved. This is usually expressed as $\lambda/d\lambda$; thus, a resolution of 10,000 would indicate that the smallest detectable difference in wavelength was 1/10,000 of the wavelength upon which the instrument was set.

For a prism spectroscope, the prism is frequently the limiting aperture, and it can be shown that when the prism is used at minimum deviation, the resolution is given by

$$\frac{\lambda}{d\lambda} = B \frac{dN}{d\lambda} \tag{6.18}$$

where B is the length of the base of the prism and $dN/d\lambda$ is the dispersion of the prism material.

A diffraction grating consists of a series of precisely ruled lines on a clear (or reflecting) base. Light can pass directly through a grating, but it is also diffracted. As with the slit aperture discussed above, at certain angles the diffracted wavelets reinforce, and maxima are produced when

$$\text{Sin } a = \frac{m\lambda}{S} + \text{Sin } I \tag{6.19}$$

where λ is the wavelength, I is the angle of incidence, S is the spacing of the grating lines and m is an integer, called the *order* of the maxima. Since a depends on the wavelength λ, such a device can be used to separate the diffracted light into its component wavelengths. When used as indicated in Fig. 6.18, the resolution of a grating is given by

$$\frac{\lambda}{d\lambda} = mN \tag{6.20}$$

where m is the order and N is the total number of lines in the grating (assuming the size of the grating to be the limiting aperture of the system).

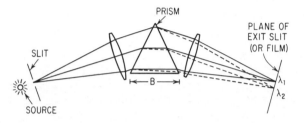

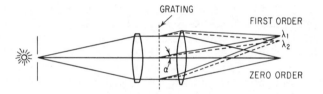

FIG. 6.18. (Upper) Prism Spectrometer. (Lower) Grating Spectrometer.

References

1. Jacobs, "Fundamentals of Optical Engineering," McGraw-Hill, 1943
2. Hardy and Perrin, "The Principles of Optics," McGraw-Hill, 1932
3. Morgan, "Introduction to Geometrical and Physical Optics," McGraw-Hill, 1953
4. Strong, "Concepts of Classical Optics," Freeman, 1958
5. Jenkins and White, "Fundamentals of Optics," McGraw-Hill, 1957

Exercises

1. Find the positions and diameters of the entrance and exit pupils of a 100 mm focal length lens with a diaphragm 20 mm

to the right of the lens, if the lens diameter is 15 mm and the diaphragm diameter is 10 mm.
Ans. Entrance pupil is 25 mm to the right and 12.5 mm in diameter. Exit pupil is 20 mm to the right and 10 mm in diameter.

2. What is the relative aperture ($f/\#$) of the lens of exercise #1 with light incident a) from the left, and b) from the right?
Ans. a) $f/8$ b) $f/10$

3. A telescope is composed of an objective lens $f = 10''$, diameter = $1''$ and an eyelens $f = 1''$, dia. = $1/2''$ which are $11''$ apart. a) locate the entrance and exit pupils and find their diameters. b) determine the object and image fields of view in radians. Assume object and image to be at infinity.
Ans. a) Entrance pupil is at the objective, diameter $1''$. Exit pupil is $1.1''$ to the right of the eyelens and is $0.1''$ diameter. b) for zero vignetting, object field is $\pm.01818$ and image field is $\pm.1818$. For complete vignetting, object field is $\pm.02727$ and image field is $\pm.2727$.

4. A $4''$ focal length $f/4$ lens is used to project an image at a magnification of four times ($m = -4$). What is the numerical aperture in object space and in image space?
Ans. N.A. = 0.1; N.A. = 0.025

5. An optical system composed of two thin elements forms an image of an object located at infinity. The front lens has a $16''$ focal length, the rear lens an $8''$ focal length and the spacing between the two is $8''$. If the exit pupil is located at the rear lens and there is no vignetting, what is the illumination at an image point $3''$ from the axis relative to the illumination on the axis?
Ans. 41%

6. A $6''$ diameter $f/5$ paraboloid mirror is part of an infrared tracker which can tolerate a blur (due to defocusing) of 0.1 milliradians. a) What tolerance must be maintained on the position of the reticle with respect to the focal point? b) What is the tolerance if the system speed is $f/2$?
Ans. a) $\pm.015''$ b) $\pm.0024''$

7. If the hyperfocal distance of a $10''$ focal length $f/10$ lens is $100''$, a) what is the diameter of the acceptable blur spot, and b) what is the closest distance at which an object is "acceptably" in focus? c) Show that the answer to b) is always one-half the hyperfocal distance.
Ans. a) $0.111''$ b) $50''$

8. Compare the image illumination produced by an $f/8$ lens at a point 45° from the axis with that from an $f/16$ lens 30° off axis.
   Ans. the $f/16$ is 56% of the $f/8$

9. Plot the illumination (in the manner of Fig. 6.16) in the diffraction pattern at the focus of a lens with a square aperture, a) along a line through the axis at 90° to a side of the aperture, and b) along a line at 45° (the diagonal) to the sides of the aperture.

10. An optical system is required to image a distant point source as a spot 0.01 mm in diameter. Assuming that all the useful energy in the image spot will be within the first dark ring, what relative aperture (f/number) must the optical system have?
    Ans. $f/7.5$

11. A pinhole camera has no lens, but uses a very small hole a distance from the film to form its image. If we assume that light travels in straight lines, then the image of a distant point source will be a blur whose diameter is the same size as the pinhole. However, diffraction will spread the light into an Airy Disc. Thus, the larger the hole, the larger the geometrical blur but the smaller the diffraction pattern. The sharpest picture will be produced when the geometrical blur is the same size as the central bright spot of the Airy Disc. What size hole should be used when the film is 10 cm. from the hole? (Hint: equate the hole diameter to the diameter of the first dark ring of the Airy Disc given by Eq. 6.15.)
    Ans. 0.037 cm. for $\lambda = 0.55$ microns

12. What is the resolution limit (at the object) for a microscope objective whose acceptance cone has a numerical aperture of, a) 0.25, b) 0.8, c) 1.2?
    Ans. a) .0013 mm, b) .00042 mm, c) .00028 mm

13. What diameter must a telescope objective have if the telescope is to resolve 11 seconds of arc? If the eye can resolve one minute of arc, what is the minimum power of the telescope?
    Ans. 0.5 inch; 5.5 ×

14. Compare the resolution of a prism and a grating. The prism has a 1″ base and its glass has a dispersion of 0.1 per micron. The grating is 1″ wide and is ruled with 15,000 lines per inch.
    Ans. prism resolution 2,540; grating resolution 15,000 1st order, 30,000 2nd order, etc.

7

Optical Materials and Coatings

7.1

To be useful as an optical material, a substance must meet certain basic requirements. It should be able to accept a smooth polish, be mechanically and chemically stable, have a homogeneous index of refraction, be free of undesirable artifacts, and of course transmit (or reflect) radiant energy in the wavelength region in which it is to be used.

The two characteristics of an optical material which are of primary interest to the optical engineer are its transmission and its index of refraction, both of which vary with wavelength. The transmission of an optical element must be considered as two separate effects. At the boundary surface between two optical media, a fraction of the incident light is reflected. For light normally incident on the boundary the fraction is given by

$$R = \frac{(N' - N)^2}{(N' + N)^2} \tag{7.1}$$

where N and N' are the indices of the two media (a more complete expression for Fresnel surface reflection is given in Section 7.9).

Within the optical element, some of the radiation may be absorbed by the material. Assume that a one millimeter thickness of a filter material transmits 25% of the incident radiation at a given wavelength (excluding surface reflections). Then two millimeters will transmit 25% of 25%. Therefore, if t is the transmission of a unit thickness of material, the transmission through a thickness of x units will be given by

$$T = t^x \tag{7.2}$$

This relationship is often stated in the following form, where a is called the absorption coefficient and is equal to $-\log_e t$.

$$T = e^{-ax} \tag{7.3}$$

Thus, it can be seen that the total transmission through an optical element is a sort of product of its surface transmissions and its internal transmission. For a plane parallel plate in air, the transmission of the first surface is given (from Eq. 7.1) as

$$T = 1 - R = 1 - \frac{(N - 1)^2}{(N + 1)^2} = \frac{4N}{(N + 1)^2} \tag{7.4}$$

Now the light transmitted through the first surface is partially transmitted by the medium and goes on to the second surface, where it is partly reflected and partly transmitted. The reflected portion passes (back) through the medium and is partly reflected and partly transmitted by the first surface, and so on. The resulting transmission can be expressed as the infinite series

$$T_{1,2} = T_1 T_2 (K + K^3 R_1 R_2 + K^5 (R_1 R_2)^2 + K^7 (R_1 R_2)^3 + \cdots\cdots)$$

$$= \frac{T_1 T_2 K}{1 - K^2 R_1 R_2} \tag{7.5}$$

where T_1 and T_2 are the transmissions of the two surfaces, R_2 and R_1 are the reflectances of the surfaces, and K is the transmittance of the medium between them. (This equation can also be used to determine the transmission of two or more elements, e.g. flat plates, by finding first $T_{1,2}$, then using $T_{1,2}$ and T_3 together, and so on.)

If we set $T_1 = T_2 = 4N/(N + 1)^2$ from Eq. 7.4 into Eq. 7.5, and assume that $K = 1$, we find that the transmission, including all internal reflections, of a completely non-absorbing plate is given by

$$T = \frac{2N}{(N^2 + 1)} \tag{7.6}$$

Similarly, the reflection is given by

$$R = 1 - T = \frac{(N - 1)^2}{(N^2 + 1)} \tag{7.7}$$

It should be emphasized that the transmission of a material, being wavelength dependent, may not be treated as a simple number over any appreciable wavelength interval. For example, suppose that a filter is found to transmit 45% of the incident energy between 1 and 2 microns. It cannot be assumed that the transmission of two such filters in series will be .45 × .45 = 20% unless they have a uniform spectral transmission (neutral density). To take an extreme example, if the filter transmits nothing from 1 to 1.5 microns and 90% from 1.5 to 2 microns, its "average" transmission will be 45%. However, two such filters, when combined, will transmit zero from 1 to 1.5 microns, and about 81% from 1.5 to 2 microns, for an "average" transmission of about 40%, rather than the 20% which two neutral density filters would transmit.

The density of a filter is the log of its opacity (the reciprocal of transmittance), thus

$$D = \log \frac{1}{T}$$

where D is the density and T is the transmittance of the material. Note that transmittance does not include surface reflection losses; thus, density is directly proportional to thickness. To a fair approximation, the density of a "stack" of filters is the sum of the individual densities.

The index of refraction of an optical material varies with wavelength as indicated in Fig. 7.1 where a long spectral range is shown. The dashed portions of the curve represent absorption bands. Notice that at each absorption band the index rises markedly, and then begins to drop with increasing wavelength. As the wavelength continues to increase, the slope of the curve levels out until the next absorption band is approached, when the slope increases again. For optical materials we need concern ourselves with only one section of the curve, since most optical materials have an absorption band in the ultraviolet and another in the infrared and their useful spectral region lies between the two.

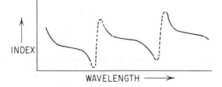

FIG. 7.1. Dispersion curve of an optical material. The dashed lines indicate absorption bands. (Anomolous dispersion).

Many investigators have attacked the problem of devising an equation to describe the irrational variation of index with wavelength. Such expressions are of value in interpolating between measured points on the dispersion curve and also in the study of the secondary spectrum characteristics of optical systems. Several of these dispersion equations are listed below.

Cauchy $$N = a + \frac{b}{\lambda^2} + \frac{c}{\lambda^4} + \cdots \cdots \qquad (7.8)$$

Hartmann $\qquad N = a + \dfrac{b}{(c - \lambda)^{1.2}}$ $\qquad$ (7.9)

Hartmann $\qquad N = a + \dfrac{b}{(c - \lambda)} + \dfrac{d}{(e - \lambda)}$ $\qquad$ (7.10)

Conrady $\qquad N = a + \dfrac{b}{\lambda} + \dfrac{c}{\lambda^{3.5}}$ $\qquad$ (7.11)

Kettler-Drude $\qquad N^2 = a + \dfrac{b}{c - \lambda^2} + \dfrac{d}{e - \lambda^2} + \cdots$ $\qquad$ (7.12)

Sellmeier $\qquad N^2 = a + \dfrac{b\lambda^2}{c - \lambda^2} + \dfrac{d\lambda^2}{e - \lambda^2} + \cdots$ $\qquad$ (7.13)

Hertzberger $\qquad N = a + b\lambda^2 + \dfrac{e}{(\lambda^2 - .035)} + \dfrac{d}{(\lambda^2 - .035)^2}$ $\qquad$ (7.14)

The constants (a, b, c, etc.) are, of course, derived for each individual material by substituting known index and wavelength values and solving the resulting simultaneous equations for the constants. The Cauchy equation obviously allows for only one absorption band at zero wavelength. The Hartmann formula is an empirical one, but does allow the absorption band to be located at wavelength c, and an additional term, as in Eq. 7.10, will accomodate the second band at wavelength e. The Hertzberger expression is an approximation of the Kettler-Drude equation and is reliable through the visible to about one micron in the near infrared. The Conrady equation is empirical and designed for the visible region. All these equations suffer from the drawback that the index approaches infinity as the absorption wavelength is approached. Since little use is made of any material close to an absorption band, this is usually of small consequence.

The dispersion of a material is the rate of change of index with respect to wavelength, that is, $dN/d\lambda$. From Fig. 7.2, it can be seen that the dispersion is large at short wavelengths and becomes less at longer wavelengths. At still longer wavelengths, the dispersion increases again as the long wavelength absorption band is approached.

For materials which are used in the visible spectrum, the refractive characteristics are conventionally specified by giving two numbers, the index of refraction for the sodium D line (0.5893 microns) and the Abbe V-number, or reciprocal relative dispersion. The V-number, or V-value, is defined as

$$V = \frac{N_D - 1}{N_F - N_C}$$ (7.15)

where N_D, N_F and N_C are the indices of refraction for the sodium D line, the hydrogen F line (0.4861 microns) and the hydrogen C line (0.6563 microns) respectively. Note that $\Delta N = N_F - N_C$ is a measure of the dispersion, and its ratio with $N_D - 1$ (which effectively indicates the basic refracting power of the material) gives the dispersion relative to the amount of bending that a light ray undergoes.

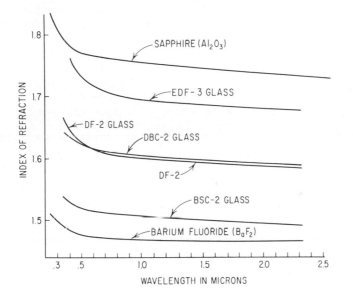

FIG. 7.2. The dispersion curves for several optical materials.

For optical glass, these two numbers describe the glass type and are conventionally written $(N_D - 1):V$. For example, a glass with an N_D of 1.517 and a V of 64.5 would be identified as 517:645.

For many purposes, the index and V-value are sufficient information about a material. For secondary spectrum work, however, it is necessary to know more, and the relative partial dispersion

$$P_C = \frac{N_D - N_C}{N_F - N_C} \qquad (7.16)$$

is frequently used for this purpose. P_C is a measure of the rate of change of the slope of the index vs wavelength curve (i.e. the second derivative).

The index of refraction values conventionally given in catalogs, handbooks, etc., are those arrived at by measuring a sample piece in air, and are thus the index relative to the index of air at the

the wavelength, temperature, humidity and pressure encountered in the measurement. Since the index is used in optical calculations as a relative number, this causes no difficulty when the index of air is assumed to be 1.0.

7.2 Optical Glass

Optical glass is almost the ideal material for use in the visual and near infrared spectral regions. It is stable, readily fabricated, homogenous, clear, and economically available in a fairly wide range of characteristics.

Figure 7.3 gives some indication of the variety of American-made optical glasses. Each point in the figure represents a glass whose N_D is plotted against its V-value; note that the V-values are conventionally plotted in reverse, i.e. descending, order. Glasses are somewhat arbitrarily divided into two groups, the crown glasses and the flint glasses, crowns having a V-value of 50 or more, flints 50 or less. The "ordinary" glasses are located along the glass line.

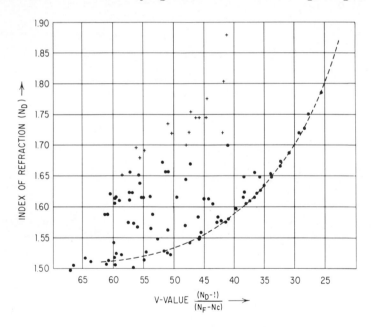

FIG. 7.3. The Glass Veil. Index (N_D) plotted against the reciprocal relative dispersion. The ordinary glasses are those along the dashed "glass line." Above the line are the barium glasses and the rare earth glasses (crosses).

The addition of lead to crown glass causes its index to rise, and its V-value to decrease, along the glass line. Immediately above the glass line are the barium crowns and flints; these are produced by the addition of barium to the glass mix. This has the effect of raising the index without markedly lowering the V-value. The rare earth glasses (Fig. 7.4) are a completely different family of glasses based on the rare earths instead of silicon dioxide (which is the major constituent of the other glasses).

Type	N_D	V	$N_F - N_C$	$N_F - N_D$	P_C	S.G.	Rel. Cost
EK-110 (RE-10)	1.6968	56.2	.01241	.00874	.296	4.1	20.
EK-210 (RE-11)	1.7340	51.0	.01439	.01016	.294	4.4	20.
EK-310	1.7450	46.4	.01605	.01138	.291	4.5	20.
EK-320	1.7445	45.8	.01625	.01153	.290	4.5	20.
EK-325 (RE-12)	1.7445	45.6	.01634	.01157	.292	4.5	25.
EK-330	1.7551	47.2	.01600	.01133	.292	4.7	25.
EK-430	1.7767	44.7	.01738	.01232	.291	4.6	25.
EK-450	1.8037	41.8	.01924	.01368	.289	4.6	25.
				$N_D - N_C$			
RE-15 (LaF-2)	1.7440	44.7	.01667	.00487	.292	4.1	—
RE-21 (LaK-7)	1.6510	58.5	.01114	.00336	.302	3.95	20
RE-24 (LaK-2, 12)	1.6780	55.5	.01221	.00365	.299	3.88	—
RE-25 (LaK-10)	1.7200	50.3	.01432	.00432	.295	3.94	—
691:548 (RE-23)	1.6910	54.8	.01261	.00370	.293	4.03	20
700:480 (RE-20)	1.7000	48.0	.01457	.00424	.291	3.99	10

FIG. 7.4. Characteristics of Rare Earth Optical Glasses.

The table of Fig. 7.5 lists the characteristics of the most common optical glass types. Each glass type in the table is available from the major glass manufacturers so that all types listed are readily obtainable. The index data given are taken from the Bausch and Lomb catalogue; the equivalent glasses from other suppliers may have slightly different nominal characteristics.

Formerly, optical glass was made by heating the ingredients in a large clay pot, or crucible, stirring the molten mass for uniformity, and carefully cooling the melt. The hardened glass was then broken into chunks. Currently the molten glass is more likely to be poured into a large mould; this gives better control over the size of the pieces of glass available. Many barium glasses and all the rare earth glasses are processed in platinum crucibles, since the molten glass tends to attack the walls of a clay pot and the dissolved pot materials affect the glass characteristics. In extremely large volume production, a continuous process is used, with the raw materials going in one end of the furnace and emerging as extruded strip or rod glass at the other end. Raw glass is frequently pressed into blanks, which are roughly the size and shape of the finished element. The final stage before the glass is ready for use is annealing. This is a slow cooling process, which may

TYPE	.3650 μ	N_h .4047 μ	N_F .4861 μ	N_D .5893 μ	N_C .6563 μ	$N_{A'}$ .7665 μ	1.014 μ	1.701 μ	2.325 μ	$N_F - N_C$	V	P_C $\frac{N_D-N_C}{N_F-N_C}$	S.G.	expan.	stain	bub.	rel. cost
BSC 2 517:645	1.53644	1.53043	1.52262	1.51700	1.51461	1.51179	1.5076	1.4989	1.4893	.00801	64.5	.298	2.53	80	1	1	1.2
C 1 523:586	1.54505	1.53819	1.52929	1.52300	1.52036	1.51729	1.5130	1.5051	1.4970	.00893	58.6	.296	2.53	94	2	2	1.0
LBC 1 541:599	1.56326	1.55633	1.54736	1.54100	1.53833	1.53522	1.5309	1.5229	1.5146	.00903	59.9	.296	2.84	87	2	2	1.5
LBC 2 573:574	1.59723	1.58951	1.57953	1.57250	1.56956	1.56619	1.5616	1.5533	1.5450	.00997	57.4	.295	3.21	83	2.5	2	1.5
DBC 1 611:588	—	1.62867	1.61832	1.61100	1.60793	1.60439	—	—	—	.01039	58.8	.295	3.58	73	4	4	2.2
DBC 2 617:549	1.64516	1.63634	1.62493	1.61700	1.61370	1.60995	1.6048	1.5963	1.5881	.01123	54.9	.294	3.66	72	3	4	2.3
DBC 620:603	—	1.63748	1.62724	1.62000	1.61696	1.61342	—	—	—	.01028	60.3	.296	3.58	76	5	3	3.3
DBC 638:555	—	1.65772	1.64611	1.63800	1.63461	1.63074	1.6254	1.6157	1.6057	.01150	55.5	.295	3.59	76	4.5	3	3.3
DBC 651:558	—	1.67097	1.65924	1.65100	1.64757	1.64362	—	—	—	.01167	55.8	.294	3.81	84	4.5	3	3.6
CF 1 529:516	—	1.54633	1.53584	1.52860	1.52560	1.52217	1.5174	1.5092	1.5009	.01024	51.6	.293	2.73	87	1	1	1.6
BF 2 605:436	—	1.62987	1.61518	1.60530	1.60130	1.59682	1.5910	1.5823	1.5746	.01388	43.6	.288	3.52	90	2	2	2.2
BF 670:472	—	1.69472	1.68004	1.67000	1.66585	1.66123	—	—	—	.01419	47.2	.293	3.78	72	1	3	4.1
LF 1 573:425	1.60811	1.59637	1.58208	1.57250	1.56861	1.56425	1.5585	1.5499	1.5413	.01347	42.5	.289	3.19	91	1.5	1	2.4
LF 2 580:410	—	1.60459	1.58957	1.57950	1.57544	1.57088	1.5650	1.5559	1.5476	.01413	41.0	.287	3.27	99	1.5	1	2.2
DF 1 605:380	—	1.63358	1.61638	1.60500	1.60045	1.59538	1.5889	1.5793	1.5710	.01593	38.0	.286	3.49	86	1	1	1.2
DF 2 617:366	1.66280	1.64740	1.62904	1.61700	1.61218	1.60684	1.6001	1.5902	1.5817	.01686	36.6	.286	3.64	89	1.5	1	1.2
DF 3 621:362	—	1.65197	1.63325	1.62100	1.61610	1.61066	1.6066	—	—	.01715	36.2	.286	3.67	87	2	1	1.3
EDF 1 649:338	—	1.68397	1.66275	1.64900	1.64355	1.63754	1.6301	1.6198	1.6112	.01920	33.8	.284	3.91	85	2	1	1.3
EDF 2 689:309	—	1.72996	1.70501	1.68900	1.68271	1.67584	1.6673	1.6562	1.6475	.02230	30.9	.282	4.24	85	3	2	1.5
EDF 3 720:293	—	1.76542	1.73766	1.72000	1.71309	1.70555	1.6963	1.6846	1.6757	.02457	29.3	.281	4.51	77	2	3	1.6
EDF 673:322	—	1.71084	1.68751	1.67250	1.66663	1.66012	1.6521	1.6414	1.6328	.02088	32.2	.281	3.97	85	1	1	1.4

FIG. 7.5. Characteristics of optical glasses (selected for availability). Abbreviations are: "S.G.", specific gravity; "expan.", thermal coefficient of expansion × 10⁻⁷ per °C; "stain", an indication of the chemical stability ranked from 1 (best) to 5 (worst); "bub.", an indication of bubble frequency, ranked from 1 (fewest bubbles) to 4 (most bubbles); "rel. cost", a rough indication of the relative cost per pound, normalized to 1.0 for C-1 (523:586).

take several days, which relieves strains in the glass, assures homogeniety of index, and brings the index up to the catalog value.

The characteristics of optical glass vary somewhat from melt to melt (because of variations in composition and processing) and also due to variations in annealing procedures. Ordinarily the lower index glasses (to $N = 1.55$) are supplied to a tolerance of $\pm.001$ on the catalogue value of N_D; the higher index glasses may vary $\pm.0015$ from the nominal index. Similarly the V-value will vary from the catalogue value. Typical tolerances on V-value are ± 0.3 for V-values below 46; ± 0.4 from 46 to 58; ± 0.5 for V-values above 58. Most glass manufacturers will supply glass to closer tolerances at an increased price.

Optical glass may be obtained in hundreds of different types; complete information is best obtained from the manufacturer's catalogue. Optical glass is made in large volume by the following:

Bausch and Lomb Company	Rochester, New York
Corning Glass Works	Corning, New York
Eastman-Kodak Company	Rochester, New York
Pittsburgh Plate Glass Co.	Pittsburgh, Pennsylvania
Schott und Gen. (Jena)	Germany
Chance Bros., Ltd.	England
Parra-Mantois et Cie	France
Ohara Glass	Japan

There are also a number of firms which do not manufacture glass but which specialize in moulding raw glass into rough blanks, ready for processing.

Figures 7.6 and 7.7 give an indication of the spectral transmission of optical glasses. In general, most optical glasses transmit well from 0.4μ to 2.0μ. The heavy flints tend to absorb more at the short wavelengths and transmit more at the long wavelengths. The rare earth glasses also absorb in the blue region. Since the transmission of a glass is affected greatly by minute impurities, the exact characteristics of any given glass will vary somewhat from batch to batch, even when made by the same manufacturer.

Most optical glasses turn brown when exposed to nuclear radiation because of increased absorption of the short (blue) wavelengths. To provide glasses which can be used in a radiation environment, the glass manufacturers have developed "protected" or "non-browning" glasses containing cerium. These glasses will tolerate radiation doses to the order of a million roentgens. Glasses equivalent to BSC-2, LBC-1, LBC-2, DF-2 and EDF-1 are available in non-browning formulations. Fused quartz glass, which is discussed in the next section, is almost pure SiO_2 and is extremely resistant to radiation browning.

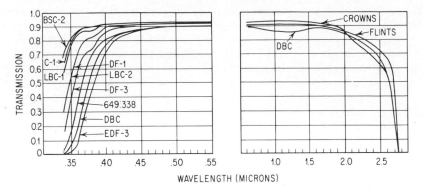

FIG. 7.6. Transmission of Glass Technology, Inc. (Hayward) optical glasses for a thickness of 0.5 inch. Data includes surface reflection losses.

Although not strictly "optical glass", ordinary window glass and plate glass are frequently used when cost is an important factor. The index of window glass ranges from about 1.514 to about 1.52, depending on the manufacturer. Ordinary window glass is slightly greenish, due primarily to modest amounts of absorption in the red and blue wavelengths; the red absorption continues to about 1.5 microns. Window glass is also available in "water white" quality, without the greenish tint. For elements with one or two plane surfaces and with modest precision requirements, window glass can often be used without further processing; the accuracy of the plane surfaces is surprisingly good. By special selection, plane parallels can be obtained which meet fairly rigorous requirements. The secret here is to

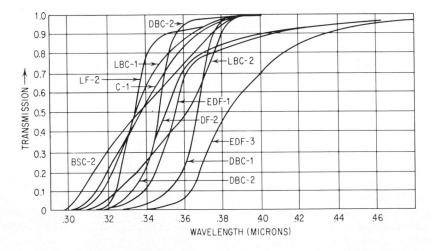

FIG. 7.7. Transmission of Bausch & Lomb optical glasses for a thickness of 10 mm. Reflection losses are not included.

avoid pieces cut from the edge of the large sheets in which this type of glass is made; the center sections are usually far more uniform in surface and thickness.

7.3 Special Glasses

Several glasses are available which differ sufficiently from the standard optical glasses to deserve special mention.

Low-Expansion Glasses: In applications where the elements of an optical system are subject to strong thermal shocks (as in projection condensers) or where extreme stability in the presence of temperature variations is necessary (such as astronomical telescope reflectors or laboratory instruments), it is desirable to use a material with a low thermal coefficient of expansion.

A number of borosilicate glasses are made with expansion coefficients which are less than half that of ordinary glass. Corning's Pyrex # 7740 and # 7760, have expansion coefficients between 30 and 40 × 10^{-7} per °C. The index of refraction of these glasses is about 1.474 and their density is about 2.2. Unfortunately they are usually afflicted with veins and striations so that they are suitable only for applications such as condensing systems when used as refracting elements. They are widely used for test plates and for mirrors. Some of these materials are yellowish, but others are available in a clear white grade.

Another low expansion glass is fused quartz, which is also called fused silica glass. This material is essentially pure (more or less, depending on the grade and manufacturer) silicon dioxide (SiO_2) and has an extremely low expansion coefficient of 5.5×10^{-7} per °C. It is usually made by fusing ground crystalline quartz. Fused quartz can be obtained in grades with homogeniety almost equal to that of good optical glass; the homogenicty seems to be limited by a slight granularity, probably due to the method of fabrication. Fused quartz has excellent spectral transmission characteristics, extending further into both the ultraviolet and infrared than ordinary optical glass. For this reason it is frequently used in spectrophotometers, infrared equipment and ultraviolet devices. The excellent thermal stability of fused quartz is responsible for its use where extremely precise reflecting surfaces are required. Large mirrors and test plates are frequently made from fused quartz for this reason. As previously mentioned, pure fused quartz is highly resistant to radiation browning. The index of refraction and transmission of fused quartz are given in Fig. 7.8. Note that the absorption bands indicated are not of the type indicated in Fig. 7.1, but are due to impurities and are thus subject to elimination, as indicated by the range of transmissions given.

A new class of materials, which are partially crystalized glasses, shows great promise for use as thermally stable mirror substrates, since they can be fabricated with a zero thermal expansion coefficient. Owens-Illinois "CER-VIT" polishes well and is available with a thermal coefficient of zero, $\pm 1 \times 10^{-7}$.

Wavelength (microns)	Index at 24 °C	Transmission 10 mm thick (incl. refl. losses)
.17		0.0 to 0.56 (depending on purity)
.1855	1.5746*	0.0 to .78
.2026	1.54725*	.3 to .84
.2573	1.50384*	.58 to .90
.2749	1.49624*	.88 to .92
.35	1.47701	.93
.40	1.47021	.93
.45	1.46564	.93
.4861 (F)	1.46320	.93
.5	1.46239	.93
.55	1.45997	.93
.5893 (D)	1.45846	.93
.60	1.45810	.93
.6563 (C)	1.45642	.93
.70	1.45535	.93
.80	1.45337	.93
1.0	1.45047	.93
1.35	Absorption Band	.76 to .93
1.5	1.44469	.93
2.0	1.43817	.93
2.2	Absorption Band	.50 to .93
2.5	1.42991	.93
2.7	Absorption Band	0 to .8
3.0	1.41937	.45 to .85
3.5	1.40601	.6 to .7
4.0		.1 to .15

$* N$ at "room temperature" $V = 67.6$ $Pc = .301$

$\Delta N = 10^{-5} \Delta t$ (°C) visible, to $0.4 \times 10^{-5} \Delta t$ at 3.5μ

Dispersion equation $N^2 = 2.978645 + \dfrac{0.008777808}{\lambda^2 - 0.010609} + \dfrac{84.06224}{\lambda^2 - 96.0}$ yields values about .00042 less than table.

FIG. 7.8. Optical characteristics of fused quartz

Infrared Transmitting Glasses:

A number of special "infrared" glasses are available. Some of these are much like extremely dense flint glasses, with index values of 1.8 to 1.9 and transmitting to 4 or 5 microns. The arsenic glasses transmit even further into the infrared. Arsenic modified selenium glass transmits from 0.8 to 18 microns, but will soften and flow at 70° C. It has the following index values:2.578

at 1.014μ; 2.481 at 5μ; 2.476 at 10μ; 2.474 at 19μ. Arsenic tri-sulfide glass transmits from 0.6 to 13 microns and is somewhat brittle and soft. Index values: 2.6365 at 0.6μ; 2.4262 at 2μ; 2.4073 at 5μ; 2.3645 at 12μ.

7.4 Crystalline Materials

The valuable optical properties of certain natural crystals have been recognized for years, but until recently the usefulness of these materials has been severely limited by the scarcity of pieces of the size and quality required for optical applications. However, many crystals are now available in synthetic form. They are grown under carefully controlled conditions to a size and clarity otherwise unavailable.

The table of Fig. 7.9 lists the salient characteristics of a number of useful crystals. The transmission range is indicated in microns for a 2 mm thick sample; the wavelengths given are the 10% transmission points. Indices are given for several wavelengths in the transmission band.

Crystal quartz and calcite are infrequently used because of their birefringence, which limits their usefulness almost entirely to polarizing prisms and the like. Sapphire is extremely hard and must be processed with diamond powder. It is used for windows, interference filter substrates, and occasionally for lens elements. The halogen salts have good transmission characteristics, but their physical properties often leave much to be desired, since they tend to be soft, fragile, and occasionally hygroscopic.

Germanium and especially silicon are widely used for re-fracting elements in infrared devices. Silicon is very much like glass in its physical characteristics, and can be processed with ordinary glass working techniques. Both are metallic in appearance, being completely opaque in the visible. Their extremely high index of refraction is a joy to the lens designer since the weak curvatures which result tend to produce designs of a quality which cannot be duplicated in comparable glass systems. Special low reflection coatings are necessary since the surface reflection (per Eq. 7.1 et seq.) is very high.

Worthy of special mention is Calcium Fluoride, or Fluorite. This material has excellent transmission characteristics in both U.V. and I.R. which make it valuable for instrumentation pur-poses. In addition, its partial dispersion characteristics are such that it can be combined with optical glass to form a lens system which is free of secondary spectrum. Its physical properties are not outstanding since it is soft, fragile, resists weathering poorly, and has a crystal structure which sometimes makes polishing difficult. In exposed applications, the Fluorite element can sometimes be sandwiched between glass elements to protect

Material	Transmission range (microns)	Index	Remarks
Crystal quartz (SiO_2)	.12 - 4.5	$N_o = 1.544$, $N_e = 1.553$	Birefringent
Calcite ($CaCO_3$)	.2 - 5.5	$N_o = 1.658$, $N_e = 1.486$	Birefringent
Rutile (TiO_2)	.43 - 6.2	$N_o = 2.62$, $N_e = 2.92$	Birefringent
Sapphire (Al_2O_3)	.14 - 6.5	1.834 @ .265, 1.755 @ 1.01, 1.586 @ 5.58	Hard, slightly birefringent
Strontium titanate ($SrTiO_3$)	.4 - 6.8	2.490 @ .486, 2.292 @ 1.36, 2.100 @ 5.3	I.R. immersion lenses
Magnesium fluoride (MgF_2)	.11 - 7.5	$N_o = 1.378$, $N_e = 1.390$	I.R. optics, low reflection coatings
Lithium fluoride (LiF)	.12 - 9.	1.439 @ .203, 1.38 @ 1.5, 1.109 @ 9.8	Prisms, windows, apochromatic lenses
Calcium fluoride (CaF_2)	.13 - 12.	See Fig. 7.10	Same as LiF
Barium fluoride (BaF_2)	.25 - 15.	1.512 @ .254, 1.468 @ 1.01, 1.414 @ 11.0	Windows
Sodium chloride (NaCl)	.2 - 26.	1.791 @ .2, 1.528 @ 1.6, 1.175 @ 27.3	Prisms, windows, hygroscopic
Silver chloride (AgCl)	.4 - 28.	2.096 @ 0.5, 2.002 @ 3., 1.907 @ 20.	Ductile, corrosive, darkens
Potassium bromide (KBr)	.25 - 40.	1.590 @ .404, 1.536 @ 3.4, 1.463 @ 25.1	Prisms, windows, soft, hygroscopic
Potassium iodide (KI)	.25 - 45.	1.922 @ .27, 1.630 @ 2.36, 1.557 @ 29	Soft, hygroscopic
Cesium bromide (CsBr)	.3 - 55.	1.709 @ 0.5, 1.667 @ 5, 1.562 @ 39	Hygroscopic, prisms and windows
Cesium iodide (CsI)	.25 - 80.	1.806 @ 0.5, 1.742 @ 5, 1.637 @ 50	Prisms and windows
Silicon (Si)	1.2 - 15.	3.498 @ 1.36, 3.432 @ 3, 3.418 @ 10	I.R. optics
Germanium (Ge)	1.8 - 23.	4.102 @ 2.06, 4.033 @ 3.42, 4.002 @ 13.	I.R. optics

Transmission range wavelengths are the 10% transmission points for a 2 mm thickness. N_o and N_e are indices for the ordinary and extraordinary rays.

FIG. 7.9. Characteristics of optical crystals

its surfaces. The table of Fig. 7.10 lists selected index and transmission values for Fluorite.

The Eastman Kodak Company produces a series of infrared transmitting materials, called Irtran 1, 2, etc., which are pressed polycrystalline compounds. The Irtrans tend to be quite rugged and are available in sizes of 5″ to 7″ in diameter. Apparently because of the manufacturing process, there is a tendency to scatter short wavelength radiation; Irtran 1 and 2, for example, are quite milky and translucent in the visible, and their useful

Wavelength (microns)	Index	Absorption coefficient (cm^{-1})
0.2	1.49531	—
0.3	1.45400	—
0.4	1.44186	—
0.4861 (F)	1.43704	—
0.5893 (D)	1.43384	—
0.6563 (C)	1.43249	—
1.014	1.42884	—
2.058	1.42360	—
3.050	1.41750	—
4.0	1.40963	—
5.0	1.39908	—
7.	—	.02
8.	—	.16
8.84	1.33075	—
9.	—	.64
10.	—	1.8

$V = 95.3 \qquad P_c = .297$

$\Delta N = -10^{-5} \Delta T \ (°C)$

FIG. 7.10. Index and transmission of Calcium Fluoride (CaF_2) for various wavelengths.

spectral range begins at about 2 microns. The following table indicates the transmission range (10% points for a thickness of .080″) and selected index values. Complete data are available from the manufacturer.

Irtran 1 (Poly. MgF_2) 0.45 to 9.2 microns

λ	1.014μ	2.153μ	4.253μ	6.238μ
N	1.3776	1.3708	1.3489	1.3122

Irtran 2 (Poly. ZnS) 0.57 to 14.7 microns

λ	1.014μ	2.153μ	4.0μ	8.0μ	13.0μ
N	2.2897	2.2616	2.2501	2.2212	2.1507

Irtran 3 (Poly. CaF_2) 0.2 to 11.5 microns

λ	$.656\mu$	1.083μ	2.058μ	4.0μ	8.25μ
N	1.4324	1.4284	1.4236	1.4096	1.3444

Irtran 4 (Poly. ZnSe) 0.48 to 21.8 microns

λ	$.489\mu$	1.0μ	2.0μ	4.0μ	$10.\mu$
N	2.786	2.491	2.448	2.434	2.408

Irtran 5 (Poly. MgO) 0.39 to 9.4 microns

λ	$.489\mu$	1.0μ	2.0μ	4.0μ	6.0μ
N	1.7461	1.7233	1.7089	1.6912	1.5961

Irtran 6 (Poly. CdTe) 0.20 to 30 microns $N - 2.673 @ 10\mu$

7.5 Plastic Optical Materials

Plastics are rarely used for precision optical elements. A great deal of effort was made to develop plastics for optical systems during the Second World War and a few systems incorporating plastics were produced. Today the use of plastics is largely restricted to novelty items such as magnifiers and toys, with the notable exceptions of low-priced camera lenses and some Schmidt aspheric corrector plates. In these applications, the fact that plastic may be conveniently molded gives it a great advantage over glass. For the toys, novelties and box camera lenses, the advantage is mass production at extremely low cost. For the Schmidt corrector plate (and the box camera lens) the advantage is the ease of production of a non-spherical surface which is essential to the design and is expensive to execute in glass.

The obvious advantages of plastic — that it is light and relatively shatterproof — are offset by a number of disadvantages. It is soft and scratches easily. Except by molding it is difficult to fabricate. Styrene plastic is frequently hazy, scatters light, and is occasionally yellowish. Plastics tend to soften at 60 to 80° C. In some plastics the index will change as much as .0005 over a period of time. Most plastics will absorb water and change dimensionally; almost all are subject to cold flow under pressure. The thermal expansion coefficient is almost ten times that of glass, being 7 or $8 \times 10^{-5}/°C$.

The density of plastics is low, usually to the order of 1.0 to 1.1. The three most widely used plastics for optical purposes are polystyrene, polycyclohexyl methacrylate and polymethyl methacrylate, frequently referred to as styrene, C. H. M., and Lucite respectively. Styrene is roughly comparable to a flint glass with $N_D = 1.591$ and $V = 31$. The methacrylates have higher V-values and lower indices; for C. H. M., $N_D = 1.506$ and $V = 56$, for Lucite $N_D = 1.491$ and $V = 61$. The index of refraction varies strongly with temperature; typically ΔN is about $-.00013$ per °C.

7.6 Absorption Filters

Absorption filters are composed of materials which transmit light selectively; that is, they transmit certain wavelengths more than others. A small percentage of the incident light is reflected, but the major portion of the energy which is not transmitted through the filter is absorbed by the filter material. Obviously, every material discussed in the preceding sections of this chapter is, in the broadest sense, an absorption filter, and occasionally these materials are introduced into optical systems as filters. However, most filters are made by the addition of metallic salts to clear glass or by dyeing a thin gelatin film to produce a more selective absorption than is available in "natural" materials.

The prime source of dyed gelatin filters is the Eastman Kodak Company, whose line of Wratten filters is widely used for applications where the versatility of dyed gelatin is required and the environmental requirements are not too severe. Gelatin filters are usually mounted between glass to protect the soft gelatin from damage.

The number of coloring materials which are suitable for use in optical filter glass is limited, and the types of filter glass available are thus not as extensive as one might desire. In the visible region, there are several main types. The red, orange and yellow glasses all transmit the red and near infrared and have a fairly sharp cut-off, as indicated in Fig. 7.11. The position of this cut-off determines the apparent color of the filter. Green filters tend to absorb both the red and blue portions of the spectrum. Their transmission curves often resemble the spectral sensitivity curve of the eye. Blue optical glass filters can be a disappointment, since they occasionally transmit not only blue light, but some green, yellow, orange, and frequently a sizeable amount of red light as well. The purple filters transmit both the red and blue ends of the spectrum, with fair suppression of the yellow and green spectral regions. Filter glass is manufactured by most optical glass companies as well as a number of establishments which make commercial colored glass (as opposed to "optical" glass which is more carefully controlled).

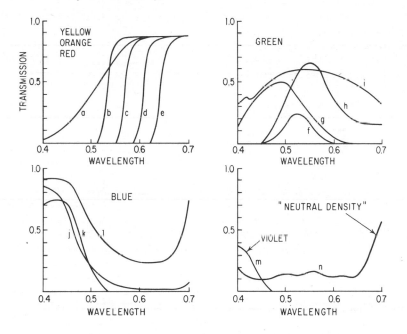

FIG. 7.11. Spectral transmission curves for several optical glass filters.

The transmission characteristics of glass filters vary from melt to melt for any given type. If a filter application requires that the transmission be accurately controlled, it is frequently necessary to adjust the finished thickness of the filter to compensate for these variations. The red filters are probably the most variable; since they are quite sensitive to heat, many red glasses cannot be re-pressed into blanks. Spectral transmission data for filters is usually given for a specific thickness and includes the losses due to Fresnel surface reflections. To determine the transmission for thicknesses other than the nominal value, the transmittance, that is the "internal" transmission of the piece without the reflection losses, must be determined. In most cases, it is sufficient to divide the transmission by Eq. 7.4 to get the transmittance. Then Eq. 7.2 or 7.3 can be used to determine the transmittance of the new thickness. This transmittance times the T of Eq. 7.4 will then give the total transmission for the filter.

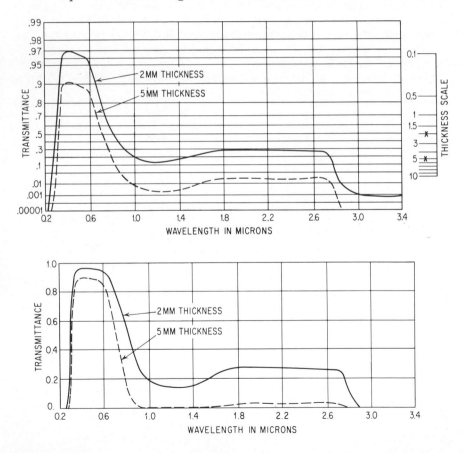

FIG. 7.12. Spectral transmittance of Schott KG 2 heat absorbing filter glass. The upper graph is plotted on a log-log scale. Note that the vertical spacing between the two plots is equal to the distance from 2 to 5 on the thickness scale at the right. The same data is plotted on a conventional linear scale in the lower figure for comparison.

This process is greatly simplified by the use of a Log-Log plot of the transmittance. The Schott catalog of Jena glass filters makes use of this type of scale. A transparent overlay makes it possible to evaluate instantly the effect of a thickness change. A study of Fig. 7.12 will indicate the utility of this type of a transmittance plot; the same filter is shown in two thicknesses on a Log-Log scale in the upper figure and on a linear scale in the lower. Against the Log-Log scale, the thickness change is effected by a simple vertical displacement of the plot. The amount of the displacement is given by the thickness scale at the right. Notice how much more information this type of plot can give (and how much more is required to prepare one!). The data plotted in this form is transmittance; to determine the total transmission of the filter, the surface reflection losses must be taken into account, either by Eq. 7.4 or 7.5.

Glass filters are also available to transmit either the ultraviolet or infrared regions of the spectrum without transmitting the visible. Typical transmission plots for these filters are shown in Fig. 7.13. Heat absorbing glasses are designed to transmit visible

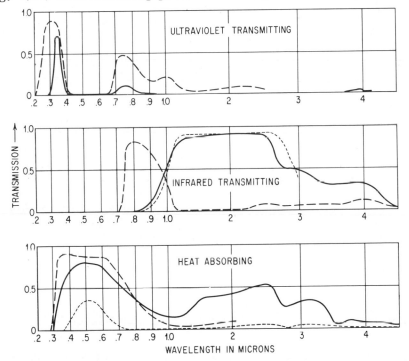

FIG. 7.13. Transmission characteristics of special purpose glass filters. U. V. transmitting: solid line, Corning 7-60; dashed, Corning 7-39. I. R. transmitting: solid, Corning 7-56 (#2540); dashed, Corning 7-69; dotted, Schott UG-8. Heat absorbing: solid, Corning 1-59 extra light Aklo; dashed, Pittsburgh Plate Glass #2043 Phosphate - 2 m.m.; dotted, Corning 1-56 dark shade Aklo.

light and absorb infrared energy. These are frequently used in projectors to protect the film from the heat of the projection lamp. Since they absorb large quantities of radiant energy, they become hot themselves and must be carefully mounted and cooled to avoid breakage from thermal expansion. From the spectral transmission characteristics given in Fig. 7.13, it is apparent that the phosphate heat absorbing glass is more efficient than the Aklo; the phosphate glass is subject to large bubbles and inclusions which do not, however, prevent its use in most applications.

7.7 Diffusing Materials

A piece of white blotting paper is an example of a (reflecting) diffusing material. Light which strikes its surface is scattered in all directions; as a result, the paper appears to have almost the same brightness regardless of the angle at which it is illuminated or the angle from which it is viewed. A perfect, or Lambertian, diffuser is one which has the same apparent brightness from any angle; thus the radiation emitted per unit area is given by $I_0 \cos \theta$, where θ is the angle to the surface normal and I_0 is the intensity of an element of area in a direction perpendicular to the surface.

There are a number of quite good reflecting diffusers with relatively high efficiencies. Matte white paper is a very convenient one and reflects 70 to 80 per cent of the incident visible light. Magnesium oxide and magnesium carbonate are frequently used in photometric work since their efficiencies are high, to the order of 97 or 98 per cent.

Transmitting diffusers are used for such applications as rear projection screens and to produce even illumination. The most commonly used are opal glass and ground glass (Fig. 7.14). Opal glass contains a suspension of minute colloidal particles and diffuses by multiple scattering from these particles. The transmitted light is slightly yellowish since the shorter wavelengths are scattered more than the longer. Opal glass is ordinarily used as flashed opal, which is a thin layer of opal glass on a supporting sheet of clear glass. The diffusion of flashed opal is quite good. When illuminated normally, the brightness at 45° from the normal is about 90% of what one would expect from a perfect diffuser. Its total transmission is quite low, about 35 or 40 per cent. It should be noted that, since good diffusion means that the incident light is scattered into 2π steradians, the axial brightness of a rear illuminated screen of good diffusion is very low when compared with a poor diffuser.

Ground glass is produced by fine grinding (or etching) the surface of a glass plate to produce a large number of very small facets which refract the incident light more or less randomly. The total transmission of ground glass is about 75%. This transmission is quite strongly directional, and ground glass is far from a perfect diffuser. Its characteristics vary somewhat, depending on the coarseness of the surface. Typically, for a normally illuminated surface, the brightness at 10° from the normal is about 50% of the normal brightness, at 20° the brightness is a few per cent of the brightness at the normal. This characteristic is of course quite useful when partial diffusion is desired. By combining two sheets of ground glass (with the ground faces in contact) the transmission is lowered about 10% but the diffusion is improved; at 20° to the normal the brightness is about 20%, at 30° about 7%. With two sheets the diffusion can be increased by spacing them apart, although this will destroy their utility as a projection screen.

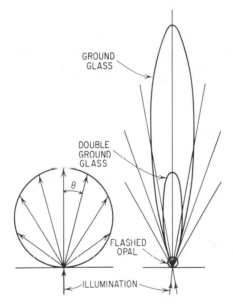

FIG. 7.14. Polar intensity plots of diffusing materials. Left: For a "perfect" diffuser, the intensity of a unit area of the surface varies with cos θ. Right: The relative intensities of single and double ground glass and flashed opal glass.

A sheet of tracing paper has diffusion characteristics quite similar to ground glass, and there are several plastic materials which are somewhat better diffusers than ground glass.

A rear projection screen, when used in a lighted room, is illuminated from both sides. The room light reduces the contrast of the projected image. This situation is sometimes alleviated by introducing a sheet of grey glass (that is, a neutral filter) between the diffusing screen and the observer. When this is done, the light from the projector is reduced by a factor of T, the transmission of the grey glass, but the room light is reduced by T^2, since the room light must pass through the grey glass twice to go from the room to the diffuser and back to the observer's eye.

7.8 Polarizing Materials

Light behaves as a transverse wave in which the waves vibrate perpendicular to the direction of propagation. If the wave motion is considered as a vector sum of two such vibrations in perpendicular planes, then plane polarized light results when one of the two components is removed from a light beam. Plane polarized light can be produced by passing the radiation from an ordinary source through a polarizing prism, several types of which are available. These prisms depend on the birefringent characteristic of calcite ($CaCO_3$) which has a different index of refraction for the two planes of polarization. Since light of one polarization is bent more strongly than the other, it is possible to separate them either by total internal reflection (as in the Nicol and Glan-Thompson prisms) or by deviation in different directions (as in the Rochon and Wollaston prisms).

Such prisms are large, heavy and expensive. Sheet polarizers, which are made by aligning microscopic crystals in a suitable base, are thin, light, relatively inexpensive, useful over a wide field of view, and simple to fabricate into an almost unlimited range of sizes and shapes. Thus, despite the fact that they are not quite as efficient as a good prism polarizer and are not effective over as large a wavelength range, they have largely supplanted prisms for the great majority of applications where polarization is required. The Polaroid Corporation of Cambridge, Massachusetts, produces a number of types of sheet polarizers. For work in the visible region, several types are available, depending on whether optimum transmission or optimum extinction is desired. Special types are available for use at high temperatures and also for use in the near infrared (0.7 to 2.2 microns). Polaroid also produces circular (as opposed to plane) polarizers in sheet form.

Since a plane polarizer will eliminate half the energy, it is obvious that the maximum transmission of a "perfect" polarizer in a beam of unpolarized light will be 50%. Practical values range from 25 to 40 per cent for sheet Polaroid, depending on the type. If two polarizers are "crossed", that is, oriented with their polarizing axes at 90°, the transmission will be zero if the polarization is complete. This can be achieved with Nicol prisms, but the sheet polarizers have a residual transmission ranging from 10^{-6} to 5×10^{-4}, again dependent on the type. The transmission characteristics of sheet polarizers are wavelength dependent as well.

When two polarizers are placed in a beam of unpolarized light, the transmission of the pair depends on the relative orientation of their polarization axes. If θ is the angle between the axes, then the transmission of the pair is given by:

$$T = K \cos^2 \theta \qquad (7.17)$$

where $K = 1.0$ for a "perfect" polarizer. For sheet polarizers, K ranges from 0.25 to 0.65, depending on the type chosen.

Reflection from the surface of a glass plate may also be used to produce plane polarized light. When light is incident on a plane surface at Brewster's angle, one plane of polarization is completely transmitted (if the glass is perfectly clean) and about 15% of the other is reflected. This occurs when the reflected and refracted rays are at 90° to each other; thus, Brewster's angle is given by:

$$I = \arctan \frac{N'}{N} \tag{7.18}$$

The reflected beam is thus completely polarized and the transmitted beam partially so. The percentage of polarized light in the transmitted beam can be increased by using a stack of thin plates all tilted to Brewster's angle.

The subject of polarized light is treated at greater length in texts devoted to physical optics, to which the reader is referred. Two additional points are worth noting: one, interference filters (Section 7.9) are frequently polarizing and are occasionally used as polarizers; and two, opal glass (Section 7.7) is an excellent depolarizer.

7.9 Dielectric Reflection and Interference Filters

The portion of the light reflected from the surface of an ordinary dielectric material (such as glass) is given by

$$R = \frac{1}{2} \left[\frac{\sin^2(I - I')}{\sin^2(I + I')} + \frac{\tan^2(I - I')}{\tan^2(I + I')} \right] \tag{7.19}$$

where I and I' are the angles of incidence and refraction respectively. The first term of Eq. 7.19 gives the reflection of the light which is polarized in the plane of incidence, and the second term the reflection for the other plane of polarization. As indicated in Section 7.1, at normal incidence this reduces to:

$$R = \frac{(N' - N)^2}{(N' + N)^2} \tag{7.20}$$

The variation of reflection from an air-glass interface as a function of the angle of incidence (I) is shown in Fig. 7.15, where the solid line is R, the dashed line is the sine term and the dotted line is the tangent term. Notice that the dotted line drops to zero reflectivity at Brewster's angle.

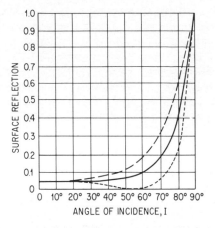

FIG. 7.15. Reflection from a single air-glass interface for index 1.523 (C-1 glass). Solid line, reflection of unpolarized light. Dashed line, reflection of light polarized in the plane of incidence. Dotted line, reflection of light polarized perpendicular to the plane of incidence.

The reflection from more than one surface can be treated as indicated by Eq. 7.5 when the separation between the surfaces is large compared to the wavelength of light. However, when the surface-to-surface separation is small, then interference between the light reflected from the various surfaces will occur and the reflectivity of the stack of surfaces will differ markedly from that given by Eq. 7.5. (At this point the reader may wish to refer to the discussion of interference effects contained in the first chapter.)

Optical coatings are thin films of various substances, notably magnesium fluoride (MgF_2) and zinc sulfide (ZnS), which are deposited in layers on an optical surface for the purpose of controlling or modifying the reflection and transmission characteristics of the surface. Such films have an optical thickness (index times mechanical thickness) which is a fraction of a wavelength, usually one-quarter or one-half wavelength. The deposition of thin films is carried out in a vacuum and is done by heating the material to be deposited to its evaporation temperature and allowing it to condense on the surface to be coated. The thickness of the film is determined by the rate of evaporation (or more precisely, condensation) and the length of time the process is allowed to continue. Since interference effects produce colors in the light reflected from thin films, just as in oil films on wet pavements, it is possible to judge the thickness of a film by the apparent color of light reflected from it. Simple coatings can be controlled visually by utilizing this effect, but coatings consisting of many layers are usually monitored photoelectrically, using nearly monochromatic light, so that the periodic rise and fall of the reflectivity can be accurately assessed and the thickness of each layer controlled to a high degree of precision.

Let us first consider a single layer film whose optical thickness (Nt) is exactly one-quarter of a wavelength. For light entering the film at normal incidence, the wave reflected from the second surface of the film will be exactly one-half wavelength out of phase with the light reflected from the first surface, resulting in destructive interference (assuming that there is no phase change by reflection). If the amount of light reflected from each surface is the same, a complete cancellation will occur and no light will be reflected. Thus, if the materials involved are non-absorbing, all the energy incident on the surface will be transmitted. This is the basis of the "quarter-wave" low-reflection coating which is almost universally used to increase the transmission of optical systems. Since low-reflection coatings reduce reflections, they tend to eliminate ghost images as well as the stray reflected light which reduces contrast in the final image. Before the invention of low-reflection coatings, optical systems which consisted of many separate elements were impractical because of the transmission losses incurred in surface reflections and the frequent ghost images. A magnesium fluoride coating has an additional benefit in that it is actually (when properly applied) a protective coating; the chemical stability of many dense barium crown glasses is enhanced by coating.

The reflectivity of one thin film is given by the equation

$$R = \frac{r_1^2 + r_2^2 + 2r_1 r_2 \cos X}{1 + r_1^2 r_2^2 + 2r_1 r_2 \cos X} \tag{7.21}$$

where

$$X = \frac{4\pi N_1 t_1 \cos I_1}{\lambda} \tag{7.22}$$

$$r_1 = \frac{-\sin(I_0 - I_1)}{\sin(I_0 + I_1)} \quad \text{or} \quad \frac{\tan(I_0 - I_1)}{\tan(I_0 + I_1)} \tag{7.23}$$

$$r_2 = \frac{-\sin(I_1 - I_2)}{\sin(I_1 + I_2)} \quad \text{or} \quad \frac{\tan(I_1 - I_2)}{\tan(I_1 + I_2)} \tag{7.24}$$

and λ is the wavelength of light, t is the thickness of the film, N_0, N_1 and N_2 are the refractive indices of the media, and I_0, I_1 and I_2 are the angles of incidence and refraction. Figure 7.16 shows a sketch of the film and indicates the physical meanings of the symbols. The sine or tangent expressions for r_1 and r_2 are chosen depending on the polarization of the incident light as in Eq. 7.19; for unpolarized light, which is composed equally of both polarizations, R is computed for each polarization and the two values are averaged. If we assume non-absorbing materials,

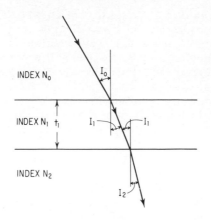

FIG. 7.16. Passage of light ray through a thin film, indicating the terms used in Equation 7.21.

the transmission T equals $(1 - R)$. At normal incidence $I_0 = I_1 = I_2 = 0$ and r_1 and r_2 reduce to

$$r_1 = \frac{N_0 - N_1}{N_0 + N_1} \tag{7.25}$$

$$r_2 = \frac{N_1 - N_2}{N_1 + N_2} \tag{7.26}$$

Using Eqs. 7.25 and 7.26 for r_1 and r_2, Eq. 7.21 can be solved for the thickness which yields a minimum reflectance. As the preceding discussion would lead one to expect, this occurs when the optical thickness of the film is one-quarter wavelength, that is

$$N_1 t_1 = \frac{\lambda}{4} \tag{7.27}$$

Using a quarter-wave film, the value for the film index to produce a zero reflection is found to be

$$N_1 = \sqrt{N_0 N_2} \tag{7.28}$$

Thus, to produce a coating which will completely eliminate reflections at an air-glass surface, a quarter-wave coating of a material whose index is the square root of the index of the glass is required. Magnesium fluoride (MgF_2) with an index of 1.38 is used for this purpose; its ability to form a hard durable film which will withstand weathering and frequent cleaning is the prime reason for its use, despite the fact that its index is higher than the optimum value for almost all optical glasses. The reflection of a low reflection coating on various index materials is shown in Fig. 7.17.

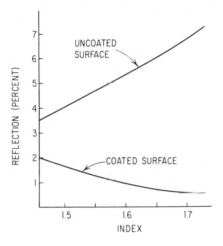

FIG. 7.17. The measured reflection of white light from an uncoated surface and from a surface coated with a quarter wave MgF$_2$ low reflection coating, as a function of the index of the base material.

From Eq. 7.21 it is apparent that the reflectivity of a coated surface will vary with wavelength. Obviously a quarter-wave coating for one wavelength will be either more or less than a quarter-wave thick for other wavelengths, and the interference effects will be modified accordingly. Thus a low-reflection coating designed for use in the visible region of the spectrum will have a minimum reflectance for yellow light and the reflectance for red and blue light will be appreciably higher. This is the cause of the characteristic purple color of single layer low-reflection coatings. Figure 7.18 indicates this variation for a typical commercial coating.

The reflection and transmission characteristics of a "stack" of several thin films can be expressed in explicit equations; however, their complexity increases rapidly with the number of films and the following recursion expressions are usually preferable. The physical thickness of each film is represented by t_j and the index by $n_j = N_j - iK_j$ (n is the complex index, N is the ordinary index of refraction and K is the absorption coefficient, which is zero for non-absorbing materials). The angle of incidence within the jth film is ϕ_j; and the "effective" refractive index is $u_j = n_j \cos \phi_j$ or $u_j = n_j/\cos \phi_j$ (for light polarized with the electric vector perpendicular to, or parallel to the plane of incidence respectively). Thus, for oblique incidence the calculations are carried out for both polarizations and the results are averaged (assuming the incident light to be unpolarized and to consist of equal parts of each polarization).

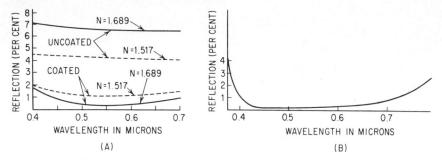

FIG. 7.18. A) The measured spectral reflectivity of a single layer low reflection coated glass surface compared with the reflectivity of uncoated glass. Coating is a quarter wave MgF_2 film on BSC-2 (517 : 645) (dashed curves) and on EDF-2 (689 : 309) (solid curves). B) A high efficiency multilayer low reflection coating on BSC-2 glass. Note the improvement over the single layer coating.

Since most calculations are carried out at normal incidence ($\phi_j = 0$) and for non-absorbing materials ($K_j = 0$), one may ordinarily use $u_j = n_j = N_j$.

The subscript notation is $j = 0$ for the substrate, $j = 1$ for the first film, $j = 2$ for the second, etc., $j = p - 1$ for the last film and $j = p$ for the final medium, which is usually air. For each film g_j, the effective optical thickness in radians, is computed from

$$g_j = \frac{2\pi n_j t_j \cos\phi_j}{\lambda} \tag{7.29}$$

where λ is the wavelength of light for which the calculation is made.

Starting with $E_1 = E_0^+ = 1.0$ and $H_1 = u_0 E_0^+ = u_0$, the following equations are applied iteratively at each surface, with the subscript j advancing from $j = 1$ to $j = p - 1$.

$$E_{j+1} = E_j \cos g_j + \frac{iH_j}{u_j} \sin g_j \tag{7.30}$$

$$H_{j+1} = iu_j E_j \sin g_j + H_j \cos g_j \tag{7.31}$$

where $i = \sqrt{-1}$ and the other terms have been defined above. Readers familiar with matrix notation may prefer to manipulate the equivalent matrix form

$$\begin{pmatrix} E_{j+1} \\ H_{j+1} \end{pmatrix} = \begin{pmatrix} \cos g_j & \dfrac{i}{u_j}\sin g_j \\ iu_j \sin g_j & \cos g_j \end{pmatrix} \begin{pmatrix} E_j \\ H_j \end{pmatrix} \tag{7.32}$$

When Eqs. 7.30 and 7.31 (or 7.32) have been applied to the entire stack, we have the values of E_p and H_p, which will generally be complex numbers of the form $z = x + iy$. These are substituted into

$$E_p^+ = \frac{1}{2}\left(E_p + \frac{H_p}{u_p}\right) = x_2 + iy_2 \tag{7.33}$$

$$E_p^- = \frac{1}{2}\left(E_p - \frac{H_p}{u_p}\right) = x_1 + iy_1 \tag{7.34}$$

and the reflectance of the thin film system is found from

$$R = \left|\frac{E_p^-}{E_p^+}\right|^2 \tag{7.35}$$

where the symbol $|z|$ indicates the modulus of a complex number z, so that

$$|z| = |x + iy| = \sqrt{x^2 + y^2}$$

and

$$R = |z|^2 = x^2 + y^2 = \left|\frac{x_1 + iy_1}{x_2 + iy_2}\right|^2 = \frac{x_1^2 + y_1^2}{x_2^2 + y_2^2}$$

If the computation has been for normal incidence through non-absorbing materials, the transmission is given by

$$T = 1 - R \tag{7.36}$$

Otherwise, the transmission is given by

$$T = \frac{n_0 \cos\phi_0}{n_p \cos\phi_p}\left|\frac{E_0^+}{E_p^+}\right|^2 \tag{7.37A}$$

or

$$T = \frac{n_0 \cos\phi_p}{n_p \cos\phi_0}\left|\frac{E_0^+}{E_p^+}\right|^2 \tag{7.37B}$$

where Eq. 7.37A is used for light polarized with the electric vector perpendicular to, and Eq. 7.37B for the electric vector parallel to, the plane of incidence.

A discussion of the design of multilayer coatings is beyond the scope of this volume; the interested reader may pursue the subject in the references listed at the end of this chapter. By suitable combinations of thin films of different indices and thicknesses a tremendous number of transmission effects can be created. Among the types of interference coatings which are readily available are long or short wavelength transmission filters, band pass filters, narrow bandpass (spike filters), achromatic extra-low-reflection coatings as well as the reflection coatings described in the next section. An extremely valuable property of thin film coatings is their spectral versatility. Once a combiantion of films has been designed to produce a desired characteristic, the wavelength region can be shifted at will by simply increasing or decreasing all the film thicknesses in proportion. For example, a spike filter designed to transmit a very narrow spectral band at one micron can be shifted to two microns by doubling the thickness of each film in the coating. This, of course, is limited by the absorption characteristics of the substrate and the film materials.

The characteristics of a number of typical interference coatings are shown in Fig. 7.19. Note that the wavelength scale is plotted

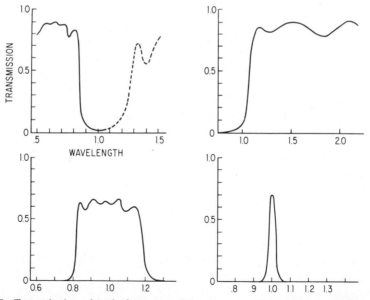

FIG. 7.19. Transmission of typical evaporated interference filters plotted against wavelength in arbitrary units. Upper left: Short pass filter (note that dashed portion of curve must be blocked by another filter if low long wavelength transmission is necessary. Upper right: Long pass filter. Lower left: Band Pass filter. Lower right: Narrow Band Pass (spike) filter.

in arbitrary units, with a central wavelength of 1, since (within quite broad limits) the characteristics can be shifted up or down the spectrum as described in the preceding paragraph. Most interference filters are very nearly 100% efficient, so that the reflection for a film is equal to one minus the transmission (except in regions where the materials used become absorbing). Since the characteristics of an interference filter depend on the thickness of the films, the characteristics will change when the angle of incidence is changed. This is in great measure due to the fact that the optical path through a film is increased when the light passes through obliquely. For *moderate* angles the effect is usually to shift the spectral characteristics to a slightly shorter wavelength.

Coatings consisting of a few layers are for the most part reasonably durable and can withstand careful cleaning. However, coatings consisting of a great number of layers (and coatings consisting of 50 or more layers are occasionally used) tend toward delicacy, and must be handled with due respect. Some multi layer coatings are quite effective polarizers (and as such, are occasionally responsible for "mysterious" happenings).

7.10 Reflectors

Although polished bulk metals are occasionally used for mirror surfaces, most optical reflectors are fabricated by evaporating one or more thin films on a polished surface, which is usually glass. Obviously the interference filters described in the preceding section can be used as special purpose reflectors in instances where their spectral characteristics are suitable. However, the workhorse reflector material for the great majority of applications is an aluminum film, deposited on a substrate by evaporation in vacuum. Aluminum has a broad spectral band of quite high reflectivity and is reasonably durable when properly applied. Almost all aluminum mirrors are "over coated" with a thin protective layer of either silicon monoxide or magnesium fluoride. This combination produces a first-surface mirror which is rugged enough to withstand ordinary handling and cleaning without scratching or other signs of wear.

The spectral reflectance characteristics of several evaporated metal films are shown in Fig. 7.20. With the exception of the curve for rhodium, the reflectivities given here can seldom be attained for practical purposes; the silver coating will tarnish and the aluminum film will oxidize, so that the reflectances tend to decrease with age, especially at shorter wavelengths. The high reflectivity of silver is only useful when the coating can be properly protected, as in an ordinary second surface mirror.

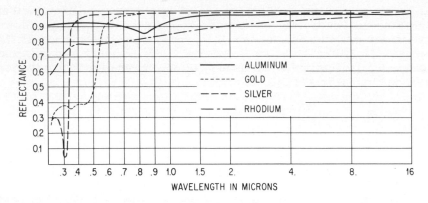

FIG. 7.20. Spectral reflectance for evaporated metal films on glass. Data represents new coatings, under ideal conditions.

Figure 7.21 indicates the variety of characteristics which are available in commercial aluminum mirrors. A run-of-the-mill protected aluminum mirror can be expected to have an average visual reflectance of about 88%. Two, four, or more interference films may be added to improve the reflectance where the additional cost can be accepted.

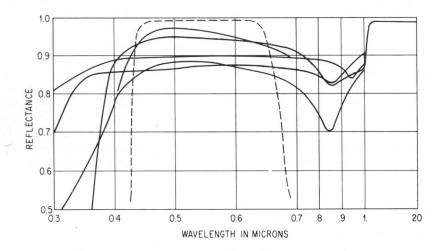

FIG. 7.21. Spectral reflectance of aluminum mirrors. The solid curves are for aluminum films with various types of thin film overcoatings - either for protection or for increased reflectivity. The dashed line is an extra high reflectance multilayer coating. All coatings shown are commercially available.

Dichroics and semi-reflecting mirrors constitute another class of reflector. Both are used to split a beam of light into two parts.

A dichroic reflector splits the light beam spectrally, in that it transmits certain wavelengths and reflects others. A semi-reflecting mirror is, nominally at least, spectrally neutral; its function is to divide a beam into two portions, each with similar spectral characteristics. Figure 7.22 shows the characteristics of a variety of these partial reflectors.

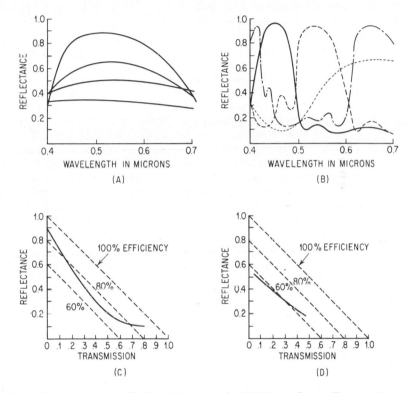

FIG. 7.22. Characteristics of Partial Reflectors. A - Multilayer "neutral" semi-reflectors (efficiency better than 99%). B - Dichroic multilayer reflectors - blue, green, red and yellow reflection. C - Visual efficiency of aluminum semi-reflectors. D - Visual efficiency of chrome semi-reflectors.

7.11 Reticles

A reticle is a pattern used at or near the focus of an optical system, such as the cross hairs in a telescope. For a simple cross hair pattern, fine wire or spider (web) hair is occasionally used, stretched across an open frame. However, a pattern which is supported on a glass (or other material) substrate offers considerably more versatility, and most reticles, scales, divided circles, and patterns are of this type.

The simplest type of reticle is produced by scribing, or scoring, the glass surface with a diamond tool. A line produced this way, while not opaque, modifies the glass sufficiently so that under the proper type of illumination the line will appear dark. Where clear lines in an opaque background are desired, the glass can be coated with an opaque coating, such as evaporated aluminum, and the lines scribed through the coating with a diamond or hardened steel tool, depending on the type of line desired. Scribing produces very fine lines.

Another old technique is to etch the substrate material. A waxy resist is coated on the substrate and the desired pattern cut through the resist. The exposed portion of the substrate is then etched (with hydrofluoric acid in the case of glass) to produce a groove in the material. The groove can be filled with titanium dioxide (white), lamp black, or evaporated metal. Etched reticles are durable and have the advantage that they can be edge lighted if illumination is necessary. Any substrate that is readily etched can be used. This process is used for many military reticles and also for accurate metrology scales on steel.

The most versatile processes for production of reticles are based on the use of a photo-resist, or photo-sensitive material. Photo-resists are exposed like a photographic emulsion, either by contact printing through a master, or by photography. However, when the photo-resist is "developed", the exposed areas are left covered with the resist and the unexposed areas are completely clear. Thus, an evaporated coating of any of a number of metals (aluminum, chrome, inconel, nichrome, copper, germanium, etc.) can be deposited over the resist. In the clear areas the coating adheres to the substrate; when the resist is removed it carries away the coating deposited upon it, leaving a durable pattern which is an exact duplicate of the master. The precision, versatility, ruggedness and suitability for mass production of this technique have earned it a prominent place in the field of reticle manufacture.

The photo-resist technique may also be combined with etching, where the material to be etched is either a metal substrate or an evaporated metal film.

Where the reticle pattern must be non-reflecting, the glue silver process or the black-print process is used. The technique is similar to that used in producing the photo-resist pattern, except that the photo-sensitive material is opaque. The clear areas are free of emulsion. Glue silver reticles are fragile but capable of very high resolution of detail. The black-print process is more durable. Occasionally an extremely high-resolution photographic emulsion is used for a reticle pattern; however, the presence of emulsion in the clear areas of the pattern is ordinarily a draw-back.

The following tabulation indicates the resolution and accuracy possible with these techniques. These figures represent the highest level of quality that reticle manufacturers are capable of at the present time; if cost is a factor, one is well advised to lower one's requirements an order of magnitude or so below the levels indicated here.

Method	Finest Line Width	Dimensional Repeatability
Scribing	.00001″	±.00001″
Etch (and fill)	.0002″-.0004″	±.0001″
Photo-resist (evaporated metal)	.0001″-.0002″	±.00005″
Glue Silver	.00003‴-.0002″	±.00005″
Black Print	.001″	±.0001″
Emulsion	.00005″	±.00005″

7.12 Cements and Liquids

Optical cements are used to fasten optical elements together. Two main purposes are served by cementing: the elements are held in accurate alignment with each other independent of their mechanical mount, and the reflections from the cemented surfaces are largely eliminated. Ordinarily the layer of cement used is extremely thin and its effect on the optical characteristics of the system can be totally neglected; some of the newer plastic cements, designed to withstand extremes of temperature, are used in thicknesses of a few thousandths of an inch (which could affect the performance of an optical system under critical conditions).

Canada balsam is made from the sap of the balsam fir. It is available in a liquid form (dissolved in xylol) and in stick, or solid form. Elements to be cemented are cleaned and placed together on a hot plate. When the elements are warm enough to melt the balsam, the stick is rubbed on the lower element. The upper element is replaced and the excess cement and any entrapped air bubbles are worked out by oscillating the upper element. The elements are then placed in an alignment fixture to cool. Balsam cement has an index of refraction of about 1.54 and a V-value of about 42. These are conveniently midway between the refractive characteristics of crown and flint glasses. Unfortunately Canada balsam will not withstand high or low temperatures. It softens when heated and splits at low temperatures and is thus unsuited for rigorous thermal environments. Balsam is rarely used today.

A great number of plastic cements have been developed to withstand extremes of both temperature and shock. For the most part, these are thermosetting (heat-curing) plastics, although a few thermoplastic (heat-softening) materials are used. Cements are available which will withstand temperatures from +180°F down to −85°F without failure when properly used. In general the thermosetting cements are supplied in two containers (sometimes refrigerated), one of which contains a catalyst which is mixed into the cement prior to use. A drop of cement is placed between the elements to be cemented, the excess cement and air bubbles are worked out, and the elements are placed in a fixture or jig for a heating cycle which cures the cement. Once the cement has set, it is exceedingly difficult to separate the components; the customary technique is to shock them apart by immersion in hot (150-200°C) castor oil. The index of refraction of plastic cements ranges from 1.47 to 1.61 depending on the type, with most cements falling between 1.53 and 1.58. Epoxies and methacrylates are widely used. Because of the variety of types and characteristics which are available, one should consult the manufacturer's literature for specific details regarding any given cement.

Optical liquids are used primarily for microscope immersion fluids and for use in index measurement (in critical-angle refractometers). For microscopy, water (N_D = 1.33), cedar oil (N_D = 1.515) and glycerin (ultraviolet N = 1.45) are frequently utilized. For refractometers alpha-bromonaphthalene (N = 1.66) is the most commonly used liquid. Methylene iodide (N = 1.74) is used for high index measurement (since the liquid index must be larger than that of the sample to avoid total internal reflection back into the sample).

References

1. Optical glass and material catalogs from manufacturers:
   Bausch & Lomb Company, Rochester, N.Y.
   Corning Glass Works, Corning, N.Y.
   Eastman-Kodak Company, Rochester, N.Y.
   Harshaw Chemical Co., Cleveland, Ohio
   Infrared Industries, Inc., Santa Barbara, Calif.
   Isomet Corp., Palisades Park, N.J.
   Liberty Mirror Div., L.O.F., Brackenridge, Pa.
   Optical Coating Laboratories, Santa Rosa, Calif.
   Pittsburg Plate Glass Co., Pittsburgh, Pa.
   Polaroid Corp., Cambridge, Mass.
   Servo Corporation, Hicksville, N.Y.
2. American Institute of Physics Handbook, McGraw-Hill, 1963.
3. Ballard, McCarthy and Wolfe, "Optical Materials for Infrared Instrumentation", University of Michigan, 1959 (Supplement 1961).
4. Handbook of Chemistry and Physics, Chemical Rubber Pub. Co.
5. Hardy and Perrin, "The Principles of Optics", McGraw-Hill, 1932.
6. Jacobs, "Fundamentals of Optical Engineering", McGraw-Hill, 1943.
7. Hertzberger, "Modern Geometrical Optics", Interscience, 1958.

8. Strong, "Concepts of Classical Optics", Freeman, 1958.
9. Jenkins and White, "Fundamentals of Optics", McGraw-Hill, 1957.
10. Hackforth, "Infrared Radiation", McGraw-Hill, 1960.
11. Jamieson et al., "Infrared Physics and Engineering", McGraw-Hill, 1963.
12. Conrady, "Applied Optics and Optical Design", Dover.
13. Vasicek, "Optics of Thin Films", North Holland, Amsterdam, 1960.
14. Heavens, "Optical Properties of Thin Solid Films", Butterworth's, London, 1955.
15. Holland, "Vacuum Deposition of Thin Films", Wiley, 1956.
16. Hass, "Physics of Thin Films, Volume I", Academic Press, 1963.

Exercises

1. a) What is the transmission of a stack of three thin plane parallel plates of glass ($N = 1.5$) at normal incidence?
   b) What percentage of the incident light is transmitted directly (i.e. without any intervening reflections)?
   Ans. a) 80%, b) $(.96)^6 = 78\%$

2. If a one centimeter thickness of a material transmits 85% and two centimeter thickness transmits 80%, a) what percentage will a three centimeter thickness transmit? b) what is the absorption coefficient of the material? (Neglect all *multiple* reflections)
   Ans. a) 75.3%; b) .06062 cm^{-1}

3. Determine the coefficients for the dispersion equations given in Section 7.1 for one of the optical glasses listed in Fig. 7.5. Evaluate the accuracy of the equations by comparing the index values given by the equations with those listed in the table (for wavelengths not used in determining the constants).

4. Using the spectral transmission curves of Fig. 7.11, plot the spectral transmission which would result from a combination of filters (c) and (f).

5. Plot, in the manner of Fig. 7.15, the curve of reflection against angle of incidence for a single surface of glass ($N = 1.52$) coated with a quarter wavelength thickness of magnesium fluoride ($N = 1.38$).

8

Radiometry and Photometry

8.1

In concept, both radiometry and photometry are quite straight-forward; however, both have been cursed with a jungle of often bewildering terminology. Radiometry deals with radiant energy (i.e. electromagnetic radiation) of any wavelength. Photometry is restricted to radiation in the visible region of the spectrum. The basic unit of power (i.e., rate of transfer of energy) in radiometry is the watt; in photometry the corresponding unit is the lumen, which is simply radiant power as modified by the relative spectral sensitivity of the eye (Fig. 5.10). Note that watts and lumens have the same dimensions, namely energy per time.

The principles of radiometry and photometry are readily understood when one thinks in terms of the basic units involved, rather than the special terminology which is conventionally used. The next five sections will discuss radiation in terms of watts; the reader should remember that the discussion is equally valid for photometry, if lumens are read for watts.

8.2 The Inverse Square Law

Consider a point (or "sufficiently" small) source of radiant energy, which is radiating uniformly in all directions. If the rate at which energy is radiated is P watts, then the source has a radiant intensity, J, of $P/4\pi$ watts per steradian*, since the solid angle into which the energy is radiated is a sphere of 4π steradians. Of course there are no truly "point" sources and no practical sources which radiate uniformly in all directions, but if a source is quite small relative to its distance, it can be treated as a point, and its radiation, in the directions in which it does radiate, can be expressed in watt. ster $^{-1}$.

---

*A steradian is the solid angle subtended (at its center) by $1/4\pi$ of the surface area of a sphere. Thus, a sphere subtends 4π (12.566) steradians from its center; a hemisphere subtends 2π steradians. The size of a solid angle in steradians is found by determining the area of that portion of the surface of a sphere which is included within the solid angle and dividing this area by the square of the radius of the sphere. For a small solid angle, the area of the included flat surface normal to the "central axis" of the angle can be divided by the square of the distance from the surface to the apex of the angle to determine its size in steradians.

If we now consider a surface which is S cm from the source, then one square cm of this surface will subtend $1/S^2$ steradians from the source (at the point where the normal to the surface from the source intersects the surface, if S is large). The irradiance, H, on this surface is the incident radiant power per unit area and is obtained by multiplying the intensity of the source by the solid angle subtended by the unit area. Thus, the irradiance is given by

$$H = J\,\frac{1}{S^2} = \frac{P}{4\pi S^2} \tag{8.1}$$

The units of irradiance are watts per square cm (watt cm^{-2}). Equation 8.1 is, of course, the "inverse square" law, which is conventionally stated: the illumination (irradiance) on a surface is inversely proportional to the square of the distance from the (point) source.

Thus, if our uniformly radiating point source emits energy at a rate of 10 watts, it will have an intensity $J = 10/4\pi = 0.8$ watt ster^{-1}, and the radiation falling on a surface 100 cm away would be 0.8×10^{-4} watt cm^{-2}, or 80 microwatts per square cm. If the surface is flat, the irradiance will, of course, be less than this at points where the radiation is incident at an angle, since the solid angle subtended by a unit of area will be reduced. From Fig. 8.1, it can be seen that the source-to-surface distance is increased to $S/\cos\theta$ and the effective area (normal to the direction of the radiation) is reduced by a $\cos\theta$ factor. Thus, the solid angle subtended, and the irradiance, are reduced by a $\cos^3\theta$ factor.

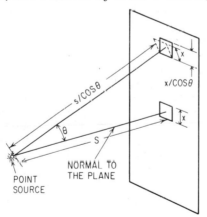

FIG. 8.1. Geometry of a point source irradiating a plane, showing that irradiance (or illumination) varies with $\cos^3\theta$.

8.3 Radiance and Lambert's Law

An extended source, that is, one whose dimensions are significant, must be treated differently than a point source. A small area of the source will radiate a certain amount of power per unit solid angle. Thus, the radiation characteristics of an extended source are expressed in terms of power per unit solid angle per unit area. This is called radiance; the usual units for radiance are watts per steradian per square cm (watt ster^{-1} cm^{-2}) and the symbol is N.

Most extended sources of radiation follow, at least approximately, what is known as Lambert's Law,

$$J_\theta = J_0 \cos \theta \qquad (8.2)$$

where J_θ is the intensity of a small incremental area of the source in a direction at an angle θ from the normal to the surface and J_0 is the intensity of the incremental area in the direction of the normal. For example, a heated metal disc with a total area of one square cm. and a radiance of one watt ster^{-1} cm^{-2} will radiate one watt ster^{-1} in a direction normal to its surface. In a direction 45° to the normal, it will radiate only 0.707 watt ster^{-1} (cos 45° = .707).

Notice that, although radiance is given in terms of watts per steradian per square cm, this should not be taken to mean that the radiation is uniform over a full steradian or over a full square cm. Consider a source consisting of a 0.1 cm square incandescent filament in a 20 cm diameter envelope. Assume that the bulb is painted so that only a 1-cm square transmits energy, and that the source radiates one-fiftieth of a watt through this square. (We assume, for convenience, that the radiation intercepted by the painted envelope is thereby totally removed from consideration). Now the filament has an area of 0.01 cm^2 and is radiating 0.02 watt into a solid angle of (approximately) 0.01 steradian. Therefore, it has a radiance of 200 watt ster^{-1} cm^{-2}, but only within the solid angle subtended by the window! Outside this angle the radiance is zero. This concept of radiance over a limited angle becomes important in dealing with the radiance of images and must be thoroughly understood.

There are several interesting consequences of Lambert's Law that are worthy of consideration, not only for their own sake but because they illustrate the basic techniques of radiometric calculations. The radiance of a surface is conventionally taken with respect to the area of a surface normal to the direction of radiation. It can be seen that, although the emitted radiation per steradian falls off with $\cos \theta$ according to Lambert's law, the "projected" surface area falls off at exactly the same rate. The result is that the radiance of a Lambertian surface is constant with respect to θ. In visual work the quantity corresponding to radiance is brightness, and the above is readily demonstrated by observing that the brightness of a diffuse source is the same regardless of the angle from which it is viewed.

8.4 Radiation into a Hemisphere

Let us determine the total power radiated from a flat diffuse source into a hemisphere. If the source has a radiance of N watt ster^{-1} cm^{-2}, one might expect that the power radiated into a hemisphere of 2π steradians would be $2\pi N$ watt cm^{-2}. That this is twice too large is readily shown. With reference to Fig. 8.2, let A represent the area of a small source with a radiance of N watt ster^{-1} cm^{-2} and an intensity of $J_\theta = J_0 \cos\theta = NA \cos\theta$ watt ster^{-1}. The incremental ring area on a hemisphere of radius R has an area of $2\pi R \sin\theta \cdot R d\theta$ and thus subtends (from A) a solid angle of $2\pi R^2 \sin\theta d\theta / R^2 = 2\pi \sin\theta d\theta$ steradians. The radiation intercepted by this ring is the product of the intensity of the source and the solid angle, or

$$dP = J_\theta \, 2\pi \sin\theta d\theta = 2\pi NA \sin\theta \cos\theta d\theta \qquad (8.3)$$

Integrating to find the total power radiated into the hemisphere from A, we get

$$P = \int_0^{\pi/2} 2\pi NA \sin\theta \cos\theta d\theta = 2\pi NA \left[\frac{\sin^2\theta}{2}\right]_0^{\frac{\pi}{2}} = \pi NA \text{ watts.} \quad (8.4)$$

Dividing by A to get watts emitted per square centimeter of source, we find the radiation into the 2πster. of the hemisphere to be πN watt cm^{-2}, *not* $2\pi N$.

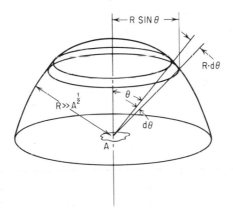

FIG. 8.2. Geometry of a Lambertian source radiating into a hemisphere.

8.5 Irradiance Produced by a Diffuse Source

It is frequently of interest to determine the irradiance produced at a point by a Lambertian source of finite size. Referring to Fig. 8.3, assume that the source is a circular disc of radius R and that we wish to determine the irradiance at some point X which is a distance S from the source and is on the normal through the center of the source. (Note that we will determine the irradiance on a plane parallel to the plane of the source.) The radiant intensity of a small element of area, dA, in the direction of point X is given by Eq. 8.2 as

$$J_\theta = J_0 \cos\theta = NdA \cos\theta$$

where N is the radiance of the source. Since the distance from dA to X is $S/\cos\theta$, and the radiation arrives at an angle θ, the incremental irradiance at X produced by dA is

$$dH = J_\theta \cos\theta \left[\frac{\cos^3\theta}{S^2}\right] = \frac{NdA \cos^4\theta}{S^2} \tag{8.5}$$

The same irradiance is produced by each incremental area making up a ring of radius r and a width dr, so that we can substitute the area of the ring, $2\pi r\, dr$, for dA in Eq. 8.5 to get the incremental irradiance from the ring,

$$dH = \frac{2\pi r\, dr\, N \cos^4\theta}{S^2} \tag{8.6}$$

To simplify the integration, we substitute

$$r = S \tan\theta$$

$$dr = S \sec^2\theta\, d\theta$$

into Eq. 8.6 to get

$$dH = \frac{2\pi S \tan\theta\, S \sec^2\theta\, d\theta\, N \cos^4\theta}{S^2}$$

$$= 2\pi N \tan\theta \cos^2\theta\, d\theta = 2\pi N \sin\theta \cos\theta\, d\theta$$

Integrating to determine the irradiance from the entire source, we get

$$H = \int_0^\theta 2\pi N \sin\theta \cos\theta\, d\theta = 2\pi N \left[\frac{\sin^2\theta}{2}\right]_0^\theta \tag{8.7}$$

$$H = \pi N \sin^2\theta \text{ watt cm}^{-2}$$

where H is the irradiance produced at a point by a circular source of radiance N watt ster^{-1} cm^{-2} which subtends an angle of 2θ from the point (when the point is on the "axis" of the source).

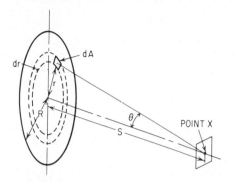

FIG. 8.3. Geometry of a circular source irradiating Point X.

Unfortunately non-circular sources do not readily yield to analysis. However, *small* non-circular sources may be approximated with a fair degree of accuracy by noting that the solid angle subtended by the source from X is

$$\Omega = 2\pi(1 - \cos\theta) = 2\pi \frac{\sin^2\theta}{(1 + \cos\theta)}$$

and for small values of θ, $\cos\theta$ equals unity and

$$\omega = \pi \sin^2\theta$$

Thus, if the angle subtended by the source is moderate, we can write

$$H = N\omega \tag{8.8}$$

If the point X does not lie on the "axis" (the normal through the center of the circular source), then the irradiance would be subject to the same factors outlined in the discussion of the "cosine-fourth" law in Section 6.7. Thus, if the line from the point X_ϕ to the center of the circle makes an angle ϕ to the normal, the irradiance at X_ϕ is given by

$$H_\phi = H_0 \cos^4\phi \tag{8.9}$$

where H_0 is the irradiance along the normal given by Eq. 8.7 or 8.8 and H_ϕ is the irradiance at X_ϕ (measured in a plane parallel to the source.)

It is apparent that Eqs. 8.8 and 8.9 may be used in combination to calculate the irradiance produced by any conceivable source configuration, to whatever degree of accuracy that time (or patience) allows.

8.6 The Radiometry of Images

When a source is imaged by an optical system, the image has a radiance, and it may be treated as a secondary source of radiation. However, one must always keep in mind that the radiance of an image differs from the radiance of an ordinary source in that the radiance of an image exists *only* within the solid angle subtended from the image by the clear aperture of the optical system. Outside of this angle, the radiance of the image is zero.

Figure 8.4 illustrates an aplanatic optical system imaging an incremental area A of a Lambertian source at A'. We will consider the radiance of the image at A' formed through a generalized incremental area P in the principal surface of the optical system. (Since the system is aplanatic, that is, free of coma and spherical aberration, the principal "planes" are spherical surfaces and are centered on the object and image.) The radiance of the source is N watt ster^{-1} cm^{-2} and the projected area of A in the direction θ is $A\cos\theta$ cm^2. The solid angle subtended by incremental area P from A is P/S^2, where S is the distance from the object to the first principal surface. Therefore, the radiant power intercepted by area P is

$$\text{power} = N\,\frac{P}{S^2}\,A\cos\theta \text{ watts}$$

This radiation is imaged by the optical system at area A', into a (projected) area $A'\cos\theta'$, through a solid angle P'/S'^2. Thus, the radiance at A' is given by

$$N' = TN\,\frac{P}{S^2}\,A\cos\theta\left[\frac{S'^2}{P'A'\cos\theta'}\right]\text{watt ster}^{-1}\text{ cm}^{-2}$$

where T is the transmission of the optical system. Now we note that the incremental areas A and A' are related by the laws of first order optics, and, if both are in media of the same index, $AS'^2 = A'S^2$. Further, the principal surfaces are unit images of each other; taking the tilts of the surfaces into account, we get $P\cos\theta = P'\cos\theta'$. Making these substitutions and clearing, we find that the radiance of the image is equal to that of the object, times the transmission of the system.

$$N' = TN \tag{8.10}$$

This fundamental relationship can be restated with slightly different emphasis: the radiance of an image cannot exceed that of the object*.

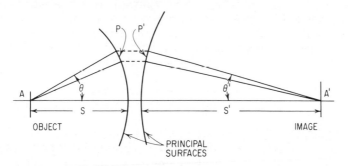

FIG. 8.4. Illustrates an aplanatic optical system imaging an incremental source area A at A'.

By the application of exactly the same integration technique used in Section 8.9, it can be shown that the irradiance produced in the plane of an image is given by:

$$H = T\pi N \sin^2 \theta \text{ watt cm}^{-2} = TN\omega \text{ (for small angles)} \tag{8.11}$$

where T is the system transmission, N (watt ster^{-1} cm^{-2}) is the object radiance and θ is the half angle subtended by the exit pupil of the optical system from the image. Small or noncircular exit pupils and cylindrical lens systems can be handled by substituting the solid angle ω for $\pi \sin^2 \theta$ (just as in Eq. 8.8); image points off

---

*This statement and Eqs. 8.10 and 8.11 are subject to the condition that both object and image lie in media of the same index of refraction. When the media have different indices, the image radiance and irradiance are multiplied by the factor $(N_i/N_0)^2$, where N_i and N_0 are the refractive indices of the image media and object media respectively. Thus, Eqs. 8.10 and 8.11 become

$$N' = TN \left(\frac{N_i}{N_0}\right)^2 \tag{8.10a}$$

$$H = TN \left(\frac{N_i}{N_0}\right)^2 \pi \sin^2 \theta \tag{8.11a}$$

$$= TN \left(\frac{N_i}{N_0}\right)^2 \omega$$

The factor $(N_i/N_0)^2$ is introduced by the use of $AN_0^2 S'^2 = A'N_i^2 S^2$ in place of $AS'^2 = A'S^2$ in the derivation of Eq. 8.10; both equalities are derived from the optical invariant relationship $hNu = h'N'u'$ (Eq. 2.40).

the optical axis are subject to the cosine-fourth law in addition to any losses due to vignetting (Eq. 8.9 and Section 6.7).

The similarity between the equations for the irradiance produced by a diffuse source and by an optical system makes it apparent that the aperture of the optical system takes on the radiance of the object it is imaging when it is viewed from the image point. This is an extremely useful concept; for radiometric purposes, a complex optical system can often be treated as if it consisted solely of a transmission loss and an exit pupil with the same radiance as the object. Similarly, when an optical system produces an image of a source, the image can be treated as a new source of the same radiance (less transmission losses). Of course the direction that radiation is emitted from the image is limited by the aperture of the system.

When an object is so small that its image is a diffraction pattern (Airy disc), then the preceding techniques, which apply to extended sources, cannot be used. Instead, the power intercepted by the optical system, reduced by transmission losses, is spread into the Airy disc. To determine the irradiance (or the radiance) of the image, we note that 84% of the power intercepted and transmitted by the lens is concentrated into the central bright spot of the Airy disc. A precise determination requires that one integrate the relative irradiance × area product over the central disc and equate this to 84% of the image power to determine the irradiance. If P is the total power in the Airy pattern, H_0 the irradiance at the center of the pattern and z the radius of the first dark ring, a numerical integration of Eq. 6.13 over the central disc yields

$$0.84P = 0.65 H_0 z^2$$

Rearranging and substituting the value of z given by Eq. 6.15, we get

$$H_0 = 1.29 \frac{P}{z^2} = 3.5P \left(\frac{N.A.}{\lambda}\right)^2$$

where λ is the wavelength and $N.A.$ is $N' \sin U'$, the numerical aperture. The irradiance for points not at the center of the pattern is then found by Eq. 6.13. Note that the preceding assumes a circular aperture; for rectangular apertures, the process would be based on Eq. 6.11.

Example A

In Fig. 8.5, A is a circular source with a radiance of 10 watt ster⁻¹ cm⁻² radiating toward plane BC. The diameter of A subtends

60° from point B. The distance AB is 100 cm and the distance BC is 100 cm. An optical system at D forms an image of the region about point C at E. Plane BC is a diffuse (Lambertian) reflector with a reflectivity of 70%. The optical system (D) has a one inch square aperture and the distance from D to E is 100 inches. The transmission of the optical system is 80%. We wish to determine the power incident on a one cm square photodetector at E.

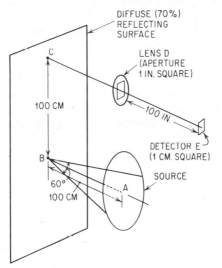

FIG. 8.5. Example A.

We begin by determining the irradiance at B, using Eq. 8.7; the source radiance is 10 watt ster^{-1} cm^{-2} and the half angle θ is 30° giving

$$H_B = \pi N \sin^2 \theta = \pi \cdot 10 \cdot \left(\frac{1}{2}\right)^2 = 7.85 \text{ watt cm}^{-2}$$

Since angle BAC is 45°, we can find the irradiance at C from Eq. 8.9, noting that cos 45° is 0.707

$$H_C = H_B \cos^4 45° = 7.85 \times (.707)^4 = 1.96 \text{ watt cm}^{-2}$$

It is now necessary to determine the radiance of the surface at C. The diffuse surface at C re-radiates 70% of the incident 1.96 watt cm^{-2} into a full hemisphere; the total power re-radiated is thus 1.37 watt cm^{-2}. In Section 8.4, it was shown that a source of radiance N radiated πN watt cm^{-2} into a hemisphere. Thus the radiance at point C is given by

$$N_C = \frac{RH}{\pi} = \frac{0.7 \times 1.96}{\pi} = \frac{1.37}{\pi} = .44 \text{ watt ster}^{-1} \text{ cm}^{-2}$$

The irradiance at E can now be determined from Eq. 8.11, noting that the solid angle subtended by the aperture of the lens system is $1/(100)^2$ or 10^{-4} ster, and substituting this for $\pi \sin^2 \theta$ in Eq. 8.11,

$$H_E = T_D \pi N_C \sin^2 \theta = T_D N_C \omega$$

$$= 0.8 \times .44 \times 10^{-4} = 0.35 \times 10^{-4} \text{ watt cm}^{-2}$$

Since the photodetector at E has an area of one square cm, the radiant power falling on it is just 0.35×10^{-4} watts, or 35 microwatts.

8.7 Spectral Radiometry

In the preceding discussion, no mention has been made of the spectral characteristics of the radiation. It is apparent that every radiant source has some sort of spectral distribution of its radiation, in that it will emit more radiation at certain wavelengths than others.

Name		Description	Units
Radiant power	P	rate of transfer of energy	watts (or joule sec^{-1})
Radiant intensity	J	power per unit solid angle from a source	watt ster^{-1}
Radiance	N	power per unit solid angle per unit area from a source	watt ster^{-1} cm^{-2}
Irradiance	H	power per unit area incident on a surface	watt cm^{-2}
Radiant energy	U		joule
Radiant emittance	W	power per unit area emitted from a surface	watt cm^{-2}

FIG. 8.6. Radiometric terminology. The names, symbols, descriptions, and preferred units for quantities in radiometric work.

For many purposes, it is necessary to treat intensity (J), irradiance (H), radiance (N), etc., (in fact, all the quantities listed in Fig. 8.6) as functions of wavelength. To do this we refer to the above quantities per unit interval of wavelength. Thus, if a source emits 5 watts of radiant power in the spectral band between 2 and 2.1 microns, it emits 50 watts per micron (watt micron^{-1}) in this region of the spectrum. The standard symbol for this type of quantity is the symbol given in Fig. 8.6 subscripted with a λ, and the name is preceded by "spectral." For example, the symbol for spectral radiance is N_λ and its units are watts per steradian per square cm per micron (watt ster^{-1} cm^{-1} micron^{-1}).

In many applications it is absolutely necessary to take the spectral characteristics of sources, detectors, optical systems, filters, and the like into account. This is accomplished by integrating the particular radiation product function over an appropriate wavelength interval. Since most spectral characteristics are not ordinary functions, the process of integration is usually numerical, and thus laborious. As a brief example, suppose that the irradiance in an image is desired. The spectral radiance of the object can be described by some function $N(\lambda)$ and the transmission of the atmosphere, the optical system, and any filters can be combined in a spectral transmission function $T(\lambda)$. Equation 8.11 will give the irradiance of the image (for any given wavelength); for use over an extended wavelength interval, we must write

$$H = \int_{\lambda_1}^{\lambda_2} T(\lambda)\, \pi N(\lambda)\, \sin^2 \theta\, d\lambda = \pi \sin^2 \theta \int_{\lambda_1}^{\lambda_2} T(\lambda)\, N(\lambda)\, d\lambda \text{ watt cm}^{-2} \quad (8.12)$$

where λ_1 and λ_2, the limits of the integration, may be zero and infinity, but are usually taken as real wavelengths which encompass the region of interest. In practice, it is usually necessary to perform the integration numerically; this process is represented (for this particular example) by the summation:

$$H = \pi \sin^2 \theta \sum_{\lambda=\lambda_1}^{\lambda_2} T(\lambda)\, N(\lambda)\, \Delta\lambda \text{ watt cm}^{-2} \quad (8.13)$$

The spectral response of a detector is included in a calculation in the same manner. For example, the effective power falling on a detector with an area of A and a relative spectral response $R(\lambda)$, when the detector is located in the image plane of the system above would be (provided that the image completely covered the detector)

$$P = A\pi \sin^2 \theta \int_{\lambda_1}^{\lambda_2} R(\lambda)\, T(\lambda)\, N(\lambda)\, d\lambda \text{ watts}$$

8.8 Black-Body Radiation

A perfect black-body is one which totally absorbs all radiation incident upon it. The radiation characteristics of a heated black-body are subject to known laws, and since it is possible to build a close approximation to a black-body, a device of this type is a very useful standard source for the calibration and testing of radiometric instruments. Further, most sources of thermal radiation, i.e. sources which radiate because they are heated, radiate energy in a manner which can be readily described in terms of a black-body emitting through a filter, making it possible to use the black-body radiation laws as a starting point for many radiometric calculations.

Planck's law describes the spectral radiant emittance of a perfect black-body as a function of its temperature and the wavelength of the emitted radiation.

$$W_\lambda = \frac{C_1}{\lambda^5 \left(e^{C_2 / \lambda T} - 1 \right)} \tag{8.14}$$

where W_λ = the radiation emitted into a hemisphere by the black-body in power per unit area per wavelength interval (watt cm^{-2} micron^{-1})

λ = the wavelength (microns)

e = the base of natural logarithms (2.718...)

T = the temperature of the black-body in degrees Kelvin (°K = °C + 273°)

C_1 = a constant = 3.7405×10^4 when area is in square centimeters and wavelength in microns.

C_2 = a constant = 1.43879×10^4 when square centimeters and microns are used.

Figure 8.7 indicates the shape of the curve of W_λ plotted against wavelength. Note that the spectral radiance (N_λ) is given by W_λ/π.

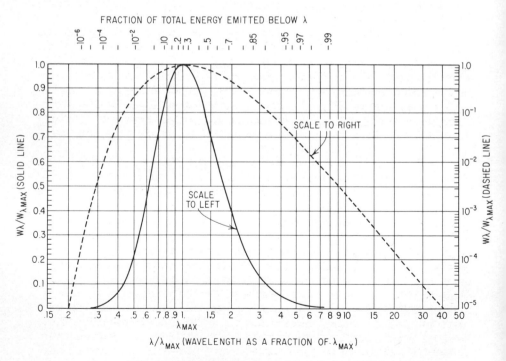

FIG. 8.7. Spectral distribution of black-body radiation.

If we integrate Eq. 8.14, we can obtain the total radiation at all wavelengths. The resulting equation is known as the Stefan-Boltzmann law,

$$W_{TOT} = 5.6697 \times 10^{-12} T^4 \text{ watt cm}^{-2} \tag{8.15}$$

and indicates that the total power radiated from a black-body varies as the fourth power of the absolute temperature.

If we differentiate Planck's equation (8.14) and set the result equal to zero, we can determine the wavelength at which the spectral emittance (W_λ) is a maximum and also the amount of W_λ at this wavelength. Wien's displacement law gives the wavelength for maximum W_λ as

$$\lambda_{max} = 2897.8 \ T^{-1} \text{ microns} \tag{8.16}$$

and W_λ at λ_{max} as:

$$W_{\lambda max} = 1.288 \times 10^{-15} T^5 \text{ watt cm}^{-2} \text{ micron}^{-1} \tag{8.17}$$

Notice that the higher the temperature, the shorter the wavelength at which the peak occurs and that W_λ at the peak varies as the fifth power of the absolute temperature.

Planck's equation is awkward to use and for this reason a number of tables, charts and slide rules are available which allow the user to simply look up the values of W_λ for the appropriate temperature and wavelength. Figure 8.7 may be used for this purpose when the precision required is relatively modest.

The use of Fig. 8.7 is quite simple: First the total energy (W_{TOT}), the peak wavelength (λmax) and the maximum spectral radiant emittance ($W_{\lambda max}$) are calculated for the desired temperature by Eqs. 8.15, 8.16 and 8.17 respectively. The graph in Fig. 8.7 is of $W_\lambda/W_{\lambda max}$ plotted against relative wavelength. Thus, if W_λ for a particular wavelength (λ) is desired, the value of $W_\lambda/W_{\lambda max}$ corresponding to the appropriate value of λ/λmax is selected and multiplied by the value of $W_{\lambda max}$ from Eq. 8.17.

Across the top of Fig. 8.7 is a scale which indicates the fraction of the total energy emitted at all wavelengths below that corresponding to the point on the scale. Note that exactly 25% of the energy from a black-body is emitted at wavelengths shorter than λmax. If it is necessary to determine the amount of power emitted in a spectral band between two wavelengths (λ_1 and λ_2), the wavelengths are converted to relative wavelengths (λ_1/λmax and λ_2/λmax) and the fractions corresponding to them are selected from the scale at the top of the figure. The total power (W_{TOT}) from Eq. 8.15 times the difference between the two fractions will give the amount of power emitted in the wavelength interval.

Example B

For a black-body at a temperature of 27°C (80.6°F), T is 273 + 27 = 300°K, and the total emitted radiation is given by Eq. 8.15

$$W_{TOT} = 5.67 \times 10^{-12} (300)^4 = 4.59 \times 10^{-2} \text{ watt cm}^{-2}$$

The wavelength at which W_λ is a maximum is given by Eq. 8.16

$$\lambda \text{max} = 2897.9 \, (300)^{-1} = 9.66 \text{ microns}$$

and the radiant emittance at this wavelength is obtained from Eq. 8.17

$$W_{\lambda \text{max}} = 1.288 \times 10^{-15} (300)^5 = 3.13 \times 10^{-3} \text{ watt cm}^{-2} \text{ micron}^{-1}$$

Suppose we wish to know the characteristics of this black-body in the wavelength region between 4 and 5 microns. We express these wavelengths in terms of λmax as $4/9.66 = 0.414$ and $5/9.66 = 0.518$. From Fig. 8.7, the corresponding values of $W_\lambda/W_{\lambda \text{max}}$ are 0.07 and 0.25; these values, multiplied by $W_{\lambda \text{max}} = 3.13 \times 10^{-3}$ watt cm^{-2} micron^{-1} give us the spectral radiant emittances for these wavelengths

$$\text{at 4 microns: } W_\lambda = 0.22 \times 10^{-3} \text{ watt cm}^{-2} \text{ micron}^{-1}$$

$$\text{at 5 microns: } W_\lambda = 0.78 \times 10^{-3} \text{ watt cm}^{-2} \text{ micron}^{-1}$$

Using the fraction scale across the top of the chart, we find that about 0.011 of the radiation is emitted below 5 microns (rel. $\lambda = 0.518$) and about 0.0015 below 4 microns. Thus, approximately 1% of the total radiation (W_{TOT}), amounting to about 4×10^{-4} watt cm^{-2}, is emitted in this spectral band. The radiance of the surface will be $4 \times 10^{-4}/\pi$ watt ster^{-1} cm^{-2} in this spectral band. If the black-body is a foot square, with an area of about 1000 square cm., it will radiate about 0.4 watts between 4 and 5 microns into a hemisphere of 2π ster.

Most thermal radiators are not perfect black-bodies. Many are what are called gray-bodies. A gray-body is one which emits radiation in exactly the same spectral distribution as a black-body at the same temperature, but with reduced intensity. The total emissivity (ϵ) of a body is the ratio of its total radiant emittance to a perfect black-body at the same temperature. Emissivity is thus a measure of the radiation and absorption efficiency of a body. For a perfect black-body $\epsilon = 1.0$, and most laboratory standard black-bodies are within a percent or two of this value. The table of Fig. 8.8 lists the emissivity of a number of common materials.

Material	Total emissivity	
Tungsten	500°K	.05
	1000°K	.11
	2000°K	.26
	3000°K	.33
	3500°K	.35
Polished silver		.03
Polished aluminum	300°K	.03
Polished aluminum	1000°K	.07
Polished copper		.15
Polished iron		.2
Polished brass		.03
Oxidized iron		.8
Black oxidized copper		.78
Aluminum oxide		.75
Water		.94
Paper		.92
Glass		.94
Lampblack		.95
Laboratory black body cavity		.98 - .99

FIG. 8.8. The total emissivity of a number of materials.

When dealing with gray-bodies, it is necessary to insert the emissivity factor ϵ into the black-body equations. Planck's law (Eq. 8.14), the Stefan-Boltzmann law (Eq. 8.15), and the Wien displacement law (Eq. 8.17) should be modified by multiplying the right hand term by the appropriate value of ϵ. For many materials the emissivity is a function of wavelength. This is apparent from the fact that many substances (glass, for example) have a negligible absorption, and consequent low emissivity, at certain wavelengths, while they are almost totally absorbent at other wavelengths. In regions of the spectrum where this occurs, emissivity becomes spectral emissivity (ϵ_λ) and is treated just as any other spectral function. For many materials, emissivity will decrease as wavelength increases. It should also be noted that most materials show a variation of emissivity with temperature, as well as wavelength, and precise work must take this into account. Emissivity usually increases with temperature.

Note that not all sources are continuous emitters. Gas discharge lamps at low pressure emit discrete spectral lines; the plot of spectral radiant emittance for such a source is a series of sharp spikes, although there is usually a low level background continuum. In high pressure arcs, the spectral lines broaden and merge into a continuous background with less pronounced spikes.

Before leaving the subject of black-body radiation, the concept of color temperature should be mentioned. The color temperature of a source of light is a colorimetric concept related to the apparent visual color of a source, not its temperature. For a black-body, the color temperature is equal to the actual temperature in °K. For other sources, the color temperature is the actual temperature of the black-body which has the same effective spectral distribution (in the visible range). Thus, exceedingly bright or dim sources may have the same color temperature, but radically different radiances or intensities. Color temperature is extremely important in colorimetry and in color photography where fidelity of color rendition is important, but is little used in radiometry.

8.9 Photometry

Photometry deals with luminous radiation, that is, radiation which the human eye can detect. The basic photometric unit of radiant power is the lumen, which is defined as the luminous flux emitted into a solid angle of one steradian by a point source whose intensity is 1/60 of that of one square centimeter of a black-body at the solidification temperature of platinum (2042°K). From the preceding section, we know that a black body radiates energy throughout the entire electromagnetic spectrum. Chapter 5 indicated that the eye was sensitive to only a small interval of this spectrum and that its response to different wavelengths within this interval varied widely. Thus, if a source of radiation has a spectral power function $P(\lambda)$ (watts micron^{-1}), the visual effect of this radiation is obtained by multiplying it by $V(\lambda)$, the visual response function which is plotted in Fig. 5.10. The effective visual power of a source is, therefore, the integral (or summation) of $P(\lambda) V(\lambda) d\lambda$ over the appropriate wavelength interval. From the definition of the lumen, it can be determined that one watt of radiant energy at the wavelength of maximum visual sensitivity (0.555 microns) is equal to 680 lumens. Therefore, the luminous flux emitted by a source with a spectral power of $P(\lambda)$ watts micron^{-1} is given by*

$$F = 680 \int V(\lambda) P(\lambda) d\lambda \text{ lumens} \qquad (8.18)$$

The unit of luminous intensity is called the candle (or "candela") and is so named because the original standard of intensity was an actual candle. A point of source of one candle power is one which emits one lumen into a solid angle of one steradian. A source of

---

*Note that $V(\lambda)$ is customarily the photopic (normal level of illumination and brightness) visual response curve. Under conditions of complete dark adaptation, the visual response for scotopic vision would be used.

one candle intensity which radiates uniformly in all directions emits 4π lumens. From the definition of the lumen, it is apparent that a one square cm. black body at 2042°K has an intensity of 60 candles.

Illumination, or illuminance, is the luminous flux per unit area incident on a surface. The most widely used unit of illumination is the foot-candle. One foot-candle is one lumen incident per square foot. The misleading name foot-candle resulted from the fact that it is the illumination produced on a surface one foot away from a source of one candle intensity. The photometric term illuminance corresponds to irradiance in radiometry.

The term brightness, or luminance, corresponds to the term radiance. Brightness is the luminous flux emitted from a surface per unit solid angle per unit of area (projected on a plane normal to the line of sight). There are several commonly used units of brightness. The candle per square centimeter is equal to one lumen emitted per steradian per square centimeter. The lambert is equal to $1/\pi$ candles per square centimeter. The foot-lambert is equal to $1/\pi$ candles per square foot. The foot-lambert is a convenient unit for illuminating engineering work, since it is the brightness which results from one foot-candle of illumination falling on a

Source	Brightness
Sun (zenith)	1.6×10^5 candles/cm^2
Sun (horizon)	6×10^2
Blue sky	0.8
Dark cloudy sky	4×10^{-3}
Night sky	5×10^{-9}
Moon	0.25
Exteriors - daylight (typical)	1.
Exteriors - night (typical)	10^{-6}
Interiors - daylight (typical)	10^{-2}
Mercury arc - high pressure	5×10^5
Mercury arc - laboratory	10.
Carbon arc	10^4 to 10^5
Tungsten melting point 3655°K	5.7×10^3
3500°K	4.2×10^3
3000°K	1.3×10^3
Tungsten filament - ordinary lamp	5×10^2
- projection lamp	3×10^3
Fluorescent lamp	0.6
Sodium lamp	6.
Flame - candle, kerosene	1.
Least perceptible brightness	5×10^{-11}
Least perceptible point source	2×10^{-8} candles at 3 meter dist.

FIG. 8.9. Typical values for the brightness of a number of sources

"perfect" diffusing surface. (Since one lumen is incident on the one square foot area under an illumination of one foot-candle, the total flux radiated into a hemisphere of 2π ster. from a perfectly diffuse (Lambertian) surface is just one lumen. As pointed out in Section 8.4 and Example A, the resulting brightness is $1/\pi$ lumen ster^{-1} foot^{-2}, not $1/2\pi$ lumen ster^{-1} foot^{-2}). The brightness of a number of sources is tabulated in Fig. 8.9.

The terminology of photometry has grown through engineering usage, and is thus far from orderly. Special terms have derived from special usages, and many such terms have survived. A tabulation of photometric units is given in Fig. 8.10.

FLUX (Symbol F)
    lumen   defined in text

INTENSITY (Symbol I)

candle ("candela")	one lumen per steradian emitted from a point source. 1/60 of the intensity of one sq. cm. of a black body at 2042° K.
carcel	9.6 candles
hefner	0.9 candles
"old candle"	1.02 candles (candela)

ILLUMINATION (Symbol E) (Also called illuminance)

foot candle	one lumen per square foot incident on a surface.
phot	one lumen per square centimeter.
lux	one lumen per square meter.
meter-candle	one lumen per square meter.

BRIGHTNESS (Symbol B) (also called luminance)

candle per sq. cm.	one lumen emitted per steradian per square cm. area projected normal to direction.
stilb	one candle per square centimeter
lambert	$1/\pi$ candles per square centimeter
foot-lambert	$1/\pi$ candles per square foot

FIG. 8.10. Photometric quantities

Photometric calculations may be carried out exactly as radiometric calculations, using the relationships presented in Sections 8.2 through 8.6. If lumens are substituted for watts in all the expressions, the computations are straightforward. When the starting and final data must be expressed in the special terminology of photometry (as opposed to what one might term the rational units of lumens, steradians, and square centimeters),then conversion factors may be necessary for each relationship. A very simple way of avoiding this difficulty is to convert the starting data to lumens, ster. and cm^2, complete the calculation and then convert the results into the desired units.

For convenience, the basic relationships are repeated here

in both radiometric (left column) and photometric (right column) form:

INTENSITY

$$J = \frac{P}{\Omega}$$

J is radiant intensity

P is the radiant power emitted into solid angle Ω

INTENSITY

$$I = \frac{F}{\Omega}$$

I is luminous intensity

F is the luminous flux emitted into solid angle Ω

IRRADIANCE

$$H = \frac{J}{S^2} = J\Omega$$

E is the irradiance incident on a surface a distance S from a point source of intensity J. Ω is the solid angle subtended by a unit area of the surface from the source.

$$H = \pi N \, Sin^2 \theta$$

H is the irradiance produced by a diffuse circular source of radiance N at a point from which the source diameter subtends 2θ.

$$H = N\omega$$

H is the irradiance produced by a diffuse source of radiance N at a point from which the area of the source subtends the solid angle ω.

$$H = T\pi N \, Sin^2 \theta$$
$$(H = TN\omega)$$

H is the irradiance at an image formed by an optical system of transmission T whose exit pupil diameter (area) subtends an angle 2θ (solid angle ω) from the image point when object radiance is N.

ILLUMINATION (Illuminance)

$$E = \frac{I}{S^2} = I\Omega$$

E is the illumination incident on a surface a distance S from a point source of intensity I. Ω is the solid angle subtended by a unit area of the surface from the source.

$$E = \pi B \, Sin^2 \theta$$

E is the illumination produced by a diffuse circular source of brightness (luminance) B at a point from which the source diameter subtends 2θ.

$$E = B\omega$$

E is the illumination produced by a diffuse source of brightness B at a point from which the area of the source subtends the solid angle ω.

$$E = T\pi B \, Sin^2 \theta$$
$$(E = TB\omega)$$

E is the illumination at an image formed by an optical system of transmission T whose exit pupil diameter (area) subtends an angle 2θ (solid angle ω) from the image point when the object brightness is B.

RADIANCE	BRIGHTNESS (Luminance)
$$N = \frac{P}{\pi A}$$	$$B = \frac{F}{\pi A}$$
N is the radiance of a diffuse source of area A which emits radiant power P into a hemisphere of 2π steradians.	B is the brightness of a diffuse source of area A which emits luminous flux F into a hemisphere of 2π steradians.

Example C

It may be instructive to repeat Example A in photometric terms and to indicate at each step in the calculation the conversions to the various photometric units. We will use Fig. 8.5 again; the only change in the starting data will be that the source A will be assumed to have a brightness of 10 lumens per steradian per square centimeter.

From Fig. 8.10, we note that the source brightness may also be expressed as 10 candles per cm^2, as 10 stilb, as 10π lamberts, or as 9290π foot lamberts.

The illumination produced at Point B is calculated from Eq. 8.7 (after rewriting it in photometric symbols)

$$H = \pi N \sin^2 \theta$$

$$E = \pi B \sin^2 \theta$$

$$= \pi (10\,L\ ster^{-1}\ cm^{-2}) \left(\frac{1}{2}\right)^2$$

$$= 7.85 \text{ lumen } cm^{-2}$$

Applying the cosine-fourth law, we find the illumination at C

$$E_C = E_B \cos^4 45°$$

$$= 7.85 \times (.707)^4$$

$$= 1.96 \text{ lumen } cm^{-2}$$

Since there are 929 cm^2 per square foot

$$E_c = 929 \times 1.96 = 1821 \text{ lumens per square foot}$$

$$= 1821 \text{ foot-candles}$$

Since the surface BC has a diffuse reflectivity of 70%, we can multiply the illumination in foot-candles by 0.7 to obtain the brightness in foot-lamberts

$$B = 0.7 \times 1821 = 1275 \text{ foot-lamberts}$$

Similarly 0.7 times the illumination in candle cm^{-2} will yield the brightness in lamberts

$$B = 0.7 \times 1.96 = 1.37 \text{ lamberts}$$

Or we can retain the lumen units, and determine that, with 1.96 lumen cm^{-2} falling on a surface 70% reflectivity, 1.37 lumen cm^{-2} will be emitted into a hemisphere, and, following our previous reasoning, compute the brightness as

$$B = \frac{1.37}{\pi}$$

$$= 0.44 \text{ lumen ster}^{-1} \text{ cm}^{-2}$$

$$= 0.44 \text{ candle cm}^{-2}$$

The illumination at E is determined from Eq. 8.11 as before

$$H = TN \sin^2 \theta$$

$$= TN\omega$$

$$E = TB\omega$$

$$= 0.8 \times 0.44 \times 10^{-4}$$

$$= 0.35 \times 10^{-4} \text{ lumen cm}^{-2}$$

$$= 929 \times 0.35 \times 10^{-4} = 0.032 \text{ foot candles.}$$

8.10   Illumination   Devices

A searchlight is one of the simpler, and at the same time one of the least understood, illuminating devices. It consists of a source of light (usually small) placed at the focal point of a lens or reflector. The image of the source is thus located at infinity. A common misconception is that the beam of light produced is a "collimated parallel bundle" which extends out to infinity with a constant diameter and a constant power density. A little consideration of the matter will reveal the fallacy: the rays from any point on the source do indeed form a collimated parallel bundle, etc. However, a geometrical point on any source of finite brightness must emit zero energy, since a point has zero area, and therefore the "collimated bundle" of rays has zero energy.

With reference to Fig.8.11, which shows a source S at the focal point of lens L, the image (S') will be located at infinity. Since S subtends an angle α from L, the image S' will also subtend α. Now the illumination at a point on the axis will be determined by the brightness of the image and the solid angle subtended by the image. Thus, for points *near the lens*, the illumination is given by

$$E = TB\omega \qquad (8.19)$$

which the reader will recognize as Eq. 8.8 rewritten in photometric symbols and with a transmission constant (T) added. B is the brightness of source S (since the brightness of an image equals the brightness of the object) and ω is the solid angle subtended by the image. (We have tacitly assumed ω to be small.) Now for a point at the lens, it is obvious that the solid angle ω subtended by the image S' is exactly equal to the solid angle subtended by the source S from the lens. Since S' is at infinity, this angle will not change as we shift our reference point a short distance along the axis away from the lens, and the illumination will remain constant in this region.

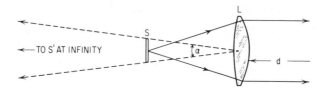

FIG. 8.11. The optics of a searchlight.

However, at a distance D = (lens diameter)/α, the source image will subtend the same angle as the diameter of the lens, and for points more distant than D, the size of the solid angle subtended by the source of illumination will be limited by the lens diameter. This solid angle will obviously be equal to (area of lens)/d^2 and the illumination beyond distance D will fall off with the square of the distance (d) to the lens. Thus, the equations governing the illumination produced by a searchlight are:

$$D = \text{(lens diameter)}/\alpha \qquad (8.20)$$

$$\text{for } d \leq D: \ E = TB\omega \ \text{(a constant)} \qquad (8.21)$$

$$\text{for } d \geq D: \ E = TB \text{ (lens area)}/d^2 \qquad (8.22)$$

The general technique used here is applicable to almost any illumination problem, and we can restate it in general terms as follows:

To determine the illumination at a point, the size and position of the source image, as seen from the point, are calculated. The pupils and windows of the system (again, as seen from the point) are determined. Then the illumination at the point is the product of the system transmission, the source brightness and the solid angle subtended by that area of the source which can be seen from the point through the pupils and windows of the system.

Note that for points (which lie within the beam) beyond the critical distance D, the searchlight acts as if it were a source of a diameter equal to that of the searchlight lens and a brightness TB. As mentioned in Section 8.6, this concept is quite useful in evaluating the illumination at an image point; here we find that it occasionally can be applied to points which are not image points.

The beam candle power of a searchlight is simply the intensity of the (point) source which would produce the same illumination at a great distance. A point source with an intensity of I candles will emit I lumens per steradian. A one square foot area placed d feet from the point source will subtend $1/d^2$ steradians from the source, and will thus be illuminated by I/d^2 lumens per square foot (foot-candles). We can determine the necessary candle power for I by equating this illumination to that produced by the searchlight according to Eq. 8.22.

$$E = \frac{I}{d^2} = \frac{TB \text{ (lens area)}}{d^2}$$

(8.23)

$$I = TB \text{ (lens area)}$$

where I is the beam candle power in lumens per steradian (or candles). Note that the lens area should be specified in the same units as the source brightness.

The second illumination device we shall consider is the projection condenser, which is schematically diagrammed in Fig. 8.12.

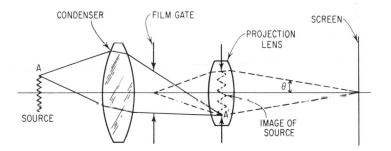

FIG. 8.12. Schematic of a projection condenser system. The condenser forms an image of the source (lamp filament) in the aperture of the projection lens.

The purpose of the projector is to produce a bright and evenly illuminated image of the film on the screen. This could be achieved by placing a sheet of diffusing material behind the film and illuminating this diffuser. The resultant image would be dim, because the maximum brightness which the image could achieve would be that of the diffuser, which would be considerably less than that of the lamp. The function of the condenser is to image the source in the pupil of the projection lens so that the lens aperture has the same brightness as the source. When this is done, the screen is illuminated according to Eq. 8.11, where the solid angle is that subtended by the source image (in the projection lens) from the screen. It is apparent that the maximum value for the screen illumination, is limited by the size of the projection lens aperture. Therefore, the maximum screen illumination is achieved when the image of the source completely fills the aperture of the lens. This is required for all points within the field of view, and the condenser diameter must be sufficiently large so that it does not vignette, if maximum illumination at the edge of the picture is required. In this regard, note that the ray from the corner of the film to the opposite edge of the lens aperture is the most demanding. The cos-fourth law will, of course, reduce the illumination at points off the axis.

When the source is irregular in shape, as in "V" filament lamps for example, the solid angle for Eq. 8.11 is determined just as one might expect, by dividing the area of the actual image of the filament by the square of the distance to the screen.

The apparent brightness of an image as seen by the eye is a function of the diameter of the pupil of the eye, since it determines the illumination of the retina, in accordance with Eq. 8.11a. When the eye is used with an optical instrument, such as a telescope, the exit pupil of the instrument enters the picture. If the exit pupil is larger than that of the eye, then the apparent brightness of the object seen through the instrument is equal to the brightness of the object (less transmission losses) since the solid angle subtended by the pupil from the retina is unchanged. When the exit pupil is smaller than that of the eye, then the apparent brightness of the object is reduced in proportion to the relative areas of the pupils. The exception to this brightness equality of object and image occurs when the object is smaller than the diffraction limit of the optical system (e.g. a star). Since this is not an extended source, all the energy in the retinal image is concentrated on a few retinal receptors, and when the magnification of a telescope is increased, its effective collection area is increased (at the objective) so that more energy is concentrated on the same retinal cells, resulting in an increase in the apparent brightness of the source. For example, if a high enough power telescope of large aperture is used, stars may be seen in daylight, since their apparent brightness is increased while that of the sky is not.

References

1. Hardy and Perrin, "The Principles of Optics," McGraw-Hill, 1932.

2. Hackforth, "Infrared Radiation", McGraw-Hill, 1960.
3. Jamieson et al., "Infrared Ohysics and Engineering", McGraw-Hill, 1963.
4. "American Institute of Physics Handbook", McGraw-Hill, 1963.
5. Walsh, "Photometry", Dover, N.Y., 1958.

Exercises

1. A point source emits 10 watts per steradian toward a 4 inch diameter optical system. How much power is collected by the optical system when its distance from the source is a) 10 feet, b) one mile?
    Ans. a) 8.73×10^{-3} watts; b) 3.13×10^{-8} watts

2. A 10 candle power point source illuminates a perfectly diffusing surface which is tilted at 45° to the line of sight to the source. What is the brightness of the surface, if it is 10 feet from the source?
    Ans. 0.0707 foot lamberts or 2.42×10^{-5} candles per cm².

3. A fluorescent lamp $10''$ long and $1''$ wide illuminates a slit, parallel to the lamp, which is $10''$ long and $10''$ from the lamp. If the lamp has a brightness of 0.5 candles per square centimeter, what is the illumination a) at the center of the slit, and b) at the ends of the slit? (Hint: divide the lamp into ten one inch square sources)
    Ans. a) .043 lumens cm⁻² or 40.2 foot candles
    b) .032 lumens cm⁻² or 29.9 foot candles

4. A 16-mm projector uses a $2''$f/1.6 projection lens and a lamp with a filament brightness of 3000 candles per cm². If the condenser fills the lens aperture with the filament image, what is the illumination produced on a screen 20 feet from the lens? Assume the transmission of the lens is 95% and the transmission of the condenser is 85%.
    Ans. 47.9 foot candles or 5.16×10^{-2} phot

5. a) What is the spectral radiance of a 1000°K black body in the region of 2 microns wavelength?
    b) If an idealized band pass filter, transmitting only between 1.95 microns and 2.05 microns is used, what is the total energy falling on a one square centimeter detector placed one meter from a one square centimeter 1000° black body? (Use Fig. 8.7)
    Ans. a) 0.89 watts per square centimeter per micron
    b) 2.83×10^{-6} watts

6. Show that, for long projection distances, the maximum lumen output of a projector is given by

$$F = \frac{\pi ABT}{4(f/\#)^2} \text{ lumens}$$

where A is the area of the film gate, B the source brightness, T the transmission of the system, and $(f/\#)$ is the relative aperture of the projection lens.

9

Basic Optical Devices

This chapter will be devoted to the first order of optics of several typical optical systems. The number of systems covered is, of necessity, limited and the emphasis is placed on those fundamental principles which are applicable to a broad range of optical systems. The rather straightforward algebraic manipulations and the considerations of image size and position which follow, are quite typical of those encountered in the rough preliminary stages of optical system design. Constructional details of the optical components have been deliberately omitted and are discussed at considerable length in later chapters.

9.1 Telescopes

The primary function of a telescope is to enlarge the apparent size of a distant object. This is accomplished by presenting to the eye an image which subtends a larger angle (from the eye) than does the object. The magnification, or power, of a telescope is simply the ratio of the angle subtended by the image to the angle subtended by the object*. Nominally, a telescope works with both its object and image located at infinity; it is referred to as an afocal instrument, since it has no focal length. In the following material, a number of basic relationships for telescopes will be presented, all based on systems with both object and image located at infinity. In practice, departures from these infinite conjugates are the rule, but for the most part they may be neglected. However, the fact that the object and/or the image are not at infinity will occasionally have a noticeable effect and must then be taken into account.

There are three major types of telescopes: astronomical (or inverting), terrestrial (or erecting), and Galilean. An astronomical telescope is composed of two positive (i.e. converging) components spaced so that the second focal point of the first component coincides with the first focal point of the second, as shown in Fig. 9.1A. The objective lens (the component nearer the object)

---

*For large angles, the magnification is the ratio of their tangents.

forms an inverted image at its focal point; the eyelens then re-images the object at infinity where it may be comfortably viewed by a relaxed eye. Since the internal image is inverted, and the eyelens does not re-invert the image, the view presented to the eye is inverted top to bottom and reversed left to right.

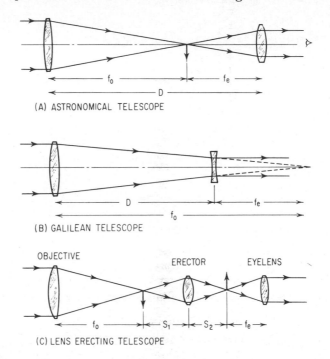

(A) ASTRONOMICAL TELESCOPE

(B) GALILEAN TELESCOPE

(C) LENS ERECTING TELESCOPE

FIG. 9.1. The three basic types of telescope.

In a Galilean telescope, Fig. 9.1B, the positive eyelens is replaced by a negative (diverging) eyelens; the spacing is the same, in that the focal points of objective and eyelens coincide. In the Galilean scope, however, the internal image is never actually formed; the object for the eyelens is a "virtual" object, no in-version occurs, and the final image presented to the eye is erect and unreversed. Since there is no real image formed in a Galilean telescope, there is no location where cross hairs or a reticle may be inserted.

Assuming the components of the telescope to be thin lenses, we can derive several important relationships which apply to *all* telescopes and which are of great utility. First, it is readily apparent that the length (D) of a simple telescope is equal to the sum of the focal lengths of the objective and eyelens.

$$D = f_o + f_e \qquad (9.1)$$

Note that in the Galilean telescope, the spacing is the difference between the absolute values of the focal lengths since f_e is negative.

The magnification, or magnifying power, of the telescope is the ratio between u_e, the angle subtended by the image, and u_o, the angle subtended by the object. The size (h) of the internal image formed by the objective will be

$$h = -u_o f_o \qquad (9.2)$$

and the angle subtended by this image from the first principal point of the eyelens will be

$$u_e = \frac{h}{f_e} \qquad (9.3)$$

Combining Eqs. 9.2 and 9.3, we get the magnification

$$\text{M.P.} = \frac{u_e}{u_o} = \frac{-f_o}{f_e} \qquad (9.4)$$

The sign convention here is that a positive magnification indicates an erect image. Thus, if objective and eyelens both have positive focal lengths, M.P. is negative and the telescope is inverting. The Galilean scope with objective and eyelens of opposite sign produces a positive M. P. and an erect image.

Note that u_o can represent the real angular field of view of the telescope and u_e the apparent angular field of view, and that Eq. 9.4 defines the relationship between the real and apparent fields for small angles. For large angles, the tangents of the half-field angles should be substituted in this expression.

From Chapter 6 we recall that the exit pupil of a system is the image (formed by the system) of the entrance pupil. In most telescopes the objective clear aperture is the entrance pupil and the exit pupil is the image of the objective as formed by the eyelens. Using the Newtonian expression relating object and image sizes $(h' = hf/x)$, and substituting CA_e (the exit pupil diameter) and CA_o (the entrance pupil diameter) for h' and h, f_e for f, and $-f_o$ for x, we get

$$\frac{CA_o}{CA_e} = \frac{-f_o}{f_e} = \text{M.P.} \qquad (9.5)$$

While the above derivation has assumed the entrance pupil to be at the objective, Eq. 9.5 is valid regardless of the pupil location, as is obvious from the rays sketched in Fig. 9.1.

Equations 9.4 and 9.5 can be combined to relate the external characteristics (magnifications, fields of view and pupils) of *any* afocal system, regardless of its internal construction

$$\text{M.P.} = \frac{u_e}{u_o} = \frac{CA_o}{CA_e} \tag{9.6}$$

The erecting telescope, Fig. 9.1C, consists of positive objective and eye-lenses with an erecting lens between the two. The erector re-images the image formed by the objective into the focal plane of the eyelens. Since it inverts the image in the process, the final image presented to the eye is erect. This is the form of telescope ordinarily used for observing terrestrial objects, where considerable confusion can result from an inverted image. (An erect image may also be obtained by the use of an erecting prism as discussed in Chapter 4.) The magnification of a terrestrial telescope is simply the magnification that the telescope would have without the erector, multiplied by the linear magnification of the erector system

$$\text{M.P.} = -\frac{f_o}{f_e} \cdot \frac{s_2}{s_1} \tag{9.7}$$

where s_2 and s_1 are the erector conjugates as indicated in Fig. 9.1C. For a scope as shown, f_o, f_e and s_2 are positive signed quantities and s_1 is negative. The resulting M.P. is thus positive, indicating an erect image.

9.2 Field Lenses and Relay Systems

In a simple two-element telescope as shown in Fig. 9.2A, the field of view is limited by the diameter of the eyelens (as discussed at greater length in Chapter 6). In the sketch, the solid rays indicate the largest field angle that a bundle may have and still pass through the telescope without vignetting; for the bundle represented by the dashed rays, only the ray through the upper rim of the objective gets through, and vignetting is effectively complete.

The function of a field lens is indicated in Fig. 9.2B. If the field lens is placed exactly at the internal image, it has no effect on the power of the telescope, but it bends the ray bundles (which would otherwise miss the eyelens) back toward the axis so that they pass through the eyelens. In this way the field of view may be increased without increasing the diameter of the eyelens. Note that the exit pupil is shifted to the left, closer to the eyelens, by the introduction of a positive field lens. The distance from the vertex of the eyelens to the exit pupil is called the eye relief (since the eye must be placed at the pupil to see the full field of view). The necessity for a positive eye relief obviously limits the strength of the field lens that can be used. In practice, field lenses are rarely located exactly at the image plane, but either ahead of or behind the image, so that imperfections in the field lens are not visible.

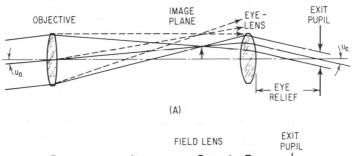

(A)

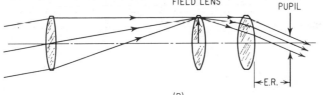

(B)

FIG. 9.2. The action of a field lens in increasing the field of view.

When it is desired to carry an image through a relatively long distance and the available space limits the diameter of the lenses which can be used, a system of relay lenses can be effective. In Fig. 9.3, the objective lens forms its image in field lens A. The image is then relayed to field lens C by lens B which functions like an erector lens. The image is then relayed again by lens D. The power of field lens A is chosen so that it forms an image of the objective at lens B; similarly field lens C forms an image of lens B in lens D. In this way, the entrance pupil (which, in this example, is at the objective) is imaged at each of the relay lenses in turn and the image is passed through the system without vignetting. The dashed rays emerging from lens A will indicate the large diameters which would otherwise be necessary to cover the same field of view.

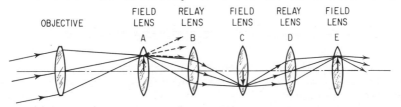

FIG. 9.3. A system of relay lenses.

9.3 Exit Pupils, The Eye, and Resolution

Since almost all telescopes are visual instruments, they must be designed to be compatible with the characteristics of the human eye. In Chapter 5, we saw that the pupil of the eye varied in diameter from 2 mm to about 8 mm, depending on the brightness of the scene being viewed. Since the pupil of the eye is, in effect, a stop of a telescopic system, its effect must be considered. For ordinary use, an exit pupil of 3 mm diameter will fill the pupil of the eye and no increase in retinal illumination will be obtained by providing a larger exit pupil. From Eq. 9.5, it is apparent that the maximum *effective* clear aperture for an ordinary telescope objective is thus limited to a diameter of about 3 mm times the magnification. This is, however, a fairly flexible situation. In surveying instruments exit pupils of 1.0 to 1.5 mm are common, since size and weight are at a premium and resolution is the most desired characteristic. In ordinary binoculars, a 5 mm pupil is usually provided; the added pupil diameter makes it much easier to align the binocular with the eyes. For the same reason, rifle scopes usually have exit pupils ranging in size from 5 to 10 mm. Telescopes and binoculars designed for use at low light levels (such as night glasses) usually have 7 or 8 mm exit pupils in order to obtain the maximum retinal illumination possible.

In Chapter 5, it was indicated that the resolution of the eye was about one minute of arc; Chapter 6 indicated that the angular resolution of a perfect optical system was $(5.5/D)$ seconds of arc when the clear aperture of the system (D) was expressed in inches. One or both of these limitations will govern the effective performance of any telescope, and for the most efficient design of a telescope, both should be taken into account. If two objects which are to be resolved are separated by an angle α, after magnification by a telescope, their images will be separated by $(M.P.)\alpha$. If $(M.P.)\alpha$ exceeds one minute of arc, the eye will be able to separate the two images; if $(M.P.)\alpha$ is less than one minute, the two objects will not be seen as separate and distinct. Thus, the magnification of a telescope should be chosen so that

$$M.P. \; > \; \frac{1}{\alpha} \; (\alpha \text{ in minutes})$$

$$> \; \frac{0.0003}{\alpha} \; (\alpha \text{ in radians})$$

(9.8)

and for critical work, a value considerably larger than indicated in Eq. 9.8 is often selected in order to minimize the visual fatigue of the viewer.

From the opposite point of view, since the resolution of a telescope (in object space) is limited to $(5.5/D)$ seconds, it is apparent that the smallest resolved detail in the image presented to the eye will subtend an angle of (M.P.) (5.5/D) seconds, and if this angle equals or exceeds one minute, the eye can discern all of the resolved details. Equating this angle to one minute (60 seconds), we find that the maximum "useful" power for a telescope is

$$M.P. = 11D \qquad (9.9)$$

(when D is in inches). Magnification in excess of this power is termed empty magnification, since it produces no increase in resolution. However, it is not unusual to utilize magnifications two or even three times this amount to minimize visual effort. The upper limit on effective magnification usually occurs at the point when the diffraction blurring of the image becomes a distraction sufficient to offset the gain in visual facility.

Example A

As numerical examples to illustrate the preceding sections, we will determine the necessary powers and spacings to produce a telescope with the following characteristics: a magnification of 4× and a length of 10″. We will do this in turn for an inverting telescope, a Galilean telescope and an erecting telescope, and will discuss the effects of arbitrarily limiting the element diameters to one inch.

For a telescope with only two components, it is apparent that Eqs. 9.1 and 9.4 together determine the powers of the objective and eyelens. Thus, we have

$$D = f_o + f_e = 10 \text{ inches}$$

$$\text{and M.P.} = \frac{-f_o}{f_e} = \pm 4X$$

where the sign of the magnification will determine whether the final image is erect (+) or inverted (−). Combining the two expressions and solving for the focal lengths, we get

$$f_o = \frac{(M.P.)D}{(M.P.) - 1} \qquad (9.10)$$

$$f_e = \frac{D}{1 - (M.P.)} \qquad (9.11)$$

For the inverting telescope, we simply substitute M.P. = −4 and $D = 10″$, to find that the required focal length for the objective is

8 inches; for the eyelens it is 2 inches. Since the lens diameters are to be one inch, the exit pupil diameter is 0.25 inch (from Eq. 9.5). The position of the exit pupil can be determined by tracing a ray from the center of the objective through the edge of the eyelens or by use of the thin-lens Eq. 2.4, as follows:

$$\frac{1}{s'} = \frac{1}{f} + \frac{.1}{s} = \frac{1}{f_e} + \frac{1}{(-D)} = \frac{1}{2} - \frac{1}{10} = 0.4$$

$s' = 2.5$ inches

Thus, the eye relief of our simple telescope is 2-1/2 inches.

The field of view of this telescope is not clearly defined, since it is determined by vignetting at the eyelens, as consideration of Fig. 9.4 will indicate. The aperture will be 50% vignetted at a field angle such that the principal (or chief) ray passes through the rim of the eyelens. Under these conditions

$$u_o = \frac{\text{dia. eyelens}}{2D} = \frac{1}{2 \times 10} = \pm 0.05 \text{ radians}$$

and the real* field of view totals 0.1 radians, or about 5.7°.

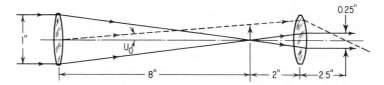

FIG. 9.4. The inverting telescope of Example A.

This is a poor representation of what the eye will see however, since the vignetted exit pupil at this angle closely approximates a semi-circle 0.25″ in diameter and can thus completely fill a 3 mm eye pupil. The field angle at which no rays get through the telescope is a somewhat more representative value for the field of view. If we visualize the size of u_o in Fig. 9.4 as being slowly increased, it is apparent that the ray from the bottom of the objective will be the first to miss the eyelens and the ray from the top of the objective will be the last to be vignetted out. For the example we have chosen, with both lenses one inch in diameter, it is apparent that the limiting diameter of the internal image will also be one inch. (For differing lens diameters, it is a simple exercise in proportion to determine the height at which this ray strikes the internal focal plane) The half field of view for 100% vignetting is

---

*The *real* field of a telescope is the (angular) field in the object space. The *apparent* field is the (angular) field in the image (i.e. eye) space.

then the quotient of the semi-diameter of the image divided by the objective focal length, or ±0.0625 radians; the total real field is 0.125 radians, or about 7.1°.

Thus, for an exit pupil of 0.25″, the field of view is totally vignetted at 0.125 rad, 50% vignetted at 0.1 rad and unvignetted at 0.075 rad. These three conditions are illustrated in Fig. 9.5, and it is apparent that the ''effective'' position of the exit pupil shifts inward as the amount of vignetting increases.

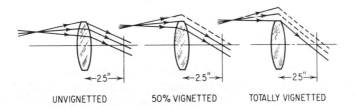

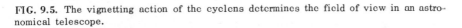

| ←2.5″→ | ←2.5″→ | ←2.5″→ |
| UNVIGNETTED | 50% VIGNETTED | TOTALLY VIGNETTED |

FIG. 9.5. The vignetting action of the eyelens determines the field of view in an astronomical telescope.

Let us now determine the minimum power for a field lens which will completely eliminate the vignetting at a field angle of ±0.0625 rad. From Fig. 9.6, it can be seen that the field lens must bend the rays from the objective so that ray B strikes no higher than the upper rim of the eyelens. The slope of ray B is equal to one inch (the difference in the heights at which it strikes the objective and the field lens) divided by minus 8 inches (the distance from field lens to objective), or −0.125. After passing through the field lens, we desire the slope to be zero (in this case) as indicated by the dashed ray B'. Using Eq. 2.31, we can solve for the power of the field lens as follows:

$$u' = u + y\phi_f$$

$$0.0 = -.125 + (0.5)\phi_f$$

$$\phi_f = +0.25$$

$$f_f = \frac{1}{\phi} = 4 \text{ inches}$$

We can now determine the new eye relief by tracing a principal ray from the center of the objective through the field and eye lenses.

$$u'_o = \frac{y_f}{(-f_o)} = -0.0625 = u_f$$

$$u'_f = u_f + y_f\phi_f = -0.0625 + 0.5(0.25) = +0.0625$$

$$y_e = y_f - u'_f f_e = 0.5 - 0.0625\,(2) = 0.375$$

$$u'_e = u'_f + y_e \phi_e = 0.0625 + 0.375\,(0.5) = 0.25$$

$$l'_e = \text{eye relief} = \frac{y_e}{u'_e} = \frac{.375}{0.25} = 1.5\,''$$

Note that u'_e and u_o are still related by the magnification, as in Eq. 9.4 where

$$\text{M.P.} = \frac{u'_e}{u_o} = \frac{+0.25}{-0.0625} = -4\times$$

since the power of the system has not been changed by the introduction of the field lens located exactly at the focal plane. If we desired to locate the field lens slightly out of the focal plane, the general approach would be the same; the distances, ray heights, etc. in the computations would, of course, be modified accordingly. The power of the telescope would be increased (and the scope shortened) if the field lens were placed to the right of the focus and vice versa.

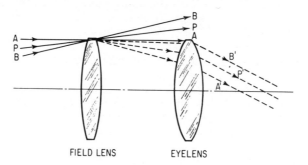

FIELD LENS   EYELENS

FIG. 9.6. Ray diagram used to determine field lens power in Example A.

For the Galilean version of our telescope, we solve for the component focal lengths by substituting +4× for the magnification in Eqs. 9.10 and 9.11 and get

$$f_o = \frac{(\text{M.P.})D}{(\text{M.P.}) - 1} = \frac{(+4)\,10}{+4 - 1} = +13.33 \text{ inches}$$

$$f_e = \frac{D}{1 - (\text{M.P.})} = \frac{10}{1 - (+4)} = -3.33 \text{ inches}$$

In a Galilean scope the aperture stop is not the objective lens, but is the pupil of the user's eye, and the exit pupil is wherever the eye is located. This is usually about 5 mm behind the eyelens.

To determine the field of view, we must trace a principal ray through the center of the pupil and passing through the edge of the objective, as indicated in Fig. 9.7. This can be done by assuming some arbitrary value for u_e and tracing the ray through, then scaling the ray data by an appropriate constant (as indicated in Chapter 6) to make the ray height at the objective equal to one-half its clear aperture. To simplify matters, we will assume here that the pupil is coincident with the eyelens; thus, u_e is equal to half the objective diameter divided by the spacing between the lenses, or 0.05 radians in this instance. Since M.P. $= u_e/u_o$ per Eq. 9.4, we can solve for $u_o = 0.05/4 = 0.0125$ radians. The total real field is 0.025 radians (about 1.5°), considerably less than that of the inverting telescope discussed above. Note that the same type of field vignetting considerations may be applied to the Galilean telescope. One must also bear in mind that the *direction* of the Galilean field of view can be changed by a lateral shift of the viewer's eye; this is not true for a telescope with a real internal image when the field stop is located at the image.

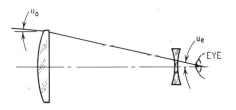

FIG. 9.7. In a Galilean telescope, the field of view is determined by the diameter of the objective lens and the location of the exit pupil, which is usually the pupil of the observer's eye.

For the erecting telescope example, we will lay out a telescopic rifle sight, with a magnification of +4×, a length of 10″ and a maximum lens diameter of one inch, as before. For small caliber (.22) rifles, a 2″ eye relief is acceptable; for heavier guns, eye reliefs of 3 to 5 inches are common. Let us assume that we desire an eye relief of 4 inches and design the telescope accordingly. The entrance pupil (at the objective) has a diameter of one inch; by Eq. 9.6, the exit pupil diameter is thus one-quarter inch. Again by Eq. 9.6, the apparent field at the eyepiece (u_e) is equal to $4u_o$, where u_o is the real field. With reference to Fig. 9.8, it is apparent that u_e is limited by the diameter of the eyelens, and that for an unvignetted pupil

$$u_e = 4u_o = \frac{1}{2R} \text{ (eyelens dia. - pupil dia.)}$$

$$u_e = \frac{1}{2 \times 4}(1 - 0.25) = \pm.09375$$

$$u_o = \pm.0234 = (\pm 1.3°)$$

To determine the spacing and powers of the components, we note that

$$L = f_o - s_1 + s_2 + f_e$$

$$M = \frac{-f_o s_2}{f_e s_1}$$

We can combine these expressions and derive equations for s_1, s_2 and f_r in terms of M, L, f_o and f_e as follows:

$$s_1 = \frac{-f_o(L - f_o - f_e)}{(Mf_e + f_o)}$$

$$s_2 = \frac{-s_1 Mf_e}{f_o} = \frac{Mf_e(L - f_o - f_e)}{(Mf_e + f_o)}$$

$$f_r = \frac{s_1 s_2}{s_1 - s_2} = \frac{Mf_e f_o(L - f_o - f_e)}{(Mf_e + f_o)^2}$$

At this point, we are faced with a situation which is very common in the layout stages of optical design. We can elect to proceed algebraically to find an expression for f_o and f_e which will yield a scope with the desired eye relief, R, or we can proceed numerically. For a one-time solution, the numerical approach is usually the better choice, especially if the system under consideration is well understood. If one is likely to design a number of systems of the same type with various parameters, or if one is "exploring" and wishes to locate all possible solutions, the often tedious labor of an algebraic solution may be well repaid.

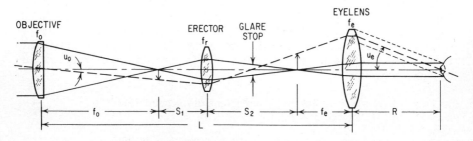

FIG. 9.8. Optics of a simple erecting telescope.

The preceding equations indicate that we have two choices (or degrees of freedom) which we can make, namely f_o and f_e, and arrive at a 4× scope of 10 inch length; we have not, however, included the eye relief in these equations. To resolve this situation numerically, we would now assume some reasonable value for f_o, then proceed to test various values of f_e, selecting the value of f_e which yields the desired value for the eye relief, R. Since R is not a critical dimension, a graphic solution (after a few values of f_e have been tried), plotting R vs f_e, would be quite adequate for our purpose. Repeating the process for several additional values of f_o would then indicate the range of solutions available.

To arrive at a solution analytically, we would proceed as follows: a principal ray, starting at the center of the objective lens with some arbitrary slope angle would be ray traced by thin-lens equations (2.31, 2.32 and 2.33), using the symbolic values for the spacings and lens powers derived from the three equations immediately preceding. The symbolic values for the powers and spacings involved would thus be:

$$\text{first air space} = f_o - s_1 = f_o + \frac{f_o(L - f_o - f_e)}{(Mf_e + f_o)}$$

$$\text{erector power } \phi_r = \frac{1}{f_r} - \frac{(Mf_e + f_o)^2}{Mf_e f_o(L - f_o - f_e)}$$

$$\text{second air space} = s_2 + f_e = f_e + \frac{Mf_e(L - f_o - f_e)}{(Mf_e + f_o)}$$

$$\text{eyelens power } \phi_e = \frac{1}{f_e}$$

The expression for the final intercept length of this ray, $l'_e = y_e/u'_e$ is then equated to the eye relief, R, and a solution for f_e expressed in terms of $f_o, M, L,$ and R is extracted. As can be imagined, the procedure is lengthy and the probability of making an error in the derivation is approximately unity for the first few attempts. Careful work and frequent checking are not only advisable, they are mandatory. When the smoke has cleared away, one finds that

$$f_e = \frac{M^2 RL - f_o(M^2 R + L)}{M^2(R + L) - f_o(M - 1)^2}$$

and that for any chosen value for f_o, (which is less than L and more than zero), a set of powers and spacings can be obtained which will satisfy our original conditions for power (M), length (L) and eye relief (R).

We are now faced, regardless of whether we have arrived via

numbers or symbols, with the problem of determining what is a suitable value for f_o upon which to base our solution. There are a number of criteria by which to judge the value of a given solution. In general, one desires to minimize the power of the components in any given system; in subsequent chapters, it will become apparent that it is often advisable to minize one or all of the following: $\Sigma|\phi|$, $\Sigma|y\phi|$, $\Sigma|y^2\phi|$ (where the symbol $|x|$ indicates the absolute value of x), ϕ is the component power and y represents the height of either the axial or principal ray on the component or the element semi-clear-aperture.

Avoiding, for a few chapters at least, the rationale behind these desiderata, we shall proceed to indicate the technique. For a number of arbitrarily chosen values of f_o, we determine the required values for f_r and f_e (as well as s_1 and s_2). Then the values of the component powers ϕ_o, ϕ_r and ϕ_e (where $\phi = 1/f$) as well as $\Sigma|\phi| = |\phi_o| + |\phi_r| + |\phi_e|$ are plotted against f_o, resulting in a graph as shown in Fig. 9.9. Note that the minimum $\Sigma|\phi|$ occurs in the region of $f_o = 3.5$; for want of a better criterion, this is a reasonable choice.

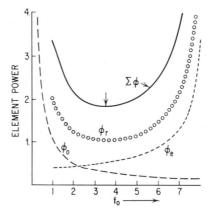

FIG. 9.9. Plot of the element powers for a 10" long erecting telescope with 4" eye relief vs the arbitrarily chosen objective focal length. ϕ_o, ϕ_r and ϕ_e are the powers of the objective, erector and eyelens respectively.

To carry the matter a bit further, we can trace an axial ray and a principal ray through each solution. The axial ray has starting data (at the objective) of $y = 0.5$ and $u = 0$; the principal ray starting data is $y_p = 0$ and $u_p = -.0234375$, chosen on the basis of eye relief and eyelens diameter considerations as discussed several paragraphs above. From these ray traces, we can determine the axial ray height, y, at each lens, y^2, and the necessary minimum clear diameter at each lens, $D = 2(|y| + |y_p|)$, to pass the full bundle of rays at the edge of the field. It turns out that, *under the conditions we have established*, the diameter for the objective and eyelens must be one inch, and the diameter of the erector lens is $0.3125''$ for all values of f_o. From this information, a graph as shown in Fig. 9.10 can be plotted. The choice of which of the four minima to select must be made on the basis of material which is contained in subsequent chapters. In general, however, a minimum $\Sigma|\phi|$ in this example would reduce the Petzval curvature of field, a minimum $\Sigma|D\phi|$ would reduce the cost of making the optics and minimum $\Sigma|D\phi|$, $\Sigma|y\phi|$ and $\Sigma|y^2\phi|$ would tend to reduce other aberrations, the choice being dependent upon which aberration one most desired to reduce.

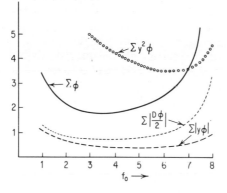

FIG. 9.10.

Assuming that we have chosen $f_o = +4$, the values of the lens powers and spacings would be determined as follows:

$$f_o = +4$$

$$f_e = \frac{4 \times 4 \times 4 \times 10 - 4(4 \times 4 \times 4 + 10)}{4 \times 4(4 + 10) - 4(4 - 1)(4 - 1)} = +1.8298$$

$$s_1 = \frac{-4(10 - 4 - 1.8298)}{(4 \times 1.8298 + 4)} = -1.4737$$

$$s_2 = \frac{-(-1.4737) \times 4 \times 1.8298}{4} = +2.6965$$

$$f_r = \frac{-(-1.4737) \times 4 \times 1.8298}{(4 \times 1.8298 + 4)} = +0.9529$$

9.4 The Simple Microscope

A microscope is an optical system which presents to the eye an enlarged image of a near object. The image is enlarged in the sense that it subtends (from the eye) a greater angle than the object does when viewed at normal viewing distance. The "normal viewing distance" is conventionally considered to be about 10"; this represents an average value for the distance at which most people see detail most clearly. (Obviously, very young people can see detail in objects a few inches from the eye and mature persons whose visual accommodation is failing, may have difficulty focusing on objects several feet away). The magnification or magnifying power of a microscope is defined as the ratio of the visual angle subtended by the image to the angle subtended by the object at a distance of 10" from the eye.

The simple microscope, or magnifying glass, consists of a lens with the object located at or within its first focal point. In Fig. 9.11, the object h, a distance s from the magnifier, is imaged at a distance s' with a height h'. As shown, the image is virtual and both s and s' are negative quantities according to our sign convention. We can readily determine the magnification by using the first order equations (2.4 and 2.7) as follows. The object and image distance equation

$$\frac{1}{s'} = \frac{1}{f} + \frac{1}{s}$$

is solved for s

$$s = \frac{f s'}{f - s'}$$

and substituted into the equation for the image height

$$h' = \frac{h s'}{s} = \frac{h(f - s')}{f}$$

Now if the eye is located at the lens, the angle subtended by the image is given by

$$\alpha' = \frac{h'}{s'} = \frac{h(f - s')}{f s'}$$

If the (unaided) eye were to view the object at a distance of (−10) inches, the angle subtended would be

$$\alpha = \frac{-h}{10''}$$

The magnifying power is the ratio between these two angles

$$\text{M.P.} = \frac{\alpha'}{\alpha} = \frac{h(f - s')}{f s'} \times \frac{(-10'')}{h}$$

$$\tag{9.10}$$

$$\text{M.P.} = \frac{10''}{f} - \frac{10''}{s'}$$

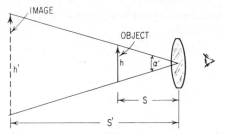

FIG. 9.11. The simple microscope, or magnifier, forms an erect, virtual image of the object.

Thus we find that the magnification produced by a simple microscope depends not only on its focal length but on the focus position chosen. If one adjusts the object distance so that the image is at infinity (i.e. $s = -f$ and $s' = \infty$) and can be viewed with a relaxed eye, then the magnification becomes simply

$$\text{M.P.} = \frac{10''}{f} \qquad\qquad (9.10\text{a})$$

If the focus is set so that the image appears to be 10 inches away (i.e. $s' = -10''$) then

$$\text{M.P.} = \frac{10''}{f} + 1 \qquad\qquad (9.10\text{b})$$

The values of M.P. given by Eqs. 9.10a and 9.10b are those conventionally used to express the power of magnifiers, eyepieces, and even compound microscopes.

9.5 The Compound Microscope

As illustrated in Fig. 9.12, a compound microscope consists of an objective lens and an eyelens. The objective lens produces a real inverted image (usually enlarged) of the object. The eyelens re-images the object at a comfortable viewing distance and magnifies the image still further. The magnifying power of the system can be determined by substituting the value of the combined focal length of the two components (as given by Eq. 2.35) into Eq. 9.10a

$$f_{eo} = \frac{f_e f_o}{f_e + f_o - d}$$

$$\text{M.P.} = \frac{10''}{f_{eo}} = \frac{(f_e + f_o - d)\,10''}{f_e f_o} \qquad\qquad (9.11)$$

The more conventional way to determine the magnification is to view it as the product of the objective magnification times the eyepiece magnification. With reference to Fig. 9.12, this approach gives

$$\text{M.P.} = M_o \times M_e = \frac{s_2}{s_1} \cdot \frac{10''}{f_e} \tag{9.12}$$

Equations 9.11 and 9.12 yield exactly the same value of magnification, as can be shown by substituting $(d - f_e)$ for s_2, determining s_1 in terms of d, f_e and f_o (from Eq. 2.4), and substituting in Eq. 9.12 to get Eq. 9.11.

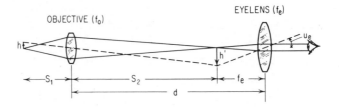

FIG. 9.12. The compound microscope.

An ordinary laboratory microscope has a tube length of 160 millimeters. The tube length is the distance from the second (i.e. internal) focal point of the objective to the first focal point of the eyepiece. Thus, by Eq. 2.6, the objective magnification is $160/f_o$, and, rewriting Eq. 9.12 for millimeter measure, we get

$$\text{M.P.} = \frac{-160}{f_o} \cdot \frac{254}{f_e} \tag{9.13}$$

Standard microscope optics are usually referred to by their power. Thus, a 16 mm focal length objective has a power of 10× and a one-half inch focal length eyepiece has a power of 20×. The combination of the two would have a magnifying power of 200×, or 200 diameters.

The resolution of a microscope is limited by both diffraction and the resolution of the eye in the same manner as in a telescope. In the case of the microscope, however, we are interested in the linear resolution, rather than angular resolution. By Rayleigh's criterion, the smallest separation between two object points that will allow them to be resolved is given by Eq. 6.15

$$Z = \frac{0.61\lambda}{\text{N.A.}}$$

where λ is wavelength and N.A. = $N \sin U$, the numerical aperture of the system. Note that the index, N, and the slope of the marginal ray, U, are those at the object. Because of the importance of the numerical aperture in this regard, microscope objectives are usually specified by power and numerical aperture; for example, a 16 mm objective is usually listed as a 10× N.A. 0.25.

At a distance of 10″, the visual resolution of one minute of arc (0.0003 radians) corresponds to a linear resolution of about 0.003″, or 0.076 mm. When the object is magnified by an optical system, the *visual* resolution at the object is thus

$$R = \frac{0.003''}{M.P.} = \frac{0.076 \text{ mm}}{M.P.} \qquad (9.14)$$

If we now equate the visual resolution R with the diffraction limit Z and solve for the magnification, we find that

$$M.P. = \frac{0.12 \text{ N.A.}}{\lambda} \qquad (9.15)$$

with λ in mm, is the magnification at which the diffraction limit and visual limit match. At this power the eye can resolve all the detail present in the image, and setting $\lambda = 0.55\mu$, any magnification beyond 225 N.A. is "empty magnification." However, as with telescopes, magnifications several times this amount are regularly used.

9.6 Rangefinders

Figure 9.13 is a schematic diagram of a simplified rangefinder. The eye views the object by two paths; directly through semi-transparent mirror M_1 and by an offset path via M_1 and fully reflecting mirror M_2. The angular position of one of the mirrors is adjusted until both images coincide. In the rudimentary instrument shown here, a pointer attached to mirror M_2 can be used to read the value of $\theta/2$; the distance to the object is found from

$$D = \frac{B}{\tan \theta} \qquad (9.16)$$

where B is the base length of the instrument. In actual rangefinders, a telescope is often combined with the mirror system to increase the accuracy of the reading, and any one of a number of devices may be used to determine θ; the distance is usually read directly from a suitable range scale so that no calculation is necessary.

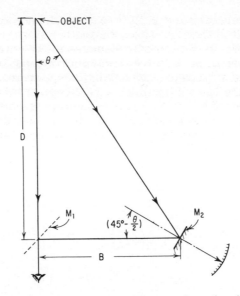

FIG. 9.13. Basic rangefinder optical system. The eye views the object directly through semi-reflector M_1 and also through movable mirror M_2. The angular setting of M_2 which brings both views into coincidence determines the range.

The accuracy of the value of D depends on how accurately θ can be measured. For large ratios of D/B, we can write

$$D = \frac{B}{\theta} \qquad (9.17)$$

and differentiating with respect to θ, we get

$$dD = -B\theta^{-2}\,d\theta \qquad (9.18a)$$

Substituting $\theta = B/D$ into Eq. 9.18a we find that the error in D due to a setting error of $d\theta$ is

$$dD = \frac{-D^2}{B}\,d\theta \qquad (9.18b)$$

Now $d\theta$ is primarily limited by how well the eye can determine when the two images are in coincidence. This is essentially the vernier acuity of the eye and is about 10 seconds of arc (0.00005 radians). If the magnification of the rangefinder optical system is M, then $d\theta$ is .00005/M radians and

$$dD = \pm\frac{5 \times 10^{-5}D^2}{MB} \qquad (9.18c)$$

Thus, the greater the base B and the greater the magnification M, the more accurate the value of the range D.

A few of the devices encountered in rangefinders are illustrated in Fig. 9.14. In Fig. 9.14a the end mirrors are replaced by penta-prisms (or "penta"-reflectors), which are constant deviation devices, bending the line of sight 90° regardless of their orientation. The reason for their use is to remove a source of error, since no change in the relative angular position of the two images is produced by misalignment of the penta-prisms as would be the case with simple 45° mirrors. A double telescope is built into the system to provide magnification; the power of each branch of the telescope must be carefully matched to avoid errors. The coincidence prism is provided to split the field of view into two halves, with a sharply focused dividing line between. In the system as shown, the final image is inverted; an erecting system, either prism or lens, is frequently included. Actual coincidence prisms are usually much more complex than that shown here.

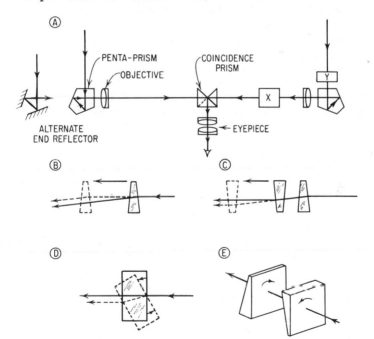

FIG. 9.14. Typical Rangefinder Optical Devices. a. A telescopic rangefinder with coincidence prism and penta-prism end reflectors. b. Sliding prism used at "x" to establish coincidence. c. Pair of sliding prisms used at "x". d. Rotating parallel plate used at "x". e. Counter rotating prisms used at "y" to establish coincidence.

A great variety of devices may be utilized to bring the two images into coincidence. Those shown in Fig. 9.14b, c, and d are located between the objective and eyelens, usually in the region marked X in Fig. 9.14a. The sliding prism of Fig. 9.14b produces a deviation at the image plane which increases with its distance from the image; it is usually an achromatic prism. Figure 9.14c shows two identical prisms with variable spacing, which displace but do not deviate the rays. The rotating block in Fig. 9.14b operates on the same principle. All of the above tend to introduce astigmatism (that is, a difference of focal position in vertically and horizontally aligned images) since they are tilted surfaces in a convergent beam. The counter-rotating wedges of Fig. 9.14e can be located in parallel light (region Y in Fig. 9.14a) and thus avoid this difficulty. Note that as one wedge turns clockwise, the other must rotate counter-clockwise through exactly the same angle; in this way the vertical deviation is maintained at zero while the horizontal deviation can be varied plus or minus twice the deviation of an individual wedge.

9.7 Radiometers and Detector Optics

A radiometer is a device for measuring the radiation from a source. In its simplest form, it consists of an objective lens (or mirror) which collects the radiation from the source and images it directly on the sensitive surface of a detector capable of converting the incident radiation into an electrical signal. A "chopper", which may be as simple as a miniature fan blade, is usually interposed in front of the detector to provide an alternating signal for the benefit of the electronic circuitry which must amplify and process the detector output.

The radiometer is widely used for the purpose its name would seem to imply, to measure radiation. However, it is also the basic instrument in many other applications. The receiver in a communications system by which one talks over a beam of light is a radiometer whose output is converted into audible form. The seeker head of an infrared homing air to air missile (e.g. the Sidewinder) is basically a radiometer whose output is arranged to indicate whether the hot exhaust pipe of an enemy jet is on or off the line of sight.

A simple radiometer is sketched in Fig. 9.15. The detector, with a diameter D, is located at the focus of an objective with a focal distance F and a diameter A. The half-field of view of the system is α, and since the detector is at the focus of the system, it is apparent that the half-field of view is given by

$$\alpha = \frac{D}{2F} \tag{9.19}$$

Now in the various applications of radiometers, the following are frequently desired

1. In order to collect a large quantity of power from the source, the diameter, A, of the system should be as large as possible,

2. In order to increase the signal to noise ratio, the size D of the detector should be as small as possible, and

3. In order to cover a practical field of view, the field angle α should be of reasonable size (and often, should be as large as possible).

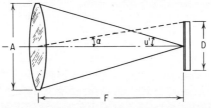

FIG. 9.15. A simple radiometer with an objective lens which forms an image of the radiation source directly on the detector cell.

The relationship between A and F, is, as we have previously noted, a limited one. If the optical system is to be aplanatic* (that is, free of spherical aberration and coma) the second principal surface (or "plane") must be spherical; for this reason, the effective diameter A cannot exceed twice the focal distance F, and the slope of the marginal ray at the image cannot exceed 90° (or the ray would be incident on the wrong side of the detector). This limits the numerical aperture of the system to N. A. $= N' \sin 90° = N'$; for systems in air with distant sources the limiting relative aperture becomes $f/0.5$. There are other limits imposed on the speed of the objective lens; the design of the system may be incapable of whatever resolution is required at large aperture ratios, or physical limitations (or predetermined relationships) may limit the acceptable speed of the objective.

We can introduce the effective $f/\#$ of the objective by multiplying both sides of Eq. 9.19 by A; setting $(f/\#) = F/A$ and rearranging, to get, for systems in air,

$$(f/\#) = \frac{D}{2A\alpha} \tag{9.20}$$

or for systems with the final image in a medium of index N'

$$\text{N.A.} = N' \sin u' = \frac{A\alpha}{D} \tag{9.21}$$

---

*The frequent assumption of aplanatic systems in the analysis of radiometric systems is based 1) on the usual need for good image quality and 2) on the fact that the image illumination (irradiance) produced by an aplanatic system cannot be exceeded, so that the assumption provides a limiting case.

Equation 9.21 can be demonstrated by setting the optical invariant (Eq. 2.39) at the objective $(I = A\alpha/2)$ equal to the invariant at the image $(I = \frac{1}{2}DN'u')$ and substituting $\sin u'$ for u' (in accordance with our requirement for aplanatism).

Since the $(f/\#)$ cannot be less than 0.5 and $\sin u'$ cannot exceed 1.0, it is apparent that the objective aperture A, half-field angle α, and detector size D, are related by

$$\left| \frac{A\alpha}{N'D} \right| \leq 1.0 \qquad (9.22)$$

It should be noted that Eq. 9.22, since it can be derived by way of the optical invariant with no assumptions as to the system between object and detector, is valid for all types of optical systems, including reflecting and refracting objectives with or without field lenses, immersion lenses, light pipes, etc. *It is thus quite futile to attempt a design with the left member of Eq. 9.22 larger than unity; in fact, it is sometimes difficult to exceed (efficiently) a value of 0.5 when good imagery is required.*

As an example of the application of Eq. 9.22, let us determine the largest field of view possible for a radiometer with a 5″ aperture and a 1 mm detector. If the detector is in air $(N' = 1.0)$ we then have, from Eq. 9.22,

$$\frac{5\alpha}{.04} \leq 1.0 \text{ or } \alpha \leq .008 \text{ radians}$$

and the maximum total field (.016 radians) is a little less than one degree (.01745 radians). An immersion lens at the detector (described below) with an index N' would increase the maximum field angle to $.016\,N'$.

An immersion lens is a means of increasing the numerical aperture of an optical system by a factor of the index, N', of the immersion lens, usually without modifying the characteristics of the system. Another way of considering the immersion lens is to think of it as a magnifier which enlarges the apparent size of the detector. The most frequently utilized form of immersion lens is a hemispherical element in optical contact with the detector. In Fig. 9.16, a concentric immersion lens of index N' has reduced the size of the image to h'/N'. Since the first surface of the immersion lens is concentric with the axial image point, rays directed toward this point are normal to this surface and are not refracted. For this reason, neither spherical aberration nor axial coma is introduced. The optical invariant at the image is $h'N'u'$, and since u' is not changed by the immersion lens, it is apparent that as N' increases, h' must decrease.

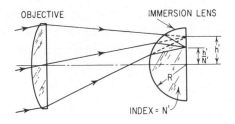

OBJECTIVE   IMMERSION LENS

INDEX = N'

FIG. 9.16. A hemispherical immersion lens concentric with the focus of an optical system reduces the linear size of the image by a factor of its index.

In the use of immersion lenses, one must beware of reflection (especially total internal reflection) at the plane surface. Ideally, the detector layer should be deposited directly on the immersion lens. Since immersion lenses are usually resorted to in cases where the angles of incidence are large, total internal reflection can occur if the immersion lens index is high and a low index layer (air, or cement for example) separates it from the detector.

In the application of radiometer type systems, it is not unusual that one wishes to use an objective of relatively low speed with a small detector and still cover a large field of view. This is readily accomplished by means of a field lens. The field lens is located at (or more frequently, near) the image plane of the objective system and redirects the rays at the edge of the field toward the detector, as indicated in Fig. 9.17. As can be seen from a brief consideration of the figure, the field lens actually images the clear aperture of the objective on the surface of the detector. The optimum arrangement is when the image of the objective aperture is the same size as the detector and

$$\frac{s_1}{s_2} = (-) \frac{A}{D}$$

This arrangement not only makes a larger field angle possible, but has the advantage of providing an even illumination over a large portion of the detector surface. Most detectors vary in sensitivity from point to point over their surface; with a field lens of focal length given by

$$f = \frac{s_1 s_2}{s_1 - s_2}$$

the same area of the detector is illuminated regardless of where the source is imaged in the field of view. Field lenses and immersion lenses are frequently combined. Note that the insertion of a field lens in a radiometer does not change the limitations of Eqs. 9.21 and 9.22; it simply permits the use of an objective system with a low numerical aperture by raising the N.A. at the detector.

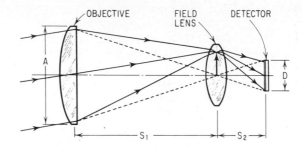

FIG. 9.17. Radiometer with field lens to increase the field of view with a small detector.

Another device to enlarge the field of view of a radiometer with a small detector is the light pipe, or cone channel condenser. In Fig. 9.18, a principal ray from the objective is shown being reflected from the walls of a tapered light pipe. Note that without the light pipe, the ray would completely miss the detector.

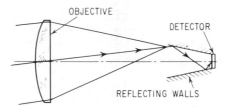
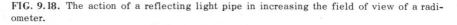

FIG. 9.18. The action of a reflecting light pipe in increasing the field of view of a radiometer.

It is instructive to consider the "unfolded" path of a ray through such a system, as indicated in Fig. 9.19. The actual reflective walls of the light pipe are shown as solid lines; the dashed lines are the images of the walls formed by reflection from each other. This layout is analogous to the unfolding technique explained in Chapter 4, and allows us to draw the path of a ray through the system as a straight line. Note that ray A in the figure undergoes three reflections before it reaches the detector end of the pipe. Ray B, entering at a greater angle, never does reach the detector, but is turned around and comes back out the large end of the pipe. This is a limit on the effectiveness of the pipe and is analogous to the $f/\#$ or N.A. limit on ordinary optical systems discussed above in the derivation of Eqs. 9.20 et seq.

A light pipe may be constructed as a hollow cone or pyramid with reflective walls in the manner indicated in Figs. 9.18 and

9.19. It is also common to construct them out of a solid piece of transparent optical material. The walls may then be reflective coated or one may rely on total internal reflection if the angles are properly chosen. Note that with a solid light pipe, total internal reflection may occur at the exit face; this can be avoided by "immersing" the detector at the exit end of the pipe. The use of a solid pipe effectively increases its acceptance angle by a factor of the index, N, of the pipe material; the effect on the system is exactly analogous to the use of an immersion lens, and the total radiometer system is still governed by Eq. 9.22 as before. Light pipes may be used with field lenses; the most common arrangement is to put a convex spherical surface on the entrance face of a solid pipe.

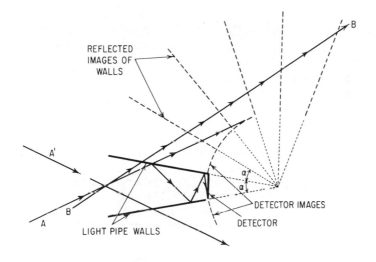

FIG. 9.19. Ray tracing through a light pipe by means of an "unfolded" diagram.

If one were to look into the large end of a pyramidal light pipe, one would see a sort of checkerboard multiple image of the exit face (or detector), as indicated in Fig. 9.19 for a two dimensional case. The checkerboard is wrapped around a sphere centered on the apex of the pyramidal pipe. This image is, of course, the effective size of the ("magnified") detector, and the cone of light from the objective, as indicated by rays A and A' is spread out over this array. This effect is occasionally useful in decorrelating the point-for-point relationship between the detector surface and the objective aperture which is established when a field lens is used. The effect is even more pronounced in a conical pipe.

The discussion in this section has been devoted to condensing radiation onto a small detector. The tables can be turned. If we replace the detector with a small source of radiation, devices such as field lenses and light pipes can be used to increase the apparent size of the source and to reduce the angle through which it radiates (or vice versa).

9.8 Fiber Optics

A long, polished cylinder of glass can transmit light from one end to the other without leakage, provided that the light strikes the walls of the cylinder with an angle of incidence greater than the critical angle for total internal reflection. The path of a meridional ray through such a cylinder is shown in Fig. 9.20. The geometric optics of meridional rays through such a device are relatively simple.

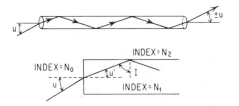

FIG. 9.20. Light is transmitted through a long polished cylinder by means of total internal reflection.

For a cylinder of length L, the path traveled by the meridional ray has a length given by

$$\text{Path length} = \frac{L}{\cos U'} \tag{9.23}$$

and the number of reflections undergone by the ray is

$$\text{No. Refl.} = \frac{\text{Path Length}}{\left(\dfrac{d}{\sin U'}\right)} = \frac{L}{d}\tan U' \pm 1 \tag{9.24}$$

where U' is slope of the ray inside the cylinder, d is the cylinder diameter and L its length. For the light to be transmitted without reflection loss, it is necessary that the angle I exceed the critical angle

$$\operatorname{Sin} I = \frac{N_2}{N_1}$$

where N_1 is the index of the cylinder and N_2 the index of the medium surrounding the cylinder. From this one can determine that the maximum external slope of a ray which is to be totally reflected is

$$\sin U = \frac{1}{N_0} \sqrt{N_1^2 - N_2^2} \qquad (9.25)$$

This "acceptance cone" of a cylinder is often specified as a numerical aperture; by rearranging Eq. 9.25, we get

$$\text{N.A.} = N_0 \sin U = \sqrt{N_1^2 - N_2^2} \qquad (9.26)$$

Again, with reference to Fig. 9.20, it is apparent that if the meridional ray had entered the cylinder well above or well below the axis, it would have emerged with a slope angle of $-U$. The path of a pair of skew rays is indicated (in an end-on view) in Fig. 9.21. Note that a skew ray is rotated with each reflection and that the amount of rotation depends on the distance of the ray from the meridional plane. Thus, a bundle of parallel rays incident on one end of a cylinder will emerge from the other end as a hollow cone of rays with an apex angle of $2U$. If the diameter of the cylinder is small, aperture effects will diffuse the hollow cone to a great extent. It is also worth noting that since the skew rays strike the surface of the cylinder at a greater angle of incidence than the meridional rays, the numerical aperture for skew rays is larger than that for meridional rays.

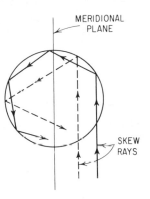

FIG. 9.21. The path of skew (non-meridional) rays through a reflecting cylinder is a sort of helix. The amount of rotation a ray undergoes in traversing a given length depends on its entrance position.

If the light transmitting cylinder is bent into a moderate curve, a certain amount of light will leak out the sides of the cylinder. However, the major portion of the light is still trapped inside the cylinder, and a curved rod (usually plastic) is occasionally a convenient device to pipe light from one location to another.

Optical fibers are extremely thin filaments of glass. Typical diameters for the fibers range from one or two microns to 50 or more microns. At these small diameters, glass is quite flexible, and a bundle of optical fibers constitutes a flexible light pipe. Figure 9.22 shows a few of the applications of fiber optics. Fig. 9.22a indicates the basic property of an oriented bundle of fibers in transmitting an image from one end of the fiber to the other. If the bundle is constrained at both ends so that each fiber occupies the same relative position at each end, then the fiber rope may literally be tied in knots without affecting its image transmitting properties. Fiber bundles with lengths of many feet are obtainable with surprisingly high transmissions. The limiting resolution (in line pairs per unit length) of a fiber bundle is approximately equal to half the reciprocal of the fiber diameter; by oscillating or scanning both ends of the fiber *simultaneously*, this resolution can be doubled. When the fibers are tightly packed, their surfaces contact each other and leakage of light from one fiber to the next will occur.

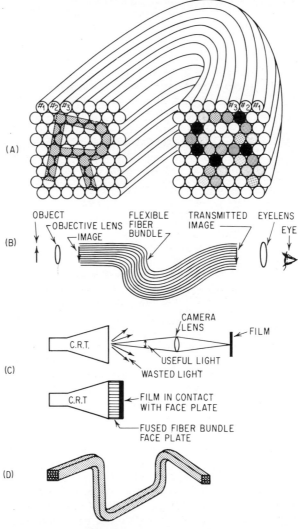

FIG. 9.22. Fiber Optics

This is prevented by coating or "cladding" each fiber with a thin layer of lower index glass. Typically, the core glass may have

N_1 = 1.72 and the cladding N_2 = 1.52, yielding a numerical aperture to the order of 0.8. Since the total internal reflection occurs at the core-cladding interface, contact between the outer surfaces does not frustrate the T.I.R. if the cladding is thick enough.

Figure 9.22b shows a flexible gastroscope. An objective lens forms an image of the object on one end of a fiber bundle; at the other end the transmitted image is viewed with the aid of an eyepiece.

Ordinary photography of a cathode ray tube face is an inefficient process. The phosphor radiates in all directions and a camera lens intercepts only a small portion of the radiated light. A tube face composed of a hermetically fused fiber array (Fig. 9.22c) can transmit all the energy radiated into a cone defined by its N.A. to a contacted photographic film with negligible loss. Fused fibers are always clad with low index glass to separate the fibers; frequently an absorbing layer is used outside the cladding to prevent contrast reduction by stray light which is emitted at angles larger than the N.A. of the fibers. Fiber optics are also available as optical conduit, that is, rigid fused bundles, for efficient transmission of light through labyrinthian paths, as shown in Fig. 9.22d.

9.9 Anamorphic Systems

An anamorphic optical system is one which has a different power or magnification in one principal meridian than in the other. Such devices usually make use of either cylinder lenses or prisms. A typical anamorphic system consists of an ordinary spherical objective lens combined with a Galilean telescope composed of cylinder lenses, as indicated in Fig. 9.23. In the upper sketch (a), it is apparent that the cylindrical afocal combination serves to shorten the focal length of the prime lens and thus widen its field of view (for a given image size). In the other meridian (Fig. 9.23b), the cylinder lenses are equivalent to plane parallel plates of glass and do not affect the focal length or coverage of the prime lens. Thus, the system has a focal length equal to that of the prime lens, F_p, in one direction and a focal length equal to the magnification of the attachment times the prime lens focal length, MF_p, in the other. In Fig. 9.23 the system is shown as a reversed Galilean telescope with a magnification of less than unity, and MF_p is less than F_p. This is the type of system used in many wide screen motion picture processes. The wide angular field is used to compress a large horizontal field of view into a normal film format. The distorted picture which results is expanded to normal proportions by projecting the film through a projection lens equipped with a similar attachment.

Cylinder lenses are also used to produce line images where a narrow slit of light is required. The image of a small light source formed by a cylinder lens is a line of light parallel to the axes of the cylindrical surfaces of the lens. The width of the line is equal to the image height given by the first order optical equations; the length of the line is limited by the length of the lens, or as shown in Fig. 9.23c, it may be controlled by another cylindrical lens oriented at 90° to the first.

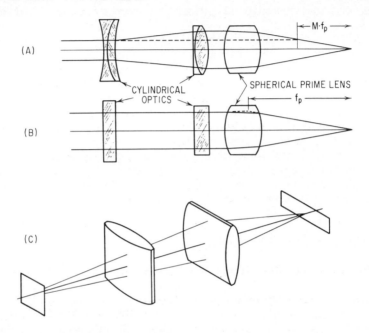

FIG. 9.23. Cylindrical Anamorphic Systems.

A prism may also be used to produce an anamorphic effect. In Section 9.1 (Eqs. 9.5 and 9.6), we saw that the magnification of an afocal optical system was given by the ratio of the diameters of its entrance and exit pupils. A refracting prism, used at other than minimum deviation, has different sized exit and entrance beams, and thus produces a magnification in the meridian in which it produces a deviation. Thus a single prism may be used as an anamorphic system. To eliminate the angular deviation, two prisms, arranged so that their deviations cancel and their magnifications combine, are usually used. Figure 9.24 illustrates the action of a single prism and also shows a compound anamorphic attachment made up of two prisms. Since the "magnification" of a prism is a function of the angle at which the beam enters the prism, a variable power anamorphic can be made by rotating the prisms in such a way that their deviations always cancel. Prism anamorphic systems are "in focus" and free of axial astigmatism only when used in parallel light. Unlike cylindrical systems, they cannot be focused by changing the space between elements. For this reason, prism anamorphics are frequently preceded by a focusable pair of spherical elements which collimate the light from the object.

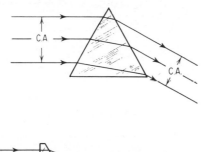

FIG. 9.24. The anamorphic action of refracting prisms.

9.10 Variable Power (Zoom) Systems

The simplest variable power system is a lens working at unit power. If the lens is shifted toward the object, the image will become larger and will move further from the object. If the lens is moved away from the object, the image will become smaller and will again move away from the object. Thus one may find any number of conjugate pairs for which the object to image distance is the same but which have magnifications which are reciprocals of each other.

Figure 9.25 indicates the relationships involved in this arrangement. The algebraic expressions shown can be derived readily by manipulation of the thin lens equation (Eq. 2.4).

The applicability of this particular zoom system is limited, since the demand for variable power systems at unit magnification is quite modest. However, by combining the moving element with one or two additional elements (usually of opposite sign), the zoom system can be made to operate at any desired set of conjugates. Several such arrangements are shown in Fig. 9.26. Note that in each system the moving lens passes through a point at which it works at unit magnification. By adding either a positive or negative eyelens or by simply adjusting the power of the last lens of the system, as indicated in the lower sketch, a telescope or afocal attachment may be made.

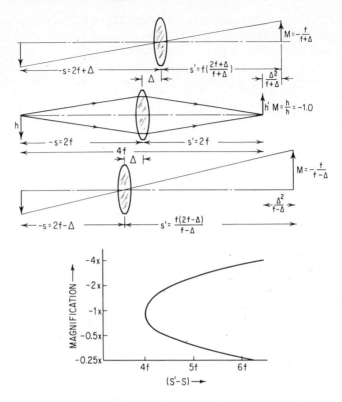

$$M = -\frac{f}{f+\Delta}$$

$$-s = 2f + \Delta$$

$$s' = f\left(\frac{2f+\Delta}{f+\Delta}\right)$$

$$\left|\frac{\Delta^2}{f+\Delta}\right|$$

$$h' \quad M = \frac{h'}{h} = -1.0$$

$$-s = 2f$$

$$s' = 2f$$

$$4f$$

$$M = -\frac{f}{f-\Delta}$$

$$\frac{\Delta^2}{f-\Delta}$$

$$-s = 2f - \Delta$$

$$s' = \frac{f(2f-\Delta)}{f-\Delta}$$

FIG. 9.25. The basic unit power zoom lens. The graph indicates the shift of the image as the lens is moved to change the magnification.

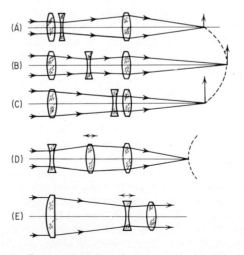

(A)

(B)

(C)

(D)

(E)

FIG. 9.26. Zoom systems based on the unit power principle.

All variable power systems with a single moving component have the same characteristic relationship between image shift and magnification (or focal length). Thus for an uncompensated "single lens" zoom system, there can be at most two magnifications at which the image is in exact focus. At all other powers, the image will be defocused. This situation can be alleviated in two ways. A "mechanically compensated" zoom system is one in which the defocusing is eliminated by introducing a compensating shift of one of the other elements of the system, as exemplified by Fig. 9.27. Since the motion of the compensating element is nonlinear, it is usually effected by a cam arrangement, hence the name "mechanically compensated."

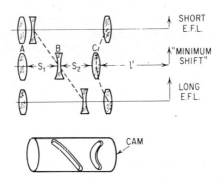

FIG. 9.27. Mechanically compensated zoom system.

Given: Φ, power (1/E.F.L.) of system at "minimum shift"
         M, ratio of power at $S_1 = 0$ to power at $S_1 = (R - 1)/R\Phi_A$
         $R = \sqrt{M}$

Choose: Φ_A, power of the first element. May be an arbitrary choice, or set
         $\Phi_A = (R - 1)/R(S_1 + S_2)$ to control the length, $(S_1 + S_2)$, at "minimum shift."

Then: $\Phi_B = -\Phi_A(R + 1)$
         $\Phi_C = (\Phi_A + \Phi)R(R + 1)/(3R - 1)$
         "minimum shift" occurs at
         $S_1 = (R - 1)/\Phi_A(R + 1) = RS_2$
         $S_2 = (R - 1)/\Phi_A R(R + 1) = S_1/R$
         $l' = (3R - 1)/\Phi R(R + 1)$
         $S_1 + S_2 + l' = \dfrac{(R - 1)}{\Phi_A R} + \dfrac{(3R - 1)}{\Phi R(R + 1)}$

Motion of lens C is computed to hold the distance from lens A to the focal point at a constant value as lens B is moved.

The formulas for a thin lens layout of this type of system are shown in Fig. 9.27 and can be derived by manipulation of the first order expressions of Chapter 2. To use the formulas, one may arbitrarily select a value for ϕ_A, the power of the first element, then determine ϕ_B, ϕ_C and the spacings for the "Minimum Shift" setting. To find the spacings for other positions of the moving lens, choose a value for one space and solve for the position of the compensating element to maintain the final focus at the same distance from the fixed element.

Zoom camera lenses are usually considered as two separate sections, a variable power afocal device (such as Fig. 9.26e) plus an image forming camera lens. An afocal device can be laid out from the formulas of Fig. 9.27 by setting Φ equal to zero.

The other technique for reducing the focus shift in a variable power system is called optical compensation. If two (or more) *alternate* lenses are linked and moved together with respect to the lenses between them, the powers and spaces can be so chosen that there are more than two magnifications at which the image is in exact focus. Two systems of this type are shown in Fig. 9.28. In the upper sketch, the first and third elements are linked and move to produce the varifocal effect. The second element, the collimator element, the prime lens and the film plane are all held in a fixed relationship with each other. The image motion produced by this type of system is a cubic curve, as shown in the upper graph. It is thus possible to arrange the powers and spaces so that the image is in exact focus for three positions in the zoom. The defocusing between these points is greatly reduced in comparison with the simpler systems described above, and if the range of powers is modest and the focal length of the system is short, a non-linear compensating motion of one of the elements is not necessary. In the second system of Fig. 9.28, the motion of the image is described by a still higher order curve, and four points of exact compensation are possible; the residual image shift is about one-twentieth of the shift of the upper system. It turns out that the maximum number of points of exact compensation is equal to the number of "active" components ahead of the first non-moving element. In Fig. 9.26 this number is two, and the image motion is parabolic with two possible points of compensation. Thus there is really no basic difference between the optics of optically and mechanically compensated zoom systems, only a matter of degree.

In zoom systems the focal lengths of the first element and of the elements following the last moving lens may be changed at will, *provided the relationship between the focal points of the elements is maintained.* Such changes modify the focal length (or power) of the over-all system and, in the case of the following elements, the amount of image shift as well. However, since a change in object position will shift the focus point of the first element with respect to the other elements, a zoom system is sensitive to object position. In order to maintain precise compensation, many zoom lenses are focused by moving the first element with respect to the rest to offset this effect.

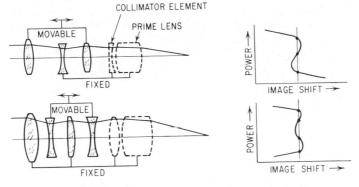

FIG. 9.28. Optically compensated zoom systems. The upper system has three "active" components and three points of compensation as indicated in the upper graph. The lower system has four "active" components and four compensation points.

References

1. Jacobs, "Fundamentals of Optical Engineering", McGraw-Hill, 1943.
2. Habell and Cox, "Engineering Optics", Pitman, 1948.
3. Strong, "Concepts of Classical Optics", (Appendix on Fiber Optics) Freeman, 1958.
4. Kingslake, "The Development of the Zoom Lens", Journal of the Society of Motion Picture and Television Engineers V69 pp. 534-544, (August, 1960).
5. Bergstein and Motz, Three papers on the optical theory of zoom lenses, Journal of the Optical Society of America, V52, pp. 363-388 (April, 1962).

Exercises

1. a) What focal lengths are required for the eyelens and objective of a 20× astronomical telescope which is 10″ long?
   b) What is the eye relief?
   c) What is the minimum objective diameter if the diffraction limit of resolution is to match the resolution of the eye?
   d) What is the maximum real field of the telescope if the eyelens is 0.5″ in diameter?

   Ans. a) $\dfrac{10''}{21}$; $\dfrac{200''}{21}$ b) 1/2″ c) 1.83″ d) ±.0296 radians

2. It is desired to add an afocal attachment in front of a 10″ f/10 camera lens to convert it to a 5″ focal length. a) What element powers are necessary for a 3 inch length reverse Galilean

telescope to accomplish this? b) What diameter must the outer element have if vignetting is not to exceed 50% for an object field of ±60? Sketch the system. Is this a reasonable diameter? Ans. a) $f_0 = -3''$; $f_e = +6''$ b) 3-1/2''

3. A microscope is required to work at a distance of 3'' from the object to the objective. If the objective and eyepiece both have 2'' focal lengths, what is the length of the microscope and what is its power?
   Ans. Length = 8''; Power = 10×

4. What is the magnification produced by a telescope made up of a 5'' focal length objective and a 5'' focal length eyepiece (and thus nominally of unit power) when it is set at minus 2 diopters (i.e. the image of an infinitely distant object is −20'' from the eyelens)?
   Ans. −1.25× (with eye at eyelens) or −0.8× (with eye at exit pupil)

5. What base length must a range finder have to measure a range of 2000 meters to an accuracy of ±0.5% if it incorporates a 20 power telescope?
   Ans. One meter

6. Determine the focal length, diameter and position (relative to the detector) for a radiometer field lens. The objective is a 5'' diameter $f/4$ paraboloid and the detector is 0.2'' square. The field to be covered is ±0.02 radians.
   Ans. $f = 0.77''$; diameter = 0.8'' minimum; $s_2 = 0.8''$

7. The entrance opening of a hollow light pipe is twice the exit opening. What is the largest angle a ray through the center of the entrance opening can make with the axis and still emerge from the small end of the pipe?
   Ans. 30° (for a long pipe) and < 90° (for a short pipe)

8. A hemi-cylindrical rod (plano-convex) with a cylindrical radius of 2.5 mm, which is 20 mm long, is located 50 mm from a 1 mm square source of light. At the "focus", what is the size of the illuminated area? (Assume the rod index is 1.5)
   Ans. 0.111 mm × 22.222 mm

9. Determine the element powers and spacings for a zoom lens of 10'' vertex length ($s_1 + s_2 = 10''$) with a zoom ratio of 4 which is to have a 10'' focal length at the "minimum shift" position. Plot the compensating motion of element C against the focal length of the lens as the element B is moved. Use Fig. 9.27.
   Ans. $M = 4$; powers: +.05, −.15, +.18; spacings: 6.67'', 3.33'';
        back focus: 8.33'';
        $M = 1/4$; powers: − 0.1, +.15, 0; spacings: 3.33'', 6.67'';
        back focus: 6.67''

10

Optical Computation

10.1

The analysis of an optical system requires a great deal of numerical computation, devoted, for the most part, to the determination of the exact paths taken by light rays as they pass through the system. As previously mentioned, a ray may be traced by the application of Snell's law at each surface. There have been a great variety of formulations devised for ray tracing. Early formulas were designed for use with logarithms, and more recently formulas which were optimized for use with mechanical desk calculators were widely used (the trigonometrical equations of Chapter 2 are of this type). Today the most widely used tool for ray tracing is the electronic computer, and the equations presented in this chapter are designed for this usage, although they can readily be used with a desk calculator. These equations do not require trigonometrical tables, nor do they require that a special computation be carried out for long radii or plane surfaces. They are further characterized by the fact that the quantities involved in them are "bounded", that is, the maximum size of each term of an equation is readily predicted in terms of the size of the optical system.

The latter sections of the chapter will present detailed directions for computing the numerical values of the aberrations discussed in Chapter 3 and also equations for determining the third order aberration contributions of surfaces and of thin lenses.

The precision required of an optical calculation is usually about six places, depending on the scale of the optical system and the application to which it is put. Trigonometric functions should be carried to six places after the decimal; this corresponds to an error of about one-fifth second of arc and is adequate for all but the most demanding applications. For moderate sized systems, linear dimensions are carried to five or six figure accuracy. Very large, diffraction-limited systems will, of course, require greater precision throughout.

The time required for an optical computation will obviously depend on the technique and equipment utilized. Tracing a meridional ray (or computing the third order aberration) through a single surface on a desk calculator is a matter of a minute or two for an experienced operator with a well thought out scheme of computation. A skew-ray trace is about an order of magnitude more time consuming. The time required on an electronic computer is a matter of seconds on the smaller machines and fractions of a second on the more powerful machines.

The problem presented by ray tracing is this: given an optical system defined by its radii, thicknesses and indices, and a ray defined by its direction and its spatial location, to find the direction and spatial location of the ray after it passes through the system.

Each set of ray tracing equations will be presented in four operational sections. First, the "opening" equations which start the ray into the system; second, the "refraction" equations which determine the ray direction after passing through a surface; third, the "transfer" equations which carry the computation to the next surface; and fourth, the "closing" equations which permit the determination of the final intercept length or height. The "refraction" and "transfer" equations are used iteratively, that is, they are repeated for each surface of the system. The "opening" and "closing" equations are only used at the start and finish of the computation.

10.2 Paraxial Rays

Although the paraxial ray tracing equations were presented in Chapter 2, they are repeated here (in slightly modified form) for completeness.

Opening: 1. given y and u at the first surface

$$\text{or } 2. \; y = lu \qquad (10.1A)$$

$$\text{or } 3. \; y = h + su \qquad (10.1B)$$

Refraction:

$$u' = \frac{cy(N'-N)}{N'} + \frac{Nu}{N'} \qquad (10.1C)$$

Transfer to the next surface:

$$y_2 = y_1 - tu'_1 \qquad (10.1D)$$

$$u_2 = u'_1 \qquad (10.1E)$$

Closing: 1. $l' = \dfrac{y}{u'}$ $\qquad (10.1F)$

$$\text{or } 2. \; h' = y - s'u' \qquad (10.1G)$$

The symbols have the following meanings:

y the height at which the ray strikes the surface; positive above the axis, negative below.

u the slope of the ray before refraction.

u' the slope of the ray after refraction; ray slopes are positive if the ray must be moved counterclockwise to reach the axis.

h the height in the object plane at which the ray originates; sign convention same as y.

h' the height at which the ray intersects the image plane.

l the distance from the first surface of the system to the axial intercept of the ray; negative if intercept point is to the left of the surface.

l' the distance from the last surface to the final axial intercept of the ray; positive if the intercept is to the right of the last surface.

s the distance from the first surface to the object plane; negative if the object plane is to the left of the surface.

s' the distance from the last surface to the image plane; positive if the image plane is to the right of the surface.

c the curvature (reciprocal radius) of the surface, equal to $1/R$; positive if the center of the curvature is to the right of the surface.

N the index of refraction preceding the surface.

N' the index of refraction following the surface.

t the vertex spacing between surfaces k and $k + 1$.

N, N' and t are positive when the ray travels from left to right, negative when the ray travels from right to left (as it does following a reflection).

The physical meanings of the symbols are indicated in Fig. 10.1.

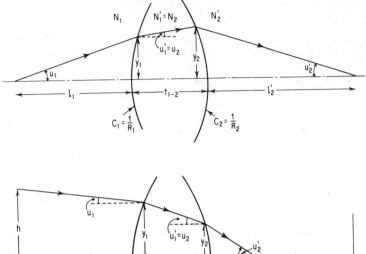

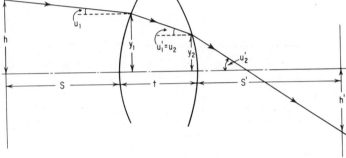

FIG. 10.1. Diagrams to illustrate the symbols used in the paraxial ray tracing equations (10.1 A through 10.1 G).

10.3 Meridional Rays

Meridional rays are those rays which are co-planar with the optical axis of the system. The plane in which both ray and axis lie is called the meridional (accent on the first i) plane and, in an axially symmetrical system, a meridional ray remains in this plane as it passes through the system. The two dimensional nature of the meridional ray makes it relatively easy to trace. For a great many optical systems, the information gained from tracing a few meridional rays (plus a Coddington trace, Section 10.6) is sufficient for most purposes.

Opening: 1. given Q and $\sin U$ at the first surface.

or   2. $Q = L \sin U$ (10.2A)

or   3. $Q = H \cos U + s \sin U$ (10.2B)

Refraction:

$$\sin I = Qc - \sin U \tag{10.2C}$$

$$\cos I = \sqrt{1 - \sin^2 I} \tag{10.2D}$$

$$\sin(U + I) = \sin U \cos I + \cos U \sin I \tag{10.2E}$$

$$\cos(U + I) = \cos U \cos I - \sin U \sin I \tag{10.2F}$$

$$\sin I' = \frac{N \sin I}{N'} \tag{10.2G}$$

$$\cos I' = \sqrt{1 - \sin^2 I'} \tag{10.2H}$$

$$\sin U' = \sin(U + I) \cos I' - \cos(U + I) \sin I' \tag{10.2I}$$

$$\cos U' = \cos(U + I) \cos I' + \sin(U + I) \sin I' \tag{10.2J}$$

$$Q' = \frac{Q(\cos U' + \cos I')}{(\cos U + \cos I)} \tag{10.2K}$$

Transfer:

$$Q_2 = Q_1' - t \sin U' \tag{10.2L}$$

$$U_2 = U_1' \tag{10.2M}$$

Closing:

$$L' = \frac{Q'}{\sin U'} \tag{10.2N}$$

or $$H' = \frac{Q' - s' \sin U'}{\cos U'} \tag{10.2P}$$

Miscellaneous:

$$y = \frac{Q[1 + \cos(U + I)]}{(\cos U + \cos I)} - \frac{Q'[1 + \cos(U + I)]}{(\cos U' + \cos I')} \tag{10.2Q}$$

$$x = \frac{Q \sin(U + I)}{(\cos U + \cos I)} \tag{10.2R}$$

$$D_{1 \text{ to } 2} = \frac{t - x_1 + x_2}{\cos U_1'} \tag{10.2S}$$

The symbols used are, for the most part, the same as those defined in Section 10.2, capitalized to differentiate them from the (lower case) paraxial symbols. Symbols new to this section are:

Q the distance from the vertex of the surface to the incident ray, perpendicular to the ray; positive if upward.

Q' the distance from the surface vertex to the refracted ray, perpendicular to the ray.

I the angle of incidence at the surface; positive if the ray must be rotated clockwise to reach the surface normal (i.e. the radius).

I' the angle of refraction.

x the longitudinal coordinate (abscissa) of the intersection of the ray with the surface; positive if the intersection is to the right of the vertex.

$D_{1\,to\,2}$ the distance along the ray between surface 1 and surface 2.

The physical meanings of the symbols are indicated in Fig. 10.2.

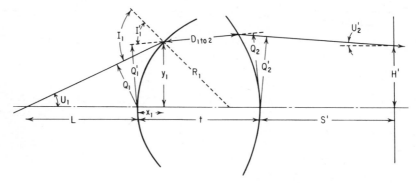

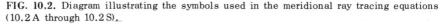

FIG. 10.2. Diagram illustrating the symbols used in the meridional ray tracing equations (10.2 A through 10.2 S).

Although this set of equations does not require the use of a table of trigonometric functions, it does involve the extraction of square roots. This can be accomplished without recourse to tables as follows: if N is the square, choose an estimated value for the root, n_0 (for Eqs. 10.2D and 10.2H, unity is a fair choice). Then a better approximation to the root is given by

$$n_1 = \frac{1}{2}\left(n_0 + \frac{N}{n_0}\right)$$
(10.3)

This process is repeated until n_k equals n_{k-1} to the number of places required.

Example A

As a numerical example, we will trace a paraxial and a meridional ray through the marginal zone of a biconvex lens with radii of 50 mm, a thickness of 15 mm and an index of 1.50. We will trace rays originating at an axial point 200 mm to the left of the first surface and determine the axial intersections for both rays after passing through the lens. We will also determine the height at which the marginal (meridional) ray intersects the paraxial focal plane. Assuming the lens to have an aperture of 40 mm, we will use a value of −0.1 for both the paraxial u and the meridional $\sin U$, so that the ray passes through the lens about 20 mm from the axis.

The following tabulation indicates both the calculation and a convenient way of arranging the ray trace data.

R		$+50.0$	-50.0	
$c = 1/R$		$+0.020$	-0.020	
t			15.0	
n	1.00		1.50	1.00

Paraxial Calculation
given: $u_1 = -0.1$
$\quad\quad l_1 = -200.0$
$\quad\quad y_1 = +20.0$ (by 10.1A)

y by 10.1D		$(+20.0)$	$+19.0$	
u by 10.1C	(-0.1)	$+0.066667$		$+0.29$
l' by 10.1F				$+65.517241$

Meridional Calculation
given: $\sin U_1 = -0.1$
$\quad\quad L_1 \quad = -200.0$
$\quad\quad Q_1 \quad = +20.0$ (by 10.2A)

Q	by 10.2L	$(+20.0)$	$+19.589064$		
$\sin I$	by 10.2C	$+.5$	$-.475278$		
$\cos I$	by 10.2D	$+.866025$	$+.879835$		
$\sin(U+I)$	by 10.2E	$+.410891$	$-.400155$		
$\cos(U+I)$	by 10.2F	$+.911684$	$+.916447$		
$\sin I'$	by 10.2G	$+.333333$	$-.712918$		
$\cos I'$	by 10.2H	$+.942809$	$+.701248$		
$\sin U'$	by 10.2I $\quad(-0.1)$	$+.083497$		$+.372744$	
$\cos U'$	by 10.2J $(.9949874)$	$+.996508$		$+.927934$	
Q'	by 10.2K	$+20.841522$		$+17.008692$	
L'	by 10.2N				$+45.631041$
$H'(s'=l')$	by 10.2P				-7.988131

10.4 Skew Rays: Spherical Surfaces

A skew ray is a perfectly general ray; however, the application of the term skew is usually restricted to rays which are not meridional rays. A skew ray must be defined in three coordinates x, y and z, instead of just x and y as in the case of meridional rays. Until the advent of the electronic computer, skew rays were rarely traced because of the lengthy computation involved. Since a skew ray takes only a bit longer to trace on an electronic computer than a meridional ray, the reverse situation is now common, and meridional rays are usually traced as special cases of general rays (to avoid having to store a separate meridional ray program in the computer memory).

The general ray tracing equations given below are those presented by D. Feder in the Journal of the Optical Society of America, V41, pp 630-636 (1951). The reader will notice that the sign convention used for the ray slope in these equations is the opposite of that used in the balance of this volume. It is left unchanged for the convenience of those who wish to refer to the original work or those who prefer this convention (which is the standard mathematical notation). Those who wish to reconcile the two notations may do so by simply reversing the sign of the offending slope of their choice wherever it occurs.

The ray is defined by the coordinates x, y and z of its intersection point with a surface, and by its direction cosines, X, Y and Z. The origin of the coordinate system is at the vertex of each surface. Figure 10.3 shows the meanings of these terms. Note that if z and Z are both zero, the ray is a meridional ray and directional cosine Y equals $(-)$ sin$U$.

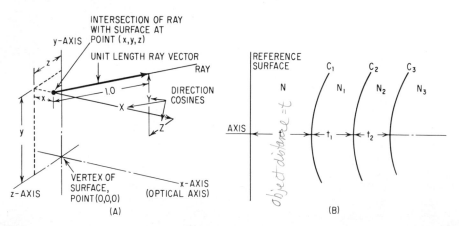

FIG. 10.3. Symbols used in skew ray tracing Equations 10.4A through 10.4P. a) The physical meanings of the spatial coordinates (x, y, z) of the ray intersection with the surface and of the ray direction cosines, X, Y and Z. b) Illustrating the system of subscript notation.

The computation is opened by determining the values for $x, y, z, X, Y,$ and Z with respect to an arbitrarily chosen reference surface, which may be plane (the usual choice) or curved. Convenient choices for the location of the reference surface are at the object (which allows the use of a curved object surface, if appropriate), at the vertex of the first surface, or at the entrance pupil.

Opening (at the reference surface):

$$c(x^2 + y^2 + z^2) - 2x = 0 \qquad (10.4A)$$

[handwritten: cos, at OBJECT SURFACE #1]

$$X^2 + Y^2 + Z^2 = 1.0 \qquad (10.4B)$$

[handwritten: direction cosines]

Transfer to the first (or next) surface:

[handwritten: Temp. storage]

$$e = tX - (xX + yY + zZ) \qquad (10.4C)$$

$$M_{1x} = x + eX - t \qquad (10.4D)$$

$$M_1^2 = x^2 + y^2 + z^2 - e^2 + t^2 - 2tx \qquad (10.4E)$$

[handwritten: angle of incidence at surface #2]

$$E_1 = \sqrt{X^2 - c_1\left(c_1 M_1^2 - 2M_{1x}\right)} \qquad (10.4F)$$

[handwritten: Ray length from surface #1 to #2]

$$L = e + \frac{\left(c_1 M_1^2 - 2M_{1x}\right)}{X + E_1} \qquad (10.4G)$$

$$x_1 = x + LX - t \qquad (10.4H)$$

$$y_1 = y + LY \qquad (10.4I)$$

$$z_1 = z + LZ \qquad (10.4J)$$

Refraction:

[handwritten: angle of refraction from surface #2]

$$E_1' = \sqrt{1 - \left(\frac{N}{N_1}\right)^2\left(1 - E_1^2\right)} \qquad (10.4K)$$

[handwritten: Temp storage?]

$$g_1 = E_1' - \frac{N}{N_1}E_1 \qquad (10.4L)$$

[handwritten: direction cosines after refraction at surface #2]

$$X_1 = \frac{N}{N_1}X - g_1 c_1 x_1 + g_1 \qquad (10.4M)$$

$$Y_1 = \frac{N}{N_1}Y - g_1 c_1 y_1 \qquad (10.4N)$$

$$Z_1 = \frac{N}{N_1} Z - g_1 c_1 z_1 \tag{10.4P}$$

Terms without subscript refer to the reference surface and the space to the right of it. Terms subscripted with 1 refer to the first surface and the following space.

The symbols have the following meanings:

x, y, z The spatial coordinates of the ray intersection with the reference surface.

x_1, y_1, z_1 The spatial coordinates of the ray intersection with surface #1.

M_1 The distance (vector) from the vertex of surface #1 to the ray, perpendicular to the ray.

M_{1x} The x component of M_1.

E_1 The cosine of the angle of incidence at surface #1.

L The distance along the ray from the reference surface (x, y, z) to surface #1 (x_1, y_1, z_1).

E_1' The cosine of the angle of refraction (I') at surface #1.

X, Y, Z The direction cosines of the ray in the space between the reference surface and surface #1 (before refraction).

X_1, Y_1, Z_1 The direction cosines after refraction by surface #1.

c The curvature (reciprocal radius = $1/R$) of the reference surface.

c_1 The curvature of surface #1.

N The index between the reference surface and surface #1.

N' The index following surface #1.

t The axial spacing between the reference surface and surface #1.

The calculation is opened by inserting c, two of the coordinates (x, y, z) and two of the direction cosines (X, Y, Z) into Eqs. 10.4A and 10.4B and solving for the third coordinate and the third direction cosine.

Then the intersection of the ray with the first surface (x_1, y_1, z_1) is determined from Eqs. 10.4C through 10.4J. Next the ray direction cosines after refraction at surface #1 (X_1, Y_1, Z_1) are found from Eqs. 10.4K through 10.4P. This completes the ray trace through the first surface; at this point Eqs. 10.4A and 10.4B (with unit subscripts) may be used to check the accuracy of the computation.

To transfer to the second surface, the subscripts of Eqs. 10.4C through 10.4J are advanced by one, and x_2, y_2, and z_2 are determined. Similarly, the direction cosines after refraction (X_2, Y_2, Z_2) at surface #2 are found by Eqs. 10.4K through 10.4P with the subscripts incremented.

This process is repeated until the intersection of the ray with the final surface of the system, which is usually the image plane, has been determined. This completes the calculation.

Note that any ray which intersects the axis is a meridional ray; thus it is only necessary to trace skew rays from off-axis object points. Further, there is no loss of generality in assuming that the object point lies in the x-y plane of the coordinate system. Therefore, any skew ray can be started with z equal to zero. When this is done, it is apparent that the two halves of the optical system, in front of, and behind the x-y plane are mirror images of each other and that any ray X_k, Y_k, Z_k passing through x_k, y_k, z_k has a mirror image X_k, Y_k, $(-Z_k)$ passing through x_k, y_k, $(-z_k)$ in the other half of the system. For this reason, it is only necessary to trace skew rays through one-half of the system aperture; rays through the other half are represented by the same data with the signs of z and Z reversed.

Example B

Using the lens of example A, we will trace a skew ray originating in the object plane (200 mm to the left of the lens) at a point 20 mm above the axis. Thus, the ray intersection coordinates in the reference plane (in this case, the object plane) are $x = 0$, $y = +20$, $z = 0$. If we set $Y = -0.1$ and $Z = +0.1$, the ray will intersect the first surface of the lens approximately in the x-z plane, about 20 mm in front of the x-axis. For the image surface we will use the paraxial focal plane as computed in example A. The calculation is shown in the table on p. 258.

10.5 Skew Rays: Aspheric Surfaces

For ray tracing purposes, an aspheric surface of rotation is conveniently represented by an equation of the form

$$x = f(y, z) = \frac{cs^2}{\left[1 + \sqrt{1 - c^2 s^2}\right]} + A_2 s^2 + A_4 s^4 + \ldots + A_j s^j \quad (10.5A)$$

where x is the longitudinal coordinate (abscissa) of a point on the surface which is a distance s from the x axis. Using the same coordinate system as Section 10.4, the distance s is related to coordinates y and z by

$$s^2 = y^2 + z^2 \qquad (10.5B)$$

The first term of the right hand side of Eq. 10.5A is the equation for a spherical surface of radius $R = 1/c$. The subsequent terms represent deformations to the spherical surface, with A_2, A_4, etc. as the constants of the second, fourth, etc. power deformation terms. Since any number of deformation terms may be included, Eq. 10.5A is quite flexible and can represent some rather extreme aspherics.

Example B – Skew Trace

	Object plane	First surface	Second surface	Image plane
R		+ 50	− 50	
c	0.0	+ 0.02	− 0.02	0.0
t	+ 200.	+ 15.	+ 65.517241	
N	1.0	1.50	1.0	
Transfer:				
e by 10.4C		+ 199.989899	+ 12.188013	+ 71.860665
M_x by 10.4D		− 2.0201011	+ 1.590643	− 3.468077
M^2 by 10.4E		+ 404.040418	+ 389.369720	+ 107.475746
E by 10.4F		+ .8588247	+ 0.8772472	+ 0.9224280
L by 10.4G		+ 206.546141	+ 6.327736	+ 75.620392
x by 10.4H	(0.0)	+ 4.470247	− 4.237125	0.000000
y by 10.4I	(+ 20.0)	− 0.654614	− 1.046031	− 7.078610
z by 10.4J	(0.0)	+ 20.654614	+ 20.116291	− 8.456088
Refraction:				
E' by 10.4K		+ 0.9398771	+ 0.6939135	
g by 10.4L		+ 0.3673272	− 0.6219573	
X by 10.4M	(+ 0.9899495)	+ 0.9944527	+ 0.9224280	
Y by 10.4N	(− 0.1)	− 0.0618575	− 0.0797745	
Z by 10.4P	(+ 0.1)	− 0.0850734	− 0.3778396	
Check:				
zero by 10.4A	(0.0)	+ .0000001	− .0000015	
1.0 by 10.4B	(1.0)	1.0000000	1.0000001	

The difficulty in tracing a ray through an aspheric surface lies in determining the point of intersection of the ray with the aspheric. In the method given here, this is accomplished by a series of approximations, which are continued until the error in the approximation is negligible.

The first step is to compute x_0, y_0 and z_0, the intersection coordinates of the ray with the spherical surface (of curvature c) which is presumed to be a fair approximation to the aspheric surface. This is done with Eqs. 10.4C through 10.4J. Note that the sign convention for the ray slope angle is the same as in Section 10.4.

Then the x coordinate of the aspheric $(\bar{x}_0)$ corresponding to this distance from the axis is found by substituting $s_0^2 = y_0^2 + z_0^2$ into the equation for the aspheric (10.5A)

$$\bar{x}_0 = f(y_0, z_0) \tag{10.5C}$$

Then compute

$$l_0 = \sqrt{1 - c^2 s_0^2} \tag{10.5D}$$

$$m_0 = -y_0 \left[c + l_0 \left(2A_2 + 4A_4 s_0^2 + \cdots + jA_j s_0^{(j-2)} \right) \right] \tag{10.5E}$$

$$n_0 = -z_0 \left[c + l_0 \left(2A_2 + 4A_4 s_0^2 + \cdots + jA_j s_0^{(j-2)} \right) \right] \tag{10.5F}$$

$$G_0 = \frac{l_0 (\bar{x}_0 - x_0)}{(Xl_0 + Ym_0 + Zn_0)} \tag{10.5G}$$

where $X, Y,$ and Z are the direction cosines of the incident ray.

Now an improved approximation to the intersection coordinates is given by

$$x_1 = G_0 X + x_0 \tag{10.5H}$$

$$y_1 = G_0 Y + y_0 \tag{10.5I}$$

$$z_1 = G_0 Z + z_0 \tag{10.5J}$$

The approximation process is now repeated (from Eq. 10.5C to 10.5J) until the error is negligible, that is until (after k times through the process)

$$x_k = \bar{x}_k \tag{10.5K}$$

within sufficient accuracy for the purposes of the computation.

The refraction at the surface is carried through with the following equations:

$$P^2 = l_k^2 + m_k^2 + n_k^2 \tag{10.5L}$$

$$F = Xl_k + Ym_k + Zn_k \tag{10.5M}$$

$$F' = \sqrt{P^2\left(1 - \frac{N^2}{N_1^2}\right) + \frac{N^2}{N_1^2}F^2} \qquad (10.5N)$$

$$g = \frac{1}{P^2}\left(F' - \frac{N}{N_1}F\right) \qquad (10.5P)$$

$$X_1 = \frac{N}{N_1}X + gl_k \qquad (10.5Q)$$

$$Y_1 = \frac{N}{N_1}Y + gm_k \qquad (10.5R)$$

$$Z_1 = \frac{N}{N_1}Z + gn_k \qquad (10.5S)$$

This completes the trace through the aspheric. The spatial intersection coordinates are x_k, y_k and z_k, and the new direction cosines are X_1, Y_1 and Z_1.

Example C

As a numerical example, let us trace the path of a ray through a paraboloidal mirror. The equation of a paraboloid with vertex at the origin is

$$x = \frac{s^2}{4f}$$

and if we choose a concave mirror with a focal length of (−) 5, the constants of Eq. 10.5A become $c = 0$, $A_2 = 1/4f = -0.05$ and A_4, A_6, etc. equal zero. Thus

$$x = -0.05s^2 = -0.05(y^2 + z^2)$$

We will place the initial reference plane at the vertex of the parabola and the final reference (image) plane at the focal point. Thus $t = 0$ and $t_1 = f = -5$ (following our usual sign convention for distance after reflections). We will trace the ray striking the reference plane at $x = 0$, $y = 0$, $z = 1.0$ at a direction of $Y = 0.1$, $Z = 0$ and (by Eq. 10.4B) $X = 0.9949874$. The index of refraction before reflection, N, equals 1.0 and the index after reflection N_1 will then be -1.0, again following the convention of reversed signs after reflection.

The computation is indicated in the following tabulation, where the applicable equation number is given in parenthesis at each step. The steps indicated by (10.5D) through (10.5C) are repeated top to bottom until $\bar{x}_k = x_k$ to (in this instance) seven places past

EXAMPLE C   261

the decimal. The fact that this example converged in only two cycles, despite the fact that $c = 0$ is a poor approximation to our paraboloid, is an indication of the rapidity of convergence of this technique.

Reference Surface: $c_0 = 0$ $t_0 = 0.0$ $N_0 = 1.0$

Aspheric: $x = -0.05\,s^2$ $c_1 = 0$ $A_2 = -0.05\,(A_4,\ \text{etc.} = 0)$ $t_1 = -5.0$

   $N_1 = -1.0$

Image Surface: $c_2 = 0$

given: $x = 0,\ y = 0,\ z = +1.0$

   $X = +0.9949874$ $Y = +0.10$ $Z = 0.0$

· Since $c = 0$ for the aspheric, it is obvious that $x_0 = x = 0$, $y_0 = y = 0$ and $z_0 = z = 1.0$. Thus, $\bar{x}_0 = -0.05\,(y^2 + z^2) = -0.05$ (by Eq. 10.5C) and $\bar{x}_0 - x_0 = -0.05$. (The same results can be obtained from Eqs. 10.4C through 10.4J)

Intersection of Ray with Aspheric:

(10.5D)	$l_0 = +1.0$	$l_1 = +1.0'$
(10.5E)	$m_0 = 0.0$	$m_1 = -0.0005025$
(10.5F)	$n_0 = +0.1$	$n_1 = +0.1$
(10.5G)	$G_0 = -0.0502519$	$G_1 = -0.0000013$
(10.5H)	$x_1 = -0.050$	$x_2 = -0.0500013$
(10.5I)	$y_1 = -0.0050252$	$y_2 = -0.0050253$
(10.5J)	$z_1 = +1.0$	$z_2 = +1.0$
(10.5C)	$\bar{x}_1 = -0.0500013$	$\bar{x}_2 = -0.0500013$
	$\bar{x}_1 - x_1 = -0.0000013$	$\bar{x}_2 - x_2 = 0.0000000$

Refraction.

(10.5L)	$P^2 = +1.0100002$
(10.5M)	$F = +0.9949372$
(10.5N)	$F' = +0.9949372$
(10.5P)	$g = +1.9701722$
(10.5Q)	$X_1 = +0.9751848$
(10.5R)	$X_1 = -0.1009900$
(10.5S)	$Z_1 = +0.1970172$
	$X_1^2 + Y_1^2 + Z_1^2 = 1.0000001$

Intersection of Ray with Image Surface:

(10.4C)	e_1	=	-5.0246880
(10.4D)	M_{2x}	=	$+0.0499993$
(10.4E)	M_2^2	=	$+0.2550229$
(10.4F)	E_2	=	$+0.9751848$
(10.4G)	L_2	=	-5.0759596
(10.4H)	x_2	=	0
(10.4I)	y_2	=	$+0.5075959$
(10.4J)	z_2	=	-0.0000513

10.6 Coddington's Equations

The tangential and sagittal curvature of field can be determined by a process which is equivalent to tracing paraxial rays along a principal ray, instead of along the axis. In Chapter 3, it was pointed out that the slope of the ray intercept curve (rim ray curve) was equal to X_t, the tangential field curvature. This slope *could* be determined by tracing two closely spaced meridional rays and computing

$$X_t = \frac{H_1' - H_2'}{\tan U_1' - \tan U_2'} = \frac{\Delta H'}{\Delta \tan U'}$$

and a similar process using close sagittal (skew) rays would yield X_s, the sagittal field curvature.

Coddington's equations are equivalent to tracing a pair of infinitely close rays, and the formulation has a marked similarity to the paraxial ray tracing equations. However, object and image distances as well as surface to surface spacings are measured along the principal ray instead of along the axis, and the surface power is modified for the obliquity of the ray.

Figure 10.4 shows a principal ray passing through a surface with sagittal and tangential ray fans originating at an object point and converging to their focii. The distance along the ray from the surface to the focus is symbolized by s and t for the object distance and by s' and t' for the image distance. The sign convention is as usual; if the focus or object point is to the left of the surface, the distance is negative; to the right, positive. In Fig. 10.4, s and t are negative, s' and t' are positive.

The computation is carried out by tracing the principal ray through the system using the meridional formulae of Section 10.3, determining the oblique power for each surface by

$$\phi = c(N' \cos I' - N \cos I) \qquad (10.6A)$$

and determining the distance (D) from surface to surface along the ray by Eq. 10.2S. The initial values of s and t are determined (Eq. 10.2S is often useful in this regard) and then the focal distances are determined by solving the following equations for s' and t'.

$$\text{(sagittal)} \quad \frac{N'}{s'} = \frac{N}{s} + \phi \tag{10.6B}$$

$$\text{(tangential)} \quad \frac{N' \cos^2 I'}{t'} = \frac{N \cos^2 I}{t} + \phi \tag{10.6C}$$

The values of s and t for the next surface are given by

$$s_2 = s'_1 - D \tag{10.6D}$$

$$t_2 = t'_1 - D \tag{10.6E}$$

where D is the value given by Eq. 10.2S.

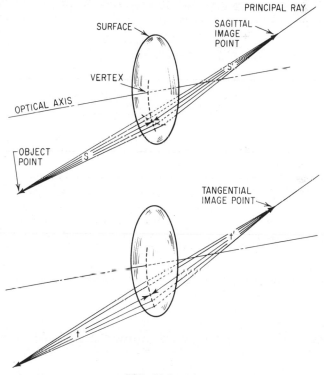

FIG. 10.4.

The calculation is repeated for each surface of the system; the final values of s' and t' represent the distances along the ray from the last surface to the final focii. The final curvature of field (with respect to a reference plane an axial distance l' from the last surface) can be found from

$$x_s = s' \cos U' + x - l' \tag{10.6F}$$

$$x_t = t' \cos U' + x - l' \tag{10.6G}$$

where x is determined for the last surface by Eq. 10.2R.

The preceding equations are ill-suited for use on an electronic computer, since s and t may be too large for the machine capacity, or too small (so that $1/s$ and $1/t$ become large). The following equations have been developed to avoid this difficulty. They make use of y_s and y_t which are fictional ray heights from the principal ray (analogous to the paraxial ray heights used in Eqs. 10.1) and equally fictional ray slope-index products P_s and P_t with respect to the principal ray.

The calculation is again begun by tracing a principal ray. The opening equations are

$$P_s = \frac{Ny_s}{s} \tag{10.6H}$$

$$P_t = \frac{Ny_t \cos^2 I}{t} \tag{10.6I}$$

where the data refers to the first surface of the system, and y_s and y_t are arbitrary numbers.

The ray "slope-index" product after refraction is determined from

$$P'_s = P_s + y_s \phi \tag{(10.6J}$$

$$P'_t = P_t + y_t \phi \tag{10.6K}$$

where ϕ is the oblique surface power given by Eq. 10.6A. The "ray height" at the next surface is given by

$$(y_s)_2 = (y_s)_1 - \frac{(P'_s)_1 D}{N'_1} \tag{10.6L}$$

$$(y_t)_2 = \frac{\cos^2 I'_1}{\cos^2 I_2} \left[(y_t)_1 - \frac{(P'_t)_1 D}{N'_1 \cos^2 I'_1} \right] \tag{10.6M}$$

At surface #2, the incident ray "slope-index" product is given by $P_2 = P_1'$.

This process is repeated for each surface of the system, and the final image distances at the last surface are found from:

$$s' = \frac{N'y_s}{P'_s} \qquad (10.6\text{N})$$

$$t' = \frac{N'y_t \cos^2 I'}{P'_t} \qquad (10.6\text{P})$$

The final curvature of field is found from Eqs. 10.6F and G.

Example D

We will use the meridional ray traced in Example A as the principal ray and trace close sagittal and tangential rays about it, assuming that the object point is at the axial intercept of the ray, that is, on the axis and 200 mm to the left of the first surface. (From a practical standpoint, this will be equivalent to determining the imagery of the lens when used with a small pinhole diaphragm located 20 mm away from the axis).

To find the initial values for s and t, we determine x at the first surface by Eq. 10.2R (using the ray trace data from Example A for the first surface). Then

$$s = t = \frac{l - x}{\cos U} = \frac{-200 - 4.415778}{.994987} = -205.445587$$

The oblique surface powers are determined from Eq. 10.6A as

$$\phi_1 = +0.02(1.5 \times 0.942809 - 1.0 \times 0.866025) = +0.0109638$$

$$\phi_2 = -0.02(1.0 \times 0.701248 - 1.5 \times 0.879835) = +0.0123701$$

Equation 10.2S gives the distance along the ray between surfaces as

$$D = \frac{15.0 - 4.415778 + (-4.177626)}{0.996508} = +6.429045$$

then for the first surface

$$\frac{1.5}{s'} = \frac{1}{-205.445} + .0109638 \qquad \text{(by 10.6B)}$$

$$s' = +246.0488$$

$$\frac{1.5(.942809)^2}{t'} = \frac{0.750}{-205.445} + .0109638$$

(by 10.6C)

$$t' = +182.3186$$

We transfer to surface 2 by Eqs. 10.6D and E to get

$$s_2 = +239.6198$$

$$t_2 = +175.8896$$

Then using Eqs. 10.6B and C for the second surface

$$\frac{1}{s'} = \frac{1.5}{239.6198} + .0123701$$

$$s_2' = +53.6768$$

$$\frac{1.491748}{t'} = \frac{1.161164}{175.8896} + .0123701$$

$$t_2' = +25.9200$$

By setting l' in Eqs. 10.6F and G equal to +45.6310 (the final intercept of the marginal ray traced in Example A), we find that, with respect to this point,

$$x_s = 49.8086 - 4.1776 - 45.6310 = 0.00$$

$$x_t = 24.0521 - 4.1776 - 45.6310 = -25.7565$$

One may gain an understanding of this rather interesting result by sketching the path of a few rays in a system of the type we have ray traced, remembering that a simple biconvex lens is afflicted with a large undercorrected spherical aberration. Alternatively, a study of the ray intercept curve for undercorrected spherical (with coordinates rotated to account for the shift of the reference plane to the focus of the marginal ray) will indicate the meaning of the value of x_t found above.

10.7 Aberration Determination

This section will briefly indicate the computational procedures involved in determining the numerical values of the various aberrations discussed in Chapter 3. Since this discussion will be somewhat condensed, the reader may wish to review Chapter 3 at this point.

We will assume that the paraxial focal distance l' (from the vertex of the last surface of the system to the paraxial image) has been determined.

Spherical Aberration

Trace a marginal meridional ray from the axial intercept of the object (through the edge of the entrance pupil of the system) and determine its final axial intercept L' and/or its intersection height, H', in the paraxial focal plane. Then the longitudinal spherical aberration is given by

$$LA' = L' - l' \qquad (10.7A)$$

and the transverse spherical aberration is given by

$$TSA' = H' = (LA') \tan U' \qquad (10.7B)$$

The spherical aberration is overcorrected if the sign of LA' is positive and undercorrected if the sign is negative.

The zonal spherical aberration is determined by tracing a second ray through the 0.707 zone (that is, a ray which strikes the entrance pupil at a distance from the axis equal to 0.707 times the distance for the marginal ray). The zonal aberration is found from Eqs. 10.7A and 10.7B. Rays may also be traced through other zones of the aperture if a more complete description of the axial correction of the system is required. The customary choice of the $0.707 = \sqrt{0.5}$ zone for zonal rays derives from the fact that, for most systems, the longitudinal spherical can be approximated by

$$LA' = aY^2 + bY^4 \qquad (10.7C)$$

where Y is the ray height and a and b are constants. Thus, if the marginal spherical, at a ray height of Y_m, is corrected to zero, the maximum zonal aberration occurs at

$$Y = \sqrt{\frac{Y_m^2}{2}} = 0.707\, Y_m$$

Coma

Three meridional rays are traced from an off-axis object point: a principal ray through the center of the entrance pupil and upper and lower rim rays through the upper and lower edges of the pupil. The final intersection heights of these rays with the paraxial focal plane are determined. Then the tangential coma is given by

$$\text{Coma}_T = \frac{H'_A + H'_B}{2} - H'_p \tag{10.7D}$$

where H'_p is the intercept for the principal ray and H'_A and H'_B are the intercepts of the rim rays.

Ordinarily, sagittal coma is very nearly equal to one third of the tangential coma (especially near the axis). Sagittal coma can be determined by tracing a skew ray through the entrance pupil at $y = 0$, $z =$ the radius of the pupil. Then the displacement of the y coordinate in the image plane from H'_p gives the sagittal coma (note that in this instance the image plane should be the plane of intersection of the upper and lower rim rays, i.e., where $H'_A = H'_B$).

The variation of coma with field angle (or image height) can be determined by repeating the process for another object height. The variation of coma with aperture is found by tracing zonal oblique rays.

O.S.C.

The offense against the sine condition is an indication of the amount of coma present in regions near the optical axis. It is determined by tracing a paraxial and a marginal ray from the axial object point and substituting their data into

$$\text{O.S.C.} = \frac{\sin U}{u} \cdot \frac{u'}{\sin U'} \cdot \frac{(l' - l'_p)}{(L' - l'_p)} - 1 \tag{10.7E}$$

where u and u' are the initial and final slopes of the paraxial ray, U and U' are the initial and final slopes for the marginal ray, l' and L' are the final intercept lengths of the paraxial and marginal rays and l'_p is the final intercept of the principal ray (thus l'_p is the distance from the last surface to the exit pupil). If the object is at infinity, the initial y and Q are substituted for u and $\sin U$ in 10.7E.

For regions near the axis

$$\text{coma}_s = H'(\text{O.S.C.})$$

$$\text{coma}_t = 3H'(\text{O.S.C.}) \tag{10.7F}$$

Distortion

Distortion is found by tracing a meridional principal ray from an off axis object point through the center of the entrance pupil and determining its intersection height H'_p in the paraxial focal plane. A paraxial ray may be traced from the same object point to determine the paraxial image height, h', or the optical invariant (I) may be used as indicated in Chapter 2.

$$\text{Distortion} = H'_p - h' \tag{10.7G}$$

Distortion is frequently expressed as a percentage of the image height, thus:

$$\text{Percent Distortion} = \frac{H'_p - h'}{h'} \times 100 \tag{10.7H}$$

The variation of distortion with image height or field angle is found by repeating the process for several object heights.

Astigmatism and Curvature of Field

Trace a principal ray from an off axis object point through the center of the entrance pupil. Then trace close skew and meridional rays by Coddington's equations (Section 10.6) and determine the final x'_s and x'_t with respect to the paraxial image plane; x'_s and x'_t are then the sagittal and tangential curvature of field for this image point.

Alternatively, a meridional ray from the object point passing through the system close to the principal ray can be traced. Then

$$X'_t = \frac{H'_p - H'}{\tan U'_p - \tan U'} \tag{10.7I}$$

will provide a close approximation to x'_t, since X'_t approaches x'_t as the two rays approach each other. A similar procedure with a close skew ray will yield X'_s.

If the variation of field curvature with image height is of interest, x'_s and x'_t may be determined for additional object heights or field angles.

Chromatic Aberration - Longitudinal

Paraxial longitudinal chromatic aberration is found by determining the paraxial image points for the longest and shortest wavelengths of light in the spectral band pass of the system. This is done by determining l' using the indices of refraction associated with one wave length and then with the other. For visual systems, the long wave length is usually taken as C-light (λ = 0.6563 microns–hydrogen line) and the short wave length as F-light (λ = 0.4861 microns–hydrogen line). The chromatic aberration is then

$$Lch A' = l'_F - l'_C \qquad (10.7J)$$

The chromatic aberrations for other zones of the aperture are found by tracing meridional rays from the axial object point for each wave length and substituting the final axial intercepts into Eq. 10.7J.

The secondary spectrum is found by tracing axial rays in at least three wave lengths, long, middle and short, and plotting their axial intercepts against wave length. A numerical value for the secondary spectrum is strictly valid only when the long and short wave length images are united at a common focus, so that

$$l'_F = l'_C$$

then

$$SS' = l'_D - l'_F = l'_D - l'_C \qquad (10.7K)$$

where the subscripts C, D and F indicate long, middle and short wave lengths. For visual work, C, F and D represent the C and F lines of hydrogen and the sodium D line at 0.5893 microns.

The spherochromatism (chromatic variation of spherical aberration) is found by determining the spherical aberration at various wave lengths. Thus, for visual work the spherochromatism would be the spherical in F light minus the spherical in C light.

Chromatic Aberration - Lateral

Lateral chromatic aberration, or chromatic difference of magnification, is determined by tracing a principal ray from an off axis object point through the center of the entrance pupil in both long and short wave lengths and finding the final intersection heights with the paraxial focal plane. Then

$$Tch A = H'_F - H'_C \qquad (10.7L)$$

for visual work. Alternatively, the paraxial lateral color can be found by tracing paraxial "principal" rays in two colors and substituting h'_F and h'_C into Eq. 10.7L.

Lateral chromatic aberration should not be confused with the transverse expression for axial (longitudinal) chromatic aberration, which is given by

$$TLch A = H'_F - H'_C = (Lch A) \tan U' \qquad (10.7M)$$

where the data are derived from meridional rays traced from an an object point *on the optical axis.*

Optical Path Difference (Wave Front Aberration)

Recalling (from Chapter 1) that a wave front which forms a "perfect" image is spherical in shape and is centered about the image point, it is apparent that the aberration of an image formed by an optical system can be expressed in terms of the departure of the wave front from an ideal sphere. The velocity of light in a medium of index N is given by c/N, where c is the speed of light in vacuum, and the time required for a point on a wave front to travel a distance D through the medium is ND/c. Thus, if a number of rays from an object point are traced through an optical system, and the distances along each ray from surface to surface are computed (by Eqs. 10.2S or 10.4G), including the distance from object point to the first surface, then the points for which $\Sigma ND/c$, or ΣND, are equal, are points through which the wave front passes at the same instant. A smooth surface through these points is the locus of the wave front.

Referring to Example D, the distance along the ray from the object point to the first surface was computed as 205.446 mm. The distance from surface 1 to surface 2 was $D = 6.429$ mm, and the distance from surface 2 to the axial intercept of the ray was $S'_2 = 53.677$.

If we now multiply each distance by the index (1.0, 1.5 and 1.0 respectively) and sum the products, we find that the Optical Path is

$$\Sigma ND = 268.766$$

The calculation can be repeated for a ray along the axis; the distances are 200 mm, 15 mm, and 45.631, and the Optical Path along the axis is

$$\Sigma ND = 268.131$$

Since the axial path is shorter by some 0.635 mm, it is apparent that when the wave front reaches this point via the axis, it is still 0.635 mm from the point along the path described by the marginal ray. If we "back up" a bit (a fraction of a nanosecond) to the time when the wave front has just emerged from the lens, and construct a reference sphere (or circle) about $L' = 45.631$, it will be apparent that the departure of the wave front from the reference sphere is equal to the difference in the Optical Paths. Thus the wave front aberration or Optical Path Difference can be found by tracing rays from the object to the surface of a reference sphere about the image point and determining

$$\text{O.P.D.} = (\Sigma ND)_A - (\Sigma ND)_B \qquad (10.7N)$$

Note that the choice of the reference image point location will have a great effect on the size of the O.P.D., since a shift of the reference point is equivalent to focusing (in the longitudinal direction) or to scanning the image plane for the point image (when shifting the reference point laterally). In the example cited, a reference sphere constructed about a point 55.57 mm from the last surface would represent a much better "fit" to the wave front, and the O.P.D. about this point would represent (approximately) the minimum obtainable for the aperture represented by this ray.

Although the example cited above showed an O.P.D. of more than 1000 wavelengths of visible light, it should be noted that O.P.D. is usually measured in wavelengths, or fractions thereof. For example, the Rayleigh criterion may be expressed as follows: An image will be "sensibly" perfect if there exists not more than one quarter wavelength difference in optical path over the wave front with reference to a sphere centered at the selected image point. The numerical precision required to obtain significant results in an O.P.D. calculation is higher than that required for ordinary ray tracing. The O.P.D. is customarily determined with respect to a spherical surface (centered about the reference point) with a radius equal to the distance from the exit pupil to the reference point.

10.8 Third Order Aberrations: Surface Contributions

If an analytic expression is derived for the transverse aberration of a general ray with respect to a reference ray (i.e. the lateral separation of their intersections in a reference plane), the expression can be broken down into orders, or powers, of the ray parameters. The parameters usually chosen are 1) the obliquity of the reference ray, and 2) the separation between the two rays at the pupil of the system; they correspond to 1) image height, and 2) system aperture. The aberrations of the first order turn out to be those which can be eliminated by locating the reference point at the paraxial image. The first order aberrations are thus defects of focus or image size which vary linearly with aperture or obliquity, such as simple focusing or paraxial chromatic aberration (transverse axial color or lateral color).

The third order terms correspond to the primary aberrations. The term in y^3 (where y is the semi-aperture, or separation of the rays) has no h (image height) component and corresponds to spherical aberration. The term in $y^2 h$ corresponds to coma. The term in yh^2 represents the astigmatism and curvature of field, and the term in h^3 is distortion. The portions of the total aberration represented by these terms are called the third order aberrations.

There will also be terms in y^5, $y^4 h$, $y^3 h^2$, $y^2 h^3$, $y h^4$, and h^5, (which are called the fifth order aberrations) as well as terms in seventh, ninth and higher exponents. (Note that in European usage, third and fifth order are frequently referred to as primary and secondary aberration). The importance of these aberration contributions diminishes rapidly as the exponent increases, just as in the series expansion for the sine of an angle

$$\sin x = x - \frac{x^3}{3!} + \frac{x^5}{5!} - \frac{x^7}{7!} + \cdots$$

The analogy here is quite good, since for optical systems in which the sines of the angles involved can be satisfactorily represented by $\sin x = x$, first order (paraxial) optics, which are based on just this approximation, are entirely adequate to describe the imagery. For systems with larger angles, more terms of the expansion are necessary to adequately describe the imaging properties, and the third (or higher) order aberration contributions must be taken into account.

Thus a knowledge of just the paraxial and third order characteristics frequently yields a fair approximation to the performance of a system which is modest in aperture and angular coverage. In systems where this approximation is poor, the third order contributions are nontheless of value. Even in systems where the fifth and higher orders are appreciable, the higher orders tend to change very slowly as the design parameters (radius, spacing, index) are varied, so that, although the first and third orders may be inadequate to fully describe the correction of the system, they are capable of indicating the changes which will be produced by moderate changes in the design parameters. For example: if a parameter change produced a change of Δx in a third order aberration, one would expect that the change in the total aberration, ΔX, (as determined by a trigonometrical ray trace) would be very nearly equal to Δx, even though the third order aberration, x, might be quite different than the trigonometrical value, X. Further, surfaces which make a large contribution to the third order aberrations also tend to make a large contribution of the same sign to the higher order aberrations, and a knowledge of the source of high order residuals is frequently useful in eliminating them.

The third order aberration contributions* can be readily calculated from the data of two paraxial rays; an axial ray (from the axial intercept of the object through the rim of the entrance pupil) and a (paraxial) principal ray (from an off axis object point through

---

*The fifth, seventh, etc., orders may also be computed from paraxial ray trace data. Buchdahl, in Reference 3, develops specific equations by which the higher order contributions may be calculated.

the center of the entrance pupil). These rays are traced by Eqs. 10.1A through 10.1G. In the following, the ray data of the axial ray will be symbolized by unsubscripted letters $(y, u, i,$ etc.) and that of the paraxial principal ray by letters with subscript $p (y_p, u_p, i_p,$ etc.).

The optical invariant I, is determined from the data of the two rays at the first surface.

$$I = y_p N u - y N u_p \tag{10.8A}$$

The final image height (i.e. the intersection point of the "principal" ray in the image plane) is determined by

$$h = \frac{I}{N'_k u'_k} \tag{10.8B}$$

where N'_k and u'_k are the index and slope (of the axial ray) after passing through the last surface of the system.

Then the following are evaluated for each surface of the system:

$$i = cy - u \tag{10.8C}$$

$$i_p = cy_p - u_p \tag{10.8D}$$

$$B = \frac{N(N' - N)}{2N'I} y(u' - i) \tag{10.8E}$$

$$B_p = \frac{N(N' - N)}{2N'I} y_p(u'_p - i_p) \tag{10.8F}$$

$$TSC = Bi^2 h \tag{10.8G}$$

$$CC = Bii_p h \tag{10.8H}$$

$$TAC = Bi_p^2 h \tag{10.8I}$$

$$TPC = \frac{(N - N')cIh}{2NN'} \tag{10.8J}$$

$$DC = h\left[B_p ii_p + \frac{1}{2}\left(u'^2_p - u^2_p\right)\right] \tag{10.8K}$$

$$TLchC = \frac{yi}{u'_k}\left(\Delta N - \frac{N}{N'}\Delta N'\right) \tag{10.8L}$$

$$TchC = \frac{yi_p}{u'_k}\left(\Delta N - \frac{N}{N'}\Delta N'\right) \tag{10.8M}$$

As previously, primed symbols refer to quantities after refraction at a surface. Most of the symbols (y, N, u, c) are defined in Section 10.2, or immediately above. Those which have not been previously defined are:

B & B_p Intermediate steps in the calculation.

i The paraxial angle of incidence.

ΔN The dispersion of the medium, equal to the difference between the index of refraction for the short wavelength and long wavelength. For visual work $\Delta N = N_F - N_C$.

The third order aberration contributions of the individual surfaces are given by Eqs. 10.8G through 10.8M where

TSC is the transverse third order spherical aberration contribution.

CC is the sagittal t. o. coma contribution.

$3CC$ is the tangential t. o. coma contribution.

TAC is the transverse t. o. astigmatism contribution.

TPC is the transverse t. o. Petzval contribution.

DC is the t. o. distortion.

$TLchC$ is the paraxial transverse axial chromatic aberration contribution.

$TchC$ is the paraxial lateral chromatic aberration contribution.

Note that $TLchC$ and $TchC$ are first order aberrations; since they are customarily computed at the same time as the third order aberrations, the equations are given at this point.

The longitudinal values of the contributions may be obtained by dividing the transverse values by u'_k, the final slope of the axial ray, thus

$$SC \quad = \quad TSC/u'_k$$

$$AC \quad = \quad TAC/u'_k$$

$$PC \quad = \quad TPC/u'_k \quad \text{slope angle} \tag{10.8N}$$

$$LchC \quad = \quad TLchC/u'_k$$

The third order aberrations at the final image are obtained by adding together the contributions of all the surfaces to get ΣTSC, ΣCC, ΣTAC, etc. These contribution sums are as follows:

ΣTSC is the third order transverse spherical aberration.

ΣSC is the t.o. longitudinal spherical aberration.

ΣCC is the t.o. sagittal coma.

$3\Sigma CC$ is the t.o. tangential coma.

ΣTAC is the t.o. transverse astigmatism.

ΣAC is the t.o. longitudinal astigmatism.

ΣTPC is the t.o. transverse Petzval sum.

ΣPC is the t.o. longitudinal Petzval sum.

ΣDC is the t.o. distortion.

$\Sigma T\,\mathrm{Lch}\,C$ is the first order transverse axial color.

$\Sigma L\mathrm{ch}\,C$ is the first order longitudinal axial color.

$\Sigma T\mathrm{ch}\,C$ is the first order lateral color.

To the extent that the first and third order aberrations approximate the complete aberration expansions, the following relationships are valid:

$$\Sigma SC \approx L' - l' \text{(spherical)}$$

$$3\Sigma CC \approx \tfrac{1}{2}(H'_A + H'_B) - H'_p \text{(tangential coma)}$$

$$x_s \approx \Sigma PC + \Sigma AC \text{ (sag. curvature of field)}$$

$$x_t \approx \Sigma PC + 3\Sigma AC \text{ (tan. curvature of field)}$$

$$\rho = \frac{h^2}{2\Sigma PC} \text{ (Petzval radius of curvature)}$$

$$\frac{100\,\Sigma DC}{h} \approx \text{ percentage distortion}$$

$$\Sigma L\mathrm{ch}\,C = l'_F - l'_C \text{ (axial color)}$$

$$\Sigma T\mathrm{ch}\,C = h'_F - h'_C \text{ (lateral color)}$$

Contributions from Aspheric Surfaces

For the purposes of computing the third order contributions, we assume that the aspheric surface is represented by the equation:

$$x = \frac{1}{2} C_e s^2 + \left(\frac{1}{8} C_e^3 + K\right) s^4 + \cdots \qquad (10.8\,\text{P})$$

in which the terms in s^6 and higher may be neglected. For aspheric surfaces given in the form of Eq. 10.5A, the equivalent curvature, C_e, and equivalent fourth order deformation constant, K, may be determined from

$$C_e = c + 2A_2 \qquad (10.8\,\text{Q})$$

$$K = A_4 - \frac{A_2}{4}\left(4A_2^2 + 6cA_2 + 3c^2\right) \qquad (10.8\,\text{R})$$

where c, A_2 and A_4 are the curvature, second and fourth order deformation terms respectively of Eq. 10.5A.

The aspheric surface contributions are determined by first computing the contributions for the equivalent spherical surface, C_e using Eqs. 10.8G through 10.8M. Then the contributions due to the equivalent fourth order deformation constant, K, are computed by the following equations and added to those of the equivalent spherical surface to obtain the total third order aberration contribution of the aspheric surface.

$$W = \frac{4K(N - N')}{I} \qquad (10.8\,\text{S})$$

$$TSC_a = Wy^4 h \qquad (10.8\,\text{T})$$

$$CC_a = Wy^3 y_p h \qquad (10.8\,\text{U})$$

$$TAC_a = Wy^2 y_p^2 h \qquad (10.8\,\text{V})$$

$$TPC_a = 0 \qquad (10.8\,\text{W})$$

$$DC_a = Wy y_p^3 h \qquad (10.8\,\text{X})$$

$$TLchC_a = 0 \qquad (10.8\,\text{Y})$$

$$TchC_a = 0 \qquad (10.8\,\text{Z})$$

Example E

We shall determine the third order surface contributions of the simple biconvex lens of Example A. We have already traced an axial paraxial ray in this example; we shall add a paraxial principal ray from an object point 20 mm below the axis and assume that the entrance pupil is at the first surface. Thus the starting data for this ray will be $y_p = 0$ and $u_p = -0.1$. We shall also assume that the lens is of crown glass with a V-value of 62.5 (and therefore $\Delta N = 0.008$.)

c		+.02		−.02		
t			15.0			
n	1.0		1.5		1.0	
y	by 10.1D	+20.0		+19.0		
u	by 10.1C −0.1		+.066667		+0.29	
i	by 10.8C	+0.5		−.446667		
				by 10.1F $l' = 65.517241$		
y_p	by 10.1D	0		+1.0		
u_p	by 10.1C −0.1		−.066667		−.09	
i_p	by 10.8D	+0.1		+.046667		
				by 10.1G $h' = 6.896552$		
	by 10.8A $I = +2.0$					
	by 10.8B $h' = 6.896552$					
B	by 10.8E	−0.722222		−2.624375		
B_p	by 10.8F	0.0		+0.025625		
TSC	by 10.8G	−1.245211		−3.610979		$\Sigma TSC = -4.856190$
SC	by 10.8N	−4.294		−12.452		$\Sigma SC = -16.745$
CC	by 10.8H	−0.249042		+0.377266		$\Sigma CC = +0.128224$
TAC	by 10.8I	−0.049808		−0.039416		$\Sigma TAC = -0.089224$
AC	by 10.8N	−0.1717		−0.1359		$\Sigma AC = -0.3077$
TPC	by 10.8J	−0.045977		−0.045977		$\Sigma TPC = -0.091954$
PC	by 10.8N	−0.1585		−0.1585		$\Sigma PC = -0.3171$
DC	by 10.8K	−0.019157		+0.008922		$\Sigma DC = -0.010235$
$T\,Lch\,C$	by 10.8L	−0.183908		−0.234115		$\Sigma T\,Lch\,C = -0.418023$
$Lch\,C$	by 10.8N	−0.6342		−0.8073		$\Sigma Lch\,C = -1.4415$
$T\,ch\,C$	by 10.8M	−0.036782		+0.024460		$\Sigma T\,ch\,C = -0.012322$

Example F

To illustrate the use of the aspheric third order contribution formulae, we shall demonstrate that the third order spherical of a paraboloidal mirror is equal to zero for an infinitely distant object. The equation for a paraboloid is simply $x = s^2/4f$, and in terms of Eq. 10.5A, $c = 0$, $A_2 = 1/4f$ and the higher order constants

$(A_4, A_6,$ etc.) are all zero. Thus, by Eqs. 10.8Q, R and S, we find that

$$C_e = \frac{1}{2f}$$

$$K = -\frac{1}{64f^3}$$

$$W = \frac{8K}{I} = -\frac{1}{8If^3}$$

remembering that for a mirror in air $N = 1.0$ and $N' = -1.0$. Then Eq. 10.8T gives the contribution of the equivalent deformation constant as

$$TSC_a = -\frac{y^4 h}{8If^3}$$

For an infinite object distance, the axial ray has a slope $u = 0$; Eq. 10.1C gives us (using C_e) $u' = y/f$ and Eq. 10.8C yields $i = y/2f$. Substituting these values into Eq. 10.8E, we get

$$B = \frac{(1.0)(-1.0 - 1.0)}{2(-1.0)I} y\left(\frac{y}{f} - \frac{y}{2f}\right)$$

$$B = \frac{y^2}{2fI}$$

Now Eq. 10.8G gives the spherical contribution of the equivalent sphere as

$$TSC = \frac{y^2}{2fI}\left(\frac{y}{2f}\right)^2 h$$

$$= \frac{y^4 h}{8If^3}$$

The contribution of the paraboloid mirror is given by the sum of TSC and TSC_a; since they are equal in magnitude and opposite in sign, the sum is zero.

Note that the demonstration did not specify that the paraboloid was concave (the more usual case); a convex paraboloid is equally free of spherical when used in this manner. And although we assumed the reflector to be in air for convenience, had we carried the indices $N' = -N$ through the calculation, the result would have been the same.

10.9 Third Order Aberrations: Thin Lenses

When the elements of an optical system are relatively thin, it is frequently convenient to assume that their thickness is zero. As we have previously noted, this assumption results in simplified approximate expressions for element focal lengths, which are none-theless quite useful for rough preliminary calculations. This approximation can be applied to third order aberration calculations; the results form a very useful tool for preliminary analytical optical system design. The following equations may be derived by application of the equations of the preceding section to a lens element of zero thickness.

The thin lens third order aberrations are found by tracing an axial and a principal ray through the system of thin lenses, in the manner outlined in Chapter 2. The equations used are

$$u' = u + y\phi \tag{10.9A}$$

$$y_2 = y_1 - du'_1 \tag{10.9B}$$

where u and u' are the ray slopes before and after refraction by the element, ϕ is the element power (reciprocal focal length), y is the height at which the ray strikes the element and d is the spacing between adjacent elements.

From Chapter 2, we also recall that the power of a thin element is given by

$$\phi = 1/f$$
$$= (N - 1)(c_1 - c_2) \tag{10.9C}$$
$$= (N - 1)c$$

where $c = c_1 - c_2$ and c_1 and c_2 are the curvatures (reciprocal radii) of the first and second surfaces of the element.

After tracing the axial and "principal" rays through the system, the following are computed for each element

$$v = \frac{u}{y} \left(\text{or } v' = \frac{u'}{y}\right) \tag{10.9D}$$

$$Q = \frac{y_p}{y} \tag{10.9E}$$

where u and y are taken from the data of the axial ray and y_p is from the principal ray data.

Then the aberration contributions may be determined from

$$SC^* \quad = SC \qquad\qquad\qquad\qquad\qquad\qquad\qquad (10.9\text{F})$$

$$CC^* \quad = CC + SC \cdot Qu'_k \qquad\qquad\qquad\qquad (10.9\text{G})$$

$$AC^* \quad = AC + CC\,\frac{2Q}{u'_k} + SC \cdot Q^2 \qquad\qquad (10.9\text{H})$$

$$PC^* \quad = PC \qquad\qquad\qquad\qquad\qquad\qquad\qquad (10.9\text{I})$$

$$DC^* \quad = (PC + 3\,AC)\,Qu'_k + CC \cdot 3 \cdot Q^2 + SC \cdot Q^3\,u'_k \qquad (10.9\text{J})$$

$$\text{Lch}\,C^* = \text{Lch}\,C \qquad\qquad\qquad\qquad\qquad\qquad (10.9\text{K})$$

$$\text{Tch}\,C \;= \text{Lch}\,C \cdot Qu'_k \qquad\qquad\qquad\qquad\qquad (10.9\text{L})$$

The starred terms are the contributions from an element which is not at the stop—that is, one for which $y_p \neq 0$. The unstarred terms are the contributions from the element when it is in contact with the stop (and $y_p = 0$) and are given by the following equations:

$$SC \;= \frac{-y^4}{u'^2_k}\left(G_1 c^3 - G_2 c^2 c_1 + G_3 c^2 v + G_4 c c_1^2 - G_5 c c_1 v + G_6 c v^2\right)$$
$$\qquad\qquad\qquad\qquad\qquad\qquad\qquad\qquad\qquad\qquad (10.9\text{M})$$

$$\;= \frac{-y^4}{u'^2_k}\left(G_1 c^3 + G_2 c^2 c_2 - G_3 c^2 v' + G_4 c c_2^2 - G_5 c c_2 v' + G_6 c v'^2\right)$$

$$CC \;- -hy^2\left(0.25\,G_5 c c_1 - G_7 c v - G_8 c^2\right)$$
$$\;= -hy^2\left(0.25\,G_5 c c_2 - G_7 c v' + G_8 c^2\right) \qquad (10.9\text{N})$$

$$AC \;= \frac{-h^2\phi}{2} \qquad\qquad\qquad\qquad\qquad\qquad (10.9\text{P})$$

$$PC \;= \frac{-h^2\phi}{2N} = \frac{AC}{N} \qquad\qquad\qquad\qquad (10.9\text{Q})$$

$$DC = 0 \qquad\qquad\qquad\qquad\qquad\qquad\qquad (10.9\text{R})$$

$$\text{Lch}\,C \;= \frac{-y^2\phi}{V u'^2_k} \qquad\qquad\qquad\qquad\qquad (10.9\text{S})$$

$$\text{Tch}\,C \;= 0 \qquad\qquad\qquad\qquad\qquad\qquad\qquad (10.9\text{T})$$

$$SSC \;= \frac{-y^2\phi P}{V u'^2_k} \qquad\qquad\qquad\qquad\qquad (10.9\text{U})$$

The symbols in the preceding have the following meanings:

u'_k is the final slope of the axial ray (at the image).

h is the image height (the intersection of the "principal" ray with the image plane).

V is the Abbe V-number of the lens material, equal to $(N_D - 1)/(N_F - N_C)$.

P is the partial dispersion of the lens material, equal to $(N_D - N_C)/(N_F - N_C)$.

G_1 through G_8 are functions of the lens material index, listed below.

SC, CC, AC, DC, PC, $LchC$, and $TchC$ have the same meanings as in Section 10.8.

SSC is the secondary spectrum contribution, equal to $l'_D - l'_C$.

$$G_1 = \frac{N^2(N - 1)}{2} \qquad G_5 = \frac{2(N + 1)(N - 1)}{N}$$

$$G_2 = \frac{(2N + 1)(N - 1)}{2} \qquad G_6 = \frac{(3N + 2)(N - 1)}{2N}$$

$$G_3 = \frac{(3N + 1)(N - 1)}{2} \qquad G_7 = \frac{(2N + 1)(N - 1)}{2N} \qquad (10.9V)$$

$$G_4 = \frac{(N + 2)(N - 1)}{2N} \qquad G_8 = \frac{N(N - 1)}{2}$$

The contributions, SC^*, CC^*, etc., are determined for each element in the system. The individual contributions are then added to get ΣSC^*, ΣCC^*, etc., and, to the extent that 1) the thin lens fiction is valid, and 2) the third order aberrations represent the total aberration of the system,

$$\Sigma SC \approx L' - l'$$

$$\Sigma CC^* \approx \text{coma}_S$$

$$\approx \frac{1}{3} \text{coma}_T$$

$$\Sigma PC^* + \Sigma AC^* \approx x_s$$

$$\Sigma PC^* + 3\Sigma AC^* \approx x_t$$

$$\frac{1}{\Sigma \dfrac{\phi}{N}} = -\rho = \text{Petzval Radius}$$

$$\frac{100\,\Sigma DC^*}{h} \approx \text{Percentage distortion}$$

$$\Sigma L\text{ch}\,C = l_F' - l_C'$$

$$\Sigma T\text{ch}\,C^* = h_F - h_C$$

$$\Sigma SSC = l_D' - l_C'$$

The thin lens third order aberration expressions (which are frequently called G-sums) can be used with the specific data of an optical system to determine the (approximate) aberration values. Another usage is in design work where the curvatures and/or spacings and powers of the elements are to be determined in such a way that the aberration values are equal to some desired set of values, as will be evident in Chapter 12.

Equations 10.9F to 10.9L are called *stop shift equations*. They may also be applied to the *surface* contributions (from Eqs. 10.8) to determine the third-order aberrations for a new, or changed, stop position by setting

$$Q = \frac{(y_p^* - y_p)}{y}$$

where y_p^* is the ray height of the "new" principal ray (i.e., after the stop is shifted) and y_p and y are as indicated in Section 10.8. Note that Q is an invariant; thus the values for y_p^*, y_p and y may be taken at *any* convenient surface. When the equations are used this way the unstarred terms (SC, CC, etc.) refer to the aberrations with the stop in the original position, while the starred terms (SC^*, CC^*, etc.) refer to the aberrations with the stop in the new position. Another consequence of the invariant nature of this definition of Q is the fact that the stop shift may be applied to either the individual surface contributions or to the contribution sums of the entire system.

Example G

We will repeat Example E, assuming that the lens is thin. Since $c_1 = +.02$ and $c_2 = -.02$, the power of the thin lens is $\phi = (1.5 - 1)\times (+.02 + .02) = +.02$. For the axial ray $u = -0.1$ and $y = 20$; Eq. 10.9A gives $u' = +0.3$, and, since there is only one element in the "system", $u' = u'_k = +0.3$. The final image distance is $20/0.3 = 66.6$ mm and the image height corresponding to an object height of -20 mm can be determined by $h' = hu/u' = +6.66$ mm, or by tracing a paraxial principal ray.

Applying Eqs. 10.9V, we find the G-functions corresponding to $N = 1.5$ to be

$$G_1 = 0.5625 \qquad\qquad G_5 = 1.666\ldots$$

$$G_2 = 1.0 \qquad\qquad G_6 = 1.08333\ldots$$

$$G_3 = 1.375 \qquad\qquad G_7 = 0.666\ldots$$

$$G_4 = 0.5833\ldots \qquad\qquad G_8 = 0.375$$

Thus we have the data (tabulated below for convenience) necessary to determine the "stop in contact" aberrations.

$$y = +20 \qquad\qquad y^2 = +400 \qquad\qquad y^4 = +160000. = 16 \times 10^4$$

$$u'_k = +0.3 \qquad\qquad u'^2_k = +0.09$$

$$c = +.04 \qquad\qquad c^2 = 16 \times 10^{-4} \qquad\qquad c^3 = 64 \times 10^{-6}$$

$$c_1 = +.02 \qquad\qquad c_1^2 = 4 \times 10^{-4}$$

$$v = -.005 \qquad\qquad v^2 = +25 \times 10^{-6}$$

$$h = +6.66\ldots \qquad\qquad h^2 = 44.44\ldots$$

$$V = 62.5$$

$$\phi = +.02$$

We will use the first surface versions of Eqs. 10.9M and N; the second surface versions (data in c_2 and v') are primarily for use in analytical work with cemented doublets where it is desirable to express the aberration of the doublet as a function of the curvature of the cemented surface.

$$SC = -\frac{16 \times 10^4}{0.09} \; [0.5625 \times 64 \times 10^{-6} - 1.0 \times 16 \times 10^{-4} \times 0.02$$

$$+ 1.375 \times 16 \times 10^{-4}(-.005) + 0.5833 \times 0.04 \times 4 \times 10^{-4}$$

$$- 1.666 \times 0.04 \times 0.02(-.005) + 1.0833 \times 0.04 \times 25 \times 10^{-6}]$$

$$SC = -1.777 \times 10^6 \, [+36 \times 10^{-6} - 32. \times 10^{-6} - 11. \times 10^{-6} + 9.33 \times 10^{-6}$$

$$+ 6.66 \times 10^{-6} + 1.0833 \times 10^{-6}]$$

$$= -1.777 \times 10^6 \, [+10.0833 \times 10^{-6}]$$

$$= -17.9259259$$

$$CC = -6.666 \times 400 \, [0.25 \times 1.666 \times 0.04 \times 0.02 - 0.666 \times 0.04 \, (-.005)$$

$$- 0.375 \times 16 \times 10^{-4}]$$

$$= -2.666 \times 10^3 \, [+3.33 \times 10^{-4} + 1.333 \times 10^{-4} - 6. \times 10^{-4}]$$

$$= -2.666 \times 10^3 \, [-1.333 \times 10^{-4}]$$

$$= +0.3555\ldots$$

$$AC = \frac{-44.44 \times 0.02}{2}$$

$$= -0.444$$

$$PC = \frac{-44.44 \times 0.02}{2 \times 1.5}$$

$$= -0.296296$$

$$DC = 0.0$$

$$LchC = \frac{-400 \times 0.02}{62.5 \times 0.09}$$

$$= -1.422$$

$$TchC = 0.0$$

The above are the third order aberrations of our thin lens with the stop (pupil) at the lens; these results may be compared with Example E (where the stop was at the first surface).

However, let us assume that the stop is 50 mm to the left of the lens. With the object height of −20 mm as before, this gives $u_p = -20/150 = -0.13333$ and $y_p = -20 - 200 \, (-0.1333) = +6.666$. Thus, Eq. 10.9E gives $Q = +0.333$ and we can determine the aberrations of the lens under these conditions from Eqs. 10.9F through 10.9L.

$$SC^* = -17.926$$

$$CC^* = +0.3555 + (-17.926)(0.333)(0.3)$$

$$= -1.4370$$

$$AC^* = -0.444 + 0.3555(2)(0.333)/0.3 + (-17.926)(0.333)(0.333)$$

$$= -1.6461$$

$$PC^* = -0.2963$$

$$DC^* = (-.2963 - 1.3333)(0.333)(0.3) + .3555(3)(0.333)(0.333)$$

$$+ (-17.926)(0.333)(0.333)(0.333)(0.3)$$

$$= -.1629629 + .1185185 - .1991767$$

$$= -0.2436$$

$$L\text{ch}\, C^* = -1.4222$$

$$T\text{ch}\, C^* = (-1.4222)(0.333)(0.3)$$

$$= -0.1422$$

Example H

As a final example for this chapter, we present a ray trace analysis of an air spaced photographic triplet lens. The constructional data shown in Fig. 10.5 are taken from K. Pestrecov's U.S. Patent No. 2,453,260 (1948). Although the data are for a focal length of 100., this lens is designed for use as an 8 or 16 mm camera objective of short focal length.

The analysis is begun by determining the size and position of the entrance pupil. The patent gives a speed of $f/2.7$; thus the pupil diameter is 37 units, and, if we assume the stop to be at R_4, the apparent position of the pupil is 25 units to the right of R_1. For an object at infinity, the paraxial rays necessary for the third order aberration calculation are represented by $u = 0$, $y = 18.5$ and $u_p = -0.25$, $y_p = -6.3$. The results are:

$EFL = 100.06$	$\Sigma CC = +.0021$	$\Sigma DC = +.057$
$BFL = 79.41$	$\Sigma TAC = +.070$	$\Sigma T\text{Lch}\, C = -.059$
$\Sigma TSC = -0.420$	$\Sigma TPC = -.272$	$\Sigma T\text{ch}\, C = +.021$

Next, meridional rays are traced for the axial bundle ($U = 0$) in C, D and F light. For the marginal ray $Q_1 = 18.5$ and for the zonal ray $Q_1 = 13.1$. The results are plotted in Fig. 10.6; plot A shows transverse measure and plot F longitudinal measure.

Principal rays are traced at several obliquities through the center of the pupil (i.e. so that $Q \sin U = l_{pr} = 25.0$) and Coddington's equations are applied to determine the field curvature, which is shown in plot G of Fig. 10.6.

To compute the data for the ray intercept curves, a fan of meridional rays was traced at each obliquity. The starting data were chosen so that one ray (principal) passed through the center of the pupil and pairs of rays passed through the rims ($Y = \pm18.5$), the 75% zones ($Y = \pm13.875$) and the 50% zones ($Y = \pm9.25$). For example, the starting values for Q for the bundle at $-14.5°$ ($\sin U_1 = -0.25$) were -6.25 for the principal ray and $+11.662548$ and -24.162548 for the rays through the pupil rim. (These three rays are shown as dashed lines in Fig. 10.5). The seven final values of H' (in the paraxial focal plane 79.4098 to the right of R_6) are plotted against $\tan U_6'$ (or $\sin U_6'$ in Fig. 10.6). Note also that the slope of the plot through the point representing the principal ray is equal to X_T ($X_T = dH'/d \tan U'$).

A sketch of the system with the rays drawn in (as in Fig. 10.5) will indicate which rays do not get through the lens; in Fig. 10.6, we indicate this by dashing the ray intercept curve. We assumed a clear aperture of 37 at R_1 and 32 at R_6.

A sagittal fan of three rays was traced with pupil intersections $x = 0$, $y = 0$ and $z = 18.5$, 13.875 and 9.25. The final values of z in the image plane are plotted in Fig. 10.6E against the final ray direction cosine, Z. The slope of this curve through the point $(0,0)$ can be obtained from $x_s = -dz/d \tan U_z$.

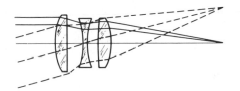

$R_1 = +40.94$			
R_2 = Plano	$t_1 = 8.74$	$N_D = 1.617$	$V = 55.0$
$R_3 = -55.65$	$t_2 = 11.05$		
$R_4 = 139.75$	$t_3 = 2.78$	$N_D = 1.649$	$V = 33.8$
$R_5 = +107.56$	$t_4 = 7.63$		
$R_6 = -43.33$	$t_5 = 9.54$	$N_D = 1.617$	$V = 55.0$

FIG. 10.5. Section drawing and constructional data for a triplet photographic objective ($f/2.7$, Focal Length 100) from U.S. Patent No. 2,453,260 (1948 - Pestrecov).

A "typical" ray trace analysis might consist of the following:

1. Paraxial trace and third order aberrations.
2. Marginal and zonal axial rays in three colors.
3. Coddington's trace at full field and 0.7 field.
4. Tangential fan of five rays (including the principal ray used in 3) at full and 0.7 field, at least one fan in three colors.

From this, one can not only obtain the aberration plots of Fig. 10.6, but a number of other relationships such as:

Variation of: Spherical with obliquity

Coma with obliquity
Coma with aperture
Distortion with obliquity
Lateral color with obliquity

The completeness with which one must analyze a system varies greatly. Systems of large aperture or field angle will require a more complete analysis. Systems of small aperture and field may not even require a "zonal" analysis. If one is familiar with the general type of system under analysis, frequently the third order aberrations plus a few carefully selected rays will yield an adequate picture of the system performance.

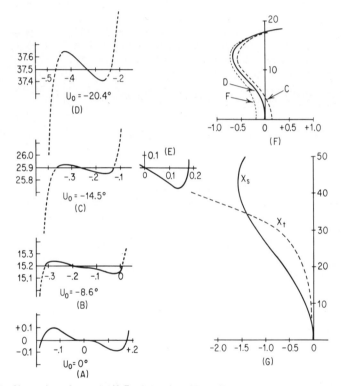

FIG. 10.6. Aberration plots of $f/2.7$ triplet (see Fig. 10.5 for constructional data). Plots A, B, C, D are meridional ray intercept curves with H' (in the paraxial focal plane) as ordinates and $\sin U_6'$ as abcissa. The dashed portions indicate rays cut off by vignetting. Plot E is one-half of a sagittal (skew) fan, with z as ordinate. Plot F shows the longitudinal spherical aberration (abscissa) as a function of the entering ray height. Plot G shows sagittal and tangential field curvature (abscissa) as a function of the final image height.

References

1. Conrady "Applied Optics and Optical Design — Part One" Oxford University Press (London) 1929 (out of print). Volumes

I and II are available in paper back form, published by Dover.

2. Hertzberger "Modern Geometrical Optics" Interscience, 1958.

3. Buchdahl "Optical Aberration Coefficients" Oxford University Press (London), 1954.

Exercises

Numerical exercises in optical computation tend to be excessively laborious, and when mistakes are made, the result is more often discouragement than enlightenment. Therefore, we suggest that the reader desirous of only a moderate amount of adventure scale the dimensional data of the numerical examples contained in this chapter by a convenient factor, say $0.5 \times$ or $2 \times$, and repeat the computations independently.

For those who wish a bit more exercise, the following problems are based on the data of Example H and Fig. 10.5. The index of refraction data are as follows:

N_D	$N_F - N_C$	N_C	N_F
1.617	.01123	1.61370	1.62493
1.649	.01920	1.64355	1.66275

1. Determine the third order aberrations. The initial ray data and answers are given in the second paragraph of Example H.

2. Trace principal rays in D, C and F light with starting data $Q = -6.25$, $\sin U = -0.25$.
   Ans. $H'_D = 25.8939$, $H'_C = 25.8867$, $H'_F = 25.9112$

3. Trace close sagittal and tangential rays from an infinitely distant object by Coddington's equations, using the D light principal ray of Exercise #2.
   Ans. $X_S = -0.9528$, $X_T = -0.4521$

4. Trace a sagittal skew ray at obliquity $\sin U = -0.25$ (direction cosine $Y = +0.25$) through the rim of the 37 diameter entrance pupil located 25 to the right of R_1.
   Ans. $X = 0.95036$ $Y = 0.25524$ $Z = -0.17791$
   $\quad\quad x = 0.0$ $\quad\quad y = 25.8808$ $\quad z = 0.0799$

5. Sketch the appearance of the ray intercept curves of Fig. 10.6 $(A, B, C, D,$ and $E)$ in a plane 79.0 from R_6 (i.e. 0.4 inside the paraxial focus). Note that this does not require additional ray tracing.

11

Image Evaluation

11.1

In the preceding chapter, we discussed the means by which ray paths are traced through an optical system and how the numerical values of the image aberrations may be determined. In this chapter, we will consider the interpretation of the results of such optical computations. The question to which we address ourselves is: "What effect does a given amount of aberration have on the performance of the optical system?"

We have seen that ray tracing yields an incomplete picture of the image forming characteristics of a system, since the image formed by a "perfect" lens or mirror is not the geometrical point that ray tracing would lead us to expect, but a finite sized diffraction pattern—the Airy disc. For small departures from perfection (that is, aberrations which cause a deformation of the wave front amounting to less than one or two wavelengths) it is thus appropriate to consider the manner in which an aberration affects the distribution of energy in the diffraction pattern. For large amounts of aberration, however, the energy distribution as described by ray tracing can yield a quite adequate representation of the performance of the system. Thus, it is convenient to divide our considerations into 1) the effects of small amounts of aberration, which we treat in terms of the wave nature of light, and 2) the effects of large amounts of aberration, which may be treated geometrically.

11.2 Optical Path Difference: Focus Shift

We will begin our discussion of small amounts of aberration by determining the optical path difference (OPD) introduced by a longitudinal shift of the reference point. Figure 11.1 shows a spherical wave front (solid line) emerging from the pupil of a "perfect" optical system with a focus at point F. We wish to determine the OPD with respect to a reference point at R, which is some arbitrary distance δ from F. If we construct a reference sphere (dashed), centered on R, which coincides with the wave front at the axis, then the

OPD for a given zone (of radius Y) is the distance* from the reference sphere to the wave front along the radius of the reference sphere, as indicated in Fig. 11.1.

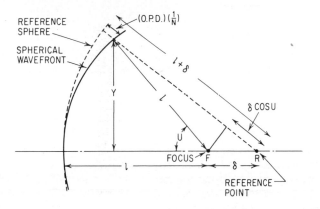

FIG. 11.1. The optical path difference (O.P.D.) introduced by a small longitudinal displacement (δ) of the reference point is equal to the index (N) times [the radius of the reference sphere ($\ell + \delta$) minus the radius of the wave front (ℓ) minus $\delta \cos U$].

From the figure we can see that, for modest amounts of OPD, the path difference is equal to the radius of the reference sphere ($l + \delta$) minus the radius of the wave front (l) all less $\delta \cos U$.

$$\frac{\text{O.P.D.}}{N} = (l + \delta - \delta \cos U - l)$$

$$= \delta(1 - \cos U)$$

To an approximation sufficient for our purposes, we can make the substitution

$$\cos U \approx 1 - \frac{1}{2} \sin^2 U$$

and the optical path difference resulting from a shift of the reference point by an amount δ is given by

$$\text{OPD} = \frac{1}{2} N \delta \sin^2 U \qquad (11.1)$$

A longitudinal shift of the reference point is equivalent to defocusing the system; by use of Rayleigh's criterion we can establish a rough allowance for the tolerable depth of focus. Setting the OPD

---

*times the index, N, of the final medium, if the final medium is not air.

equal to a quarter wavelength of light and solving for the permissible focus shift,

$$\text{Depth of focus } \delta = \pm \frac{\lambda}{2N \sin^2 U_m} \qquad (11.2)$$

where λ is the wavelength of light, N is the index of the final medium, and U_m is the final slope of the marginal ray through the system. Note that U_m is used because the maximum amount of OPD occurs at the edge of the wave front.

11.3 Optical Path Difference: Spherical Aberration

We begin by determining the OPD with respect to a reference sphere centered at the paraxial focus. In Fig. 11.2, the deformed wave front is shown as a solid line and the ray (normal to the wave front) from zone Y intersects the axis at point M. The reference sphere, centered at P, is shown dashed and the OPD is, as before, the radial distance between the two surfaces times the index. Since the wave front is shown lagging behind the reference sphere, the sign of the OPD is shown negative, to be consistent with Section 11.2.

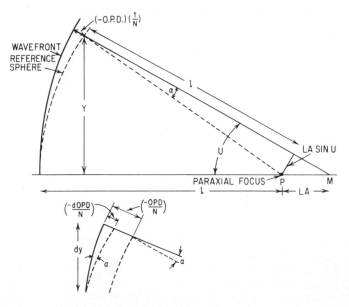

FIG. 11.2. The OPD (with reference to the paraxial focal point) produced by spherical aberration. The small diagram indicates the relationship $\alpha = (-1/N)\, d\text{OPD}/dy$. In the upper sketch, it is apparent that $\alpha = LA \sin U/l$.

The angle α between the surface normals is also the angle between the surfaces, and, as indicated in the lower sketch, the change in OPD corresponding to a small change in height, dY, is given by the relation

$$\alpha = \frac{(-d\,\text{OPD})}{N\,dY}$$

But the angular aberration, α, is also related to the spherical aberration by

$$\alpha = \frac{(LA)\,\sin U}{l} = \frac{(LA)\,Y}{l^2}$$

Combining and solving for $d\,\text{OPD}$ we get

$$d\,\text{OPD} = \frac{-YN(LA)\,dY}{l^2}$$

Now the longitudinal spherical aberration is a function of Y and can be represented by the series

$$LA = aY^2 + bY^4 + cY^6 + \cdots \qquad (11.3)$$

Making this substitution and integrating

$$
\begin{aligned}
\text{OPD} &= -\int_0^Y \frac{NY}{l^2}(aY^2 + bY^4 + cY^6 + \cdots)\,dY \\[2mm]
&= -\frac{N}{l^2}\left(\frac{aY^4}{4} + \frac{bY^6}{6} + \frac{cY^8}{8} + \cdots\right)\Big|_0^Y \\[2mm]
&= \frac{-NY^2}{2\,l^2}\left(\frac{aY^2}{2} + \frac{bY^4}{3} + \frac{cY^6}{4} + \cdots\right)
\end{aligned}
$$

$$\text{OPD} = -\frac{1}{2}N\sin^2 U\left(\frac{aY^2}{2} + \frac{bY^4}{3} + \frac{cY^6}{4} + \cdots\right) \qquad (11.4)$$

Now Eq. 11.4 is the OPD with respect to the paraxial focus of the system. It is reasonable to expect that a more desirable reference point than the paraxial focus exists. Thus, by combining Eqs. 11.1 and 11.4, we get

$$\text{OPD} = \frac{1}{2}N\sin^2 U\left[\delta - \left(\frac{aY^2}{2} + \frac{bY^4}{3} + \frac{cY^6}{4} + \cdots\right)\right] \qquad (11.5)$$

which is the OPD with respect to an axial point a distance δ from the paraxial focus.

Third Order Spherical Aberration: In a great many optical systems, the spherical aberration is almost entirely third order; this is true for almost all systems composed of simple positive elements, and very nearly true for many other systems. Under such circumstances, Eq. 11.3 reduces to

$$LA = aY^2 \tag{11.6}$$

and Eq. 11.5 reduces to

$$OPD = \frac{1}{2} N \sin^2 U \left[\delta - \frac{1}{2} aY^2 \right] \tag{11.7}$$

Now at the edge of the aperture $Y = Y_m$ and $LA = LA_m$; substituting these values into Eq. 11.6, we find that (for third order spherical)

$$a = \frac{LA_m}{Y_m^2}$$

and that

$$OPD = \frac{1}{2} N \sin^2 U \left[\delta - \frac{1}{2} LA_m \left(\frac{Y}{Y_m} \right)^2 \right] \tag{11.8}$$

To determine the value of δ which will result in the smallest amount of OPD, we can try several values of δ in Eq. 11.8 and plot the OPD for each as a function of Y. This has been done for shifts of $\delta = 0$, $\frac{1}{2} LA_m$ and LA_m; the results are plotted in Fig. 11.3.

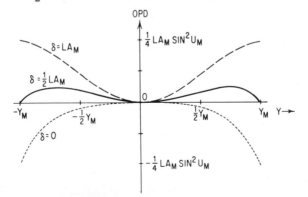

FIG. 11.3. The OPD of a system with third order spherical aberration, plotted as a function of Y for three positions of the reference point.

It is apparent that the smallest departure from the spherical reference surface occurs when the OPD is zero at the margin. The corresponding shift of the reference point is $LA_m/2$. Therefore, from the standpoint of wave front aberration, the best focus is midway between the marginal and paraxial focal points.

If we now substitute $\delta = LA_m/2$ into Eq. 11.8, we find (by differentiating with respect to Y and setting the result equal to zero) that the maximum OPD occurs at $Y = Y_m \sqrt{0.5} = 0.707 \, Y_m$ and is given by

$$OPD = \frac{LA_m}{16} N \sin^2 U_m$$

Applying Rayleigh's criterion by setting the OPD equal to one-quarter wavelength, we find the amount of marginal spherical aberration corresponding to this OPD is

$$LA_m = \frac{4\lambda}{N \sin^2 U_m} \qquad (11.9)$$

Fifth Order Spherical Aberration: When the spherical aberration consists of third and fifth order (and this includes the vast majority of all optical systems), we can write

$$LA = aY^2 + bY^4$$

Substituting $LA = LA_m$ at $Y = Y_m$ and $LA = LA_z$ at $Y = 0.707 \, Y_m$, we find that the constants a and b are related to the marginal and zonal spherical by the following expressions:

$$LA_m = aY_m^2 + bY_m^4$$

$$LA_z = \frac{aY_m^2}{2} + \frac{bY_m^4}{4}$$

$$a = \frac{4LA_z - LA_m}{Y_m^2}$$

$$b = \frac{2LA_m - 4LA_z}{Y_m^2}$$

The OPD is represented by truncating Eq. 11.5

$$OPD = \frac{1}{2} N \sin^2 \left[\delta - \frac{aY^2}{2} - \frac{bY^4}{3} \right]$$

and the graph of OPD vs. Y is a curve of the type shown in the upper plot of Fig. 11.4. The exact shape of the curve is, of course, dependent on the values of a, b and δ.

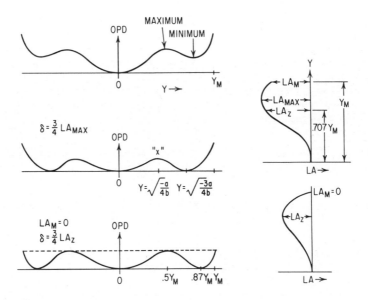

FIG. 11.4. OPD vs. Y in the presence of third and fifth order spherical aberration. Upper: OPD is a sixth order function of Y, its shape depending on the aberration coefficients, a and b, and the position of the reference point (δ). Middle: OPD vs. Y when $\delta = (3/4) LA_{max}$ Lower: OPD is minimized when $LA_m = 0$ and $\delta = (3/4) LA_z$.

The best focus occurs when

$$\delta = \frac{-3a^2}{16b} = \frac{-3(4LA_z - LA_m)^2}{32(LA_m - 2LA_z)} = \frac{3}{4} LA_{max} \qquad (11.10)$$

since at this point the OPD is zero for three values of Y as shown in the middle plot of Fig. 11.4. At the margin, the OPD is

$$OPD_m = \frac{1}{2} N \sin^2 U_m \left[\frac{-3a^2}{16b} \frac{-aY_m^2}{2} \frac{-bY_m^4}{3} \right] \qquad (11.11)$$

and at the maximum, (point X), which occurs at

$$Y = \sqrt{-\frac{a}{4b}}$$

the OPD is given by

$$\text{OPD}_X = \frac{Na^3 \sin^2 U_m}{96\,b^2\,Y_m^{\;2}} \tag{11.12}$$

If the marginal spherical aberration of the system is corrected (so that $LA_m = 0$) then the values of OPD at the margin and at point X are equal, as indicated in the lower plot of Fig. 11.4. This is the condition for minimum OPD in the presence of fifth order spherical. Then the shift of the reference point is given by

$$\delta = \frac{3}{4} LA_z$$

indicating that the best focus is three fourths of the way from the paraxial focus to the zonal focus. The residual OPD is given by

$$\text{OPD} = \frac{N LA_z \sin^2 U_m}{24} \tag{11.13}$$

and equating this to one quarter wavelength, we find that the Rayleigh criterion allows a residual zonal spherical of

$$LA_z = \frac{6\lambda}{N \sin^2 U_m} \tag{11.14}$$

11.4 Aberration Tolerances

The preceding sections form a basis for the establishment of what are usually referred to as aberration tolerances. We should note, however, that the use of the word tolerance in this connection does not carry the same go no-go connotation that it does in matters mechanical, where parts may suddenly cease to fit or function when tolerances are exceeded. *Any* amount of aberration degrades the image; a larger amount simply degrades it still more. Thus, it would be more accurate to call this section Aberration Allowances.

The Rayleigh Criterion, or Limit, allows not more than one quarter wavelength of OPD over the wave front with respect to a reference sphere about a selected image point in order that the image may be "sensibly" perfect. For convenience, we will use the term one Rayleigh Limit to mean an OPD of one-quarter wavelength. We have previously noted that the image formed by a perfect lens is a diffraction pattern which contains 84% of its energy in a central disc, the remaining 16% being distributed in the rings of the pattern. When the OPD is less than two Rayleigh Limits (one half wavelength) the size of the central disc is basically unchanged, but a noticeable shift of energy from the central disc to the rings takes place. For smaller amounts of OPD, the energy distribution is as follows:

Perfect lens (OPD = 0)	84% in disc	16% in rings
1/4 Rayleigh limit (OPD = $\lambda/16$)	83% in disc	17% in rings
1/2 Rayleigh limit (OPD = $\lambda/8$)	80% in disc	20% in rings
1 Rayleigh limit (OPD = $\lambda/4$)	68% in disc	32% in rings

Thus it is apparent that an amount of aberration corresponding to one Rayleigh Limit does cause a small but appreciable change in the characteristics of the image. For most systems, however, one may assume that, if the aberrations are reduced to the Rayleigh Limit, the performance will be first class and that it will take a determined investigator a considerable amount of effort to detect the resultant difference in performance. An occasional system does require correction to a fraction of the Rayleigh Limit, but such cases are rare. Microscopes and telescopes are usually corrected to meet the Rayleigh Criterion, on the axis at least; photographic lenses approach this level of correction only infrequently.

The following tabulation indicates the amount of aberration corresponding to one Rayleigh Limit (OPD = $\lambda/4$).

$$\text{Out of focus} = \pm \frac{\lambda}{2N \sin^2 U_m} \tag{11.15}$$

$$\text{Marginal spherical (third order)} = \pm \frac{4\lambda}{N \sin^2 U_m} \tag{11.16}$$

$$\text{Zonal spherical } (LA_m = 0) = \pm \frac{6\lambda}{N \sin^2 U_m} \tag{11.17}$$

$$\text{Sagittal coma} = \pm \frac{\lambda}{2N \sin U_m} \tag{11.18}$$

$$\text{O.S.C.} = \pm \frac{\lambda}{2NH \sin U_m} \tag{11.19}$$

$$\text{Axial chromatic aberration} = \pm \frac{\lambda}{N \sin^2 U_m} \qquad (11.20)$$

Note that, with the exception of coma and O.S.C., the aberrations are given for longitudinal measure, not transverse. The symbols are: λ, the wavelength of light; N, the index of the medium in which the image is formed; U_m, the slope angle of the marginal axial ray at the axial image; H, the image height. The allowance for longitudinal color is derived from the out of focus allowance; if the reference point is midway between the long and short wavelength focal points, it is apparent that they may be separated by twice the "out of focus" allowance before the Rayleigh Limit is exceeded. Because of the reduced sensitivity of the eye at the ends of the visual spectrum, this allowance is frequently exceeded by a factor of two or three in practice. When the chromatic aberration is in the form of secondary spectrum, the allowance may be increased somewhat, say 25 or 50%, because of the concentration of focus in the wavelengths to which the eye is most sensitive.

The allowance for coma or OSC is frequently exceeded, since it is extremely difficult to correct a system to this level of quality over an appreciable field. Conrady has suggested that for telescopes an OSC of 0.0025 be considered acceptable, and for camera lenses 0.001. These allowances, used with discretion, are adequate for most situations.

The "out of focus" allowance is, of course, applicable to curvature of field, and values of x_s and x_t should (ideally, at least) be less than twice this amount. However, it is a rare system that can be corrected to this level, and most optical systems exceed this allowance many times over.

Example A

For a visual optical system with a relative aperture of $f/5$, $\sin U_m = 0.10$ and $\lambda = 0.55$ microns = 0.00055 mm. The aberration allowances corresponding to one-quarter wave OPD are thus given by:

$$\text{Out of focus} = \pm \frac{0.00055}{2(0.1)^2} = \pm 0.0275 \text{ mm}$$

$$\text{Marginal spherical} = \pm \frac{4(0.00055)}{(0.1)^2} = 0.22 \text{ mm}$$

$$\text{Zonal spherical} = \pm \frac{6(0.00055)}{(0.1)^2} = \pm 0.33 \text{ mm} \ (LA_m = 0)$$

$$\text{Sagittal coma} = \pm \frac{0.00055}{2(0.1)} = \pm 0.00275 \text{ mm}$$

$$\text{Axial chromatic} = \pm \frac{0.00055}{(0.1)^2} = \pm 0.055 \text{ mm}$$

11.5 Image Energy Distribution (Geometrical)

When the aberrations exceed the Rayleigh Limit by several times, the results of geometrical ray tracing may be used to predict the appearance of a point image with a fair degree of accuracy. This is done by dividing the entrance pupil of the optical system into a large number of equal areas and tracing a ray from the object point through the center of each of the small areas. The intersection of each ray with the selected image plane is plotted, and since each ray represents the same fraction of the total energy in the image, the density of the points in the plot is a measure of the energy density (irradiance, illuminance) in the image. Obviously the more rays that are traced, the more accurate the representation of the geometrical image becomes. A ray intercept plot of this type is called a spot diagram. Fig. 11.5 indicates several methods of placing the rays in the entrance pupil and shows a typical example of a spot diagram.

The preparation of a spot diagram obviously entails a great amount of ray tracing. As pointed out in Section 10.4, the rays on each side of the meridional plane are mirror images of each other; this reduces the necessary ray tracing by almost 50% (almost, because the meridional rays cannot be halved). The number of rays to be traced can be reduced markedly by an interpolation process. To produce a spot diagram which faithfully reproduces the image, several hundred ray intersections are required. However, if 20 or 30 rays are traced, it is possible to fit an interpolation equation to their intercept coordinates so that the required (larger) number of points can be computed from the equation. For an optical system of any complexity, this process may represent a considerable saving in time.

An interpolation polynomial can be as simple or complex as required, and is usually a power series (in terms of the ray coordinates at either the entrance pupil of the first surface of the system) which gives the displacement of the final image plane ray intercept from the intercept of one chosen ray. If the chosen ray is a principal (meridional) ray, the displacements are $\Delta y' = y' - y'_p$ and z'. If y and z are the coordinates of the ray in the entrance pupil, then the polynomials take the following form:

$$\Delta y' = (k + kz^2 + kz^4 + kz^6 + \cdots) + y(k + kz^2 + kz^4 + kz^6 + \cdots)$$
$$+ y^2(k + kz^2 + kz^4 + kz^6 + \cdots) + y^3(k + kz^2 + kz^4 + kz^6 + \cdots)$$
$$+ y^4(k + kz^2 + kz^4 + kz^6 + \cdots) + \text{etc.} \tag{11.21}$$

$$z' = (kz + kz^3 + kz^5 + kz^7 + \cdots) + y(kz + kz^3 + kz^5 + kz^7 + \cdots)$$
$$+ y^2(kz + kz^3 + kz^5 + kz^7 + \cdots) + y^3(kz + kz^3 + kz^5 + kz^7 + \cdots)$$
$$+ \text{etc.} \tag{11.22}$$

The symbol k used above is intended to indicate a *series* of constants, *not* a single constant. Note that the expression for $\Delta y'$ contains only even powers of z; this results from the symmetry of the system about the meridional plane, since $\Delta y'$ must be the same whether z is plus or minus. The expression for z' contains only odd powers of z, since (again for reasons of symmetry) when z changes sign, z' must change sign.

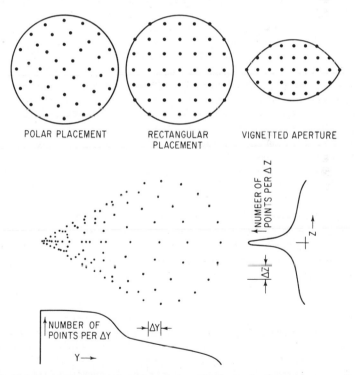

FIG. 11.5. The upper sketches show the placement of rays in the entrance pupil so that each ray "represents" an equal area. Shown below are a spot diagram (for a system with pure coma) and the line spread functions (below and to the right) obtained by counting the number of points between parallel lines separated by a small distance, ΔY or ΔZ.

In practice, the polynomials are truncated at some suitable point, typically where the highest order terms $y^m z^n$ have $(m + n)$ equal to six or seven. Then the number of rays necessary to determine the values of the constants k is equal to the number of the constants plus one (for the principal ray).

Another interpolation equation can be based on the aberration polynomial,

$$\Delta y' = A_1 s^3 \cos\theta + A_2 s^2 h (2 + \cos 2\theta) + (3A_3 + A_4) sh^2 \cos\theta + A_5 h^3$$

$$+ B_1 s^5 \cos\theta + (B_2 + B_3 \cos 2\theta) s^4 h + (B_4 + B_6 \cos^2\theta) s^3 h^2 \cos\theta$$

$$+ (B_7 + B_8 \cos 2\theta) s^2 h^3 + B_{10} sh^4 \cos\theta + B_{12} h^5 + C_1 s^7 \cos\theta + \text{etc.}$$

$$\tag{11.23}$$

$$z' = A_1 s^3 \sin\theta + A_2 s^2 h \sin 2\theta + (A_3 + A_4) sh^2 \sin\theta + B_1 s^5 \sin\theta$$

$$+ B_3 s^4 h \sin 2\theta + (B_5 + B_6 \cos^2\theta) s^3 h^2 \sin\theta + B_9 s^2 h^3 \sin 2\theta$$

$$+ B_{11} sh^4 \sin\theta + C_1 s^7 \sin\theta + \text{etc.} \tag{11.24}$$

In these equations s and θ are the polar coordinates of the ray in the entrance pupil and h is the image height. The A constants are the third-order aberration coefficients (A_1, spherical; A_2, coma; A_3, astigmatism; A_4, Petzval; A_5, distortion). The B constants are the fifth-order aberration coefficients, the C constants, seventh-order and so on.

For an accurate analysis, the effects of wavelength on the energy distribution must also be included. This is accomplished by tracing additional rays at different wavelengths; the variation of system sensitivity with wavelength may be taken into account by tracing fewer rays in the less sensitive wavelengths. For devices with appreciable fields of view, spot diagrams must also be prepared for several obliquities.

Focusing must also be taken into account. Since it is difficult to predict in advance the exact position of the plane of best focus, spot diagrams are often prepared for several positions of the image plane and the best is selected. One way of accomplishing this is to hold the final ray data (intercepts and directions) in the computer memory and to calculate a new set of intercepts (for each focus shift) from which a new interpolation polynomial is prepared.

Almost all spot diagrams are prepared with the aid of an electronic computer, and even so, the process tends to be long and costly. Indeed, for systems of moderate size it is frequently less expensive to have a precise sample of the system fabricated and tested than to completely analyze the performance by means of spot diagrams. One is well advised to weigh carefully the economics of the situation before plunging into a full dress ray trace-spot diagram-modulation transfer analysis of the optical system.

Spread Functions: The image of a point (i.e. a spot diagram) can be considered from a three dimensional point of view to be a sort of energy mountain, as sketched in Fig. 11.6. The point spread function can be described two dimensionally by a series of cross sections through the three dimensional solid. The solid corresponding to a line image is also shown in Fig. 11.6. The cross section of the line solid is called the line spread function, and can be obtained by integrating the point solid along sections parallel to the direction of the line, since the line image is simply the summation of an infinite number of point images along its length.

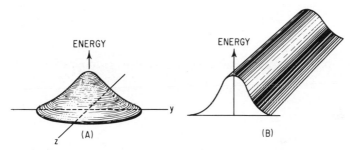

FIG. 11.6. The energy distribution in the image of a point (A) and a line (B). The line image (B) is generated by summing an infinite number of point images (A) along its length. The line spread function is the cross section of (B).

11.6 Analytical Point Spread Functions

For points on axis, it is possible to derive the point spread function analytically, when the spherical aberration and OSC are known functions of the aperture. This is done by computing the ratio dA/dA', where dA is a small unit area in the entrance pupil and dA' is the corresponding small unit area in the image plane. When the illumination is uniform at the entrance pupil, dA/dA' is the relative illumination (produced by dA) in the image plane. We shall briefly sketch the procedure for third order spherical aberration and O.S.C.

With reference to Fig. 11.7, the incremental area in the pupil is given by

$$dA = y\,dy\,d\theta$$

and similarly, in the image plane

$$dA' = y'\,dy'\,d\theta'$$

and since $d\theta = d\theta'$, the relative illumination in the image is given by

$$\frac{dA}{dA'} = \frac{y\,dy}{y'\,dy'}$$

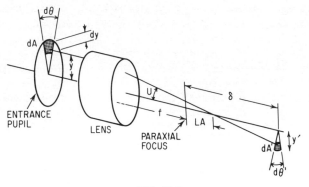

FIG. 11.7.

We assume that the image plane is to be located a distance δ from the paraxial focus; a ray through the center of the incremental area strikes the reference plane at a height given by

$$y' = (LA - \delta)\tan U$$

where LA is the spherical aberration for a ray with final slope angle U. Since we assume that the aberrations are third order, we can express LA and U as follows:

$$LA = LA_m\left(\frac{y}{y_m}\right)^2$$

$$\sin U = \frac{y}{f}\left[1 + OSC_m\left(\frac{y}{y_m}\right)^2\right]$$

By differentiating the expression for y', substituting, etc., the following result can be obtained

$$\frac{dA}{dA'} = \frac{fy^2\cos^4 U}{(LA - \delta)\sin U\left[y(LA - \delta)\left(1 + 3OSC_m\dfrac{y^2}{y_m^2}\right) + 2fLA\sin U\cos^2 U\right]}$$

$$(11.25)$$

(and as y approaches zero, dA/dA' approaches f^2/δ^2)

To obtain the spread function, the relative illumination dA/dA' is plotted against the corresponding value of y', and since this may be a multi-valued function, it is summed for each value of y' to produce a single valued plot which is the distribution of illumination in the image.

The relative illumination found in this manner goes to infinity wherever the selected image plane intersects the caustic or a focus.

These regions can be handled by approximate visual truncation of the plot or by computing two values of y' corresponding to two closely spaced values of y and determining the ratio of the finite areas involved. An image unit to the order of the size of the central Airy disc is an obviously suitable choice for the difference between the two y' values in this situation.

The principle involved here is, of course, the same as in the spot diagram, and although its application is convenient only in relatively simple cases, this sort of approach is useful when the services of a high speed computer are not available.

11.7 Geometrical Spot Size Due to Spherical Aberration

The meridional spread of an image can, of course, be read directly from a ray intercept curve (H' vs. $\tan U'$). For points on the axis, the image blur is symmetrical and it is possible to obtain simple expressions for the size of the blur spot.

Figure 11.8 shows the ray paths near the image plane of a system afflicted with third-order spherical aberration. It is apparent that the minimum diameter blur spot for this system occurs at a point between the marginal focus and the paraxial focus. This point is three quarters of the way from the paraxial focus to the marginal focus and the diameter of the spot at this point is given by:

$$B = \frac{1}{2} LA_m \tan U_m \qquad (11.26)$$

When the spherical aberration consists of both third and fifth orders, the situation is more complex. From a geometrical standpoint, the minimum spot size can be shown to occur when the marginal spherical is equal to two thirds of the (0.707) zonal spherical, or

$$LA_z = 1.5 \, LA_m$$

Then the "best" focus occurs at

$$\delta = 1.25 \, LA_m = 0.83 \, LA_z$$

and the size of the blur spot is

$$B = \frac{1}{2} LA_m \tan U_m \qquad (11.27)$$

However, if the marginal spherical is corrected to zero, then the "best" geometrical focus is at

$$\delta = 0.42 \, LA_z$$

and for small values of U, the minimum blur spot size is

$$B = 0.84\, LA_z \tan U_m \qquad (11.28)$$

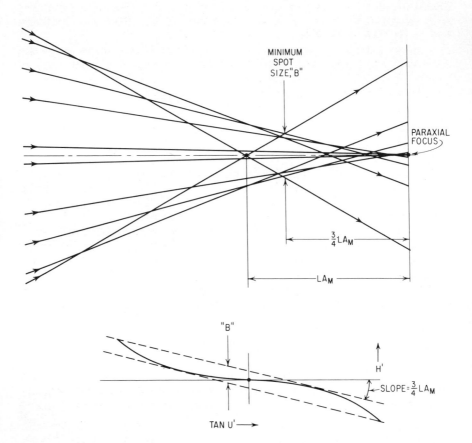

FIG. 11.8. The upper figure shows the ray paths near the focus of a system with third order spherical aberration. The smallest blur spot occurs at $0.75\,LA_m$ from the paraxial focus. The lower figure is a ray intercept curve (H' vs. $\tan U'$) for the same case; the slope of the dashed lines ($dH'/d\tan U'$) equals $0.75\,LA_m$ and their separation indicates the diameter of the blur spot.

The "best" focus positions described above are not necessarily those one would select visually, and the reader may have noticed that they differ from those selected on the basis of OPD. Figure 11.9 shows a ray intercept curve for fifth-order spherical with the marginal spherical corrected to zero. The slope of the two solid lines indicates the amount of focus shift required to minimize the blur spot. (Remember that $\Delta H/\Delta \tan U$ is equivalent to a focus shift, and that the separation of the lines indicates the size of the blur.) However, the dashed pair of lines (which enclose the ray intercepts from about 80% of the aperture) indicate a focus position at which there is a much higher concentration of energy within a much smaller spot, and this is usually the preferred focus, even though the *total* spread of the image is greater. The graph indicates the relationships between spot size and focus shift for the two cases, and is more accurate than Eq. 11.28 for large values of U.

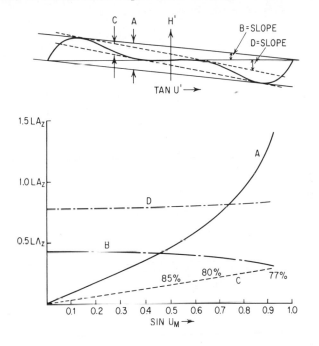

FIG. 11.9. Image blur spot size for third and fifth order spherical aberration ($LA_m = 0$). Upper: H' vs. Tan U' plot. Lower: Curve A - Diameter of smallest blur spot containing 100% of energy in image. Curve B - Distance from paraxial focus to plane of smallest 100% energy spot. Curve C - Diameter of bright core of image. Figures along curve indicate percentage of total energy within core. Curve D - Distance from paraxial focus to plane for Curve C.

The concept of minimum blur size is little used in optical systems for visual or photographic work, since the minimum blur position is seldom, if ever, chosen as the focus. However, in systems which use photodetectors, one frequently wishes to determine the smallest detector that will collect all the energy in the image. Under such circumstances, the blur spot sizes given by Eqs. 11.26, 11.27 and 11.28 are extremely useful; in Chapter 13, a number of very convenient equations will be presented which make use of this concept (in combination with the third-order aberration contribution equations) to predict the performance of several simple optical systems which are very frequently used in conjunction with photodetectors.

Example B

A visual system, working at $f/5$ ($\sin U_m = 0.1$), which has an undercorrected third-order longitudinal spherical aberration of 0.22 mm, will have its minimum diameter blur spot $0.75 \times 0.22 = 0.165$ mm ahead of the paraxial focus, and by Eq. 11.26, the size of this blur spot will be equal to

$$B = \frac{1}{2} \times 0.22 \times 0.1005 = 0.011 \text{ mm}$$

It is interesting to note that on the basis of the OPD analysis, the best focus should occur $0.5 \times 0.22 = 0.11$ mm ahead of the paraxial focus and that the diameter of the central disc of the Airy pattern is equal to

$$\frac{1.22 \lambda}{N \sin U} = \frac{1.22(0.00055)}{0.1} = 0.0066 \text{ mm}$$

This central disc should contain about 68% of the energy in the image, since a marginal spherical of 0.22 mm is equal to just one Rayleigh Limit (as shown in Example A).

If an $f/5$ system has third and fifth order spherical with a zonal residual of 0.33 mm (again in longitudinal measure), the smallest geometrical spot size would be found at about $0.42 \times 0.33 = 0.14$ mm from the paraxial focus and the spot size would be

$$B = 0.84 \times 0.33 \times 0.1005 = 0.028 \text{ mm}$$

Here the comparison with the OPD analysis is less fortuitous. The zonal spherical of 0.33 mm is again equivalent to one Rayleigh Limit; we would expect the central disc of the diffraction pattern to be 0.0066 mm as above, and the best focus to be about $0.75 \times 0.33 = 0.25$ mm from the paraxial focus. The agreement with geometry is somewhat better if we use the focus indicated by the dashed lines of Fig. 11.9; the position of "best focus" is almost exactly the same as the OPD best focus and the diameter of the intense center spot of the geometrical pattern is to the order of 0.01 mm.

11.8 The Modulation Transfer Function

A type of target commonly used to test the performance of an optical system consists of a series of alternating light and dark bars of equal width, as indicated in Fig. 11.10A. Several sets of patterns of different spacings are usually imaged by the system under test and the finest set in which the line structure can be discerned is considered to be the limit of resolution of the system, which is expressed as a certain number of lines per millimeter.*

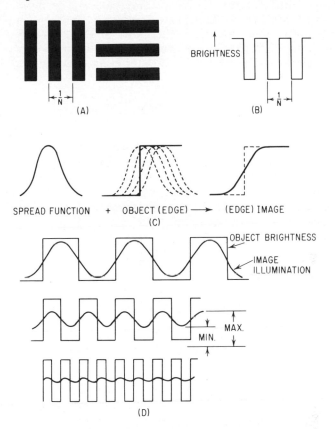

FIG. 11.10. The imagery of a bar target. A. A typical bar target used in testing optical systems consists of alternating light and dark bars. If the pattern has a frequency of *N* lines per millimeter, then it has a period of 1/*N* millimeters, as indicated. B. A plot of the brightness of A is a square wave. C. When an image is formed, each point is imaged as a blur, with an illumination distribution described by the spread function. The image then consists of the summation of all the spread functions. D. As the test pattern is made finer, the contrast between the light and dark areas of the image is reduced.

When a pattern of this sort is imaged by an optical system, each geometric line (i.e. of infinitesimal width) in the object is imaged as a blurred line, whose cross section is the line spread function. Figure 11.10B indicates a cross section of the brightness of the bar object and Fig. 11.10C shows how the image spread function

---

*Note that in optical work the convention is to consider a "line" to consist of one light bar and one dark bar, that is, one cycle. In television parlance, both light and dark lines are counted. Thus, ten "optical" lines indicate ten light and ten dark lines, whereas ten "television" lines indicate five light and five dark lines. To avoid confusion, "optical" lines are frequently referred to as line pairs, e.g. ten line pairs per millimeter.

"rounds off" the "corners" of the image. In Fig. 11.10D, the effect of the image blur on progressively finer patterns is indicated. It is apparent that when the illumination contrast in the image is less than the smallest amount that the system (e.g. the eye, film, or photodetector) can detect, the pattern can no longer be "resolved".

If we express the contrast in the image as a "modulation", given by the equation

$$\text{Modulation} = \frac{\text{max.} - \text{min.}}{\text{max.} + \text{min.}}$$

(where max. and min. are as indicated in Fig. 11.10D), we can plot the modulation as a function of the number of lines per millimeter, as indicated in Fig. 11.11A. The intersection of the modulation function line with a horizontal line representing the smallest amount of modulation which the system can detect will give the limiting resolution of the system.

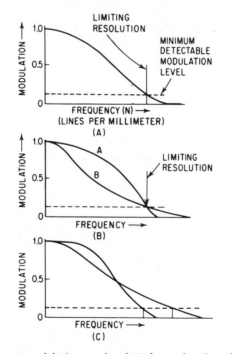

FIG. 11.11. A) The image modulation can be plotted as a function of the frequency of the test pattern. When the modulation drops below the minimum that can be detected, the target is not resolved. B) The system represented by A will produce a superior image, although both A and B have the same limiting resolution.

It should be apparent that the limiting resolution does not fully describe the performance of the system. Fig. 11.11B shows two modulation plots with the same limiting resolution, but with quite different performances. The plot with the greater modulation at the lower frequencies is obviously superior, since it will produce crisper, more contrasty images. Unfortunately, the type of choice one is usually faced with in deciding between two systems is less obvious. Consider Fig. 11.11C, where one system shows extremely high limiting resolution and the other shows high contrast at low target frequencies. In cases of this type, the decision must be based on the relative importance of contrast vs. resolution in the function of the system.*

The preceding discussion has been based on patterns whose brightness distribution is a "square wave" (Fig. 11.10B) and whose image illumination distribution is distorted by characteristics of the optical system, as indicated in Fig. 11.10D. However, if the object pattern is in the form of a sine wave, the distribution in the image is also described by a sine wave, regardless of the shape of the spread function. This fact has led to the widespread use of the Optical Modulation Transfer Function to describe the performance of a lens system. The Modulation Transfer Function is the ratio of the modulation in the image to that in the object as a function of the frequency (cycles per unit of length) of the sine wave pattern.

$$\text{MTF}\,(v)\;=\;\frac{M_i}{M_o}$$

A plot of MTF against frequency v is thus an almost universally applicable measure of the performance of an image forming system, and can be applied not only to lenses, but to films, phosphors, image tubes, the eye, and even to complete systems such as camera carrying aircraft and the like.

One particular advantage of the MTF is that it can be cascaded by simply multiplying the MTF's of two or more components to obtain the MTF of the combination. For example, if a camera lens with an MTF of 0.5 at 20 cycles per millimeter is used with a film with an MTF of 0.7 at this frequency, the combination will have an MTF of $0.5 \times 0.7 - 0.35$. If the object photographed with this camera has a contrast (modulation) of 0.1, then the image modulation is $0.1 \times 0.35 = 0.035$, close to the limit of visual detection.

One should note, however, that MTF's do not cascade between

---

*The Strehl Definition is the ratio of the light intensity at the peak of the diffraction pattern of an aberrated image to that at the peak of an aberration free image, and is one of the many criteria that have been proposed for image evaluation. It can be computed by calculating the volume under the (three dimensional) Modulation Transfer Function and dividing by the volume under the curve for an aberration free lens (Section 11.10). A similar criterion for quick general evaluation of image quality is the normalized area under the Modulation Transfer Curve.

optical components which are directly "connected", that is, lenses which are not separated by a diffuser of some sort. This is because the aberrations of one component may compensate for the aberrations in another, and thus produce an image quality for the combination which is superior to that of either component. Any "corrected" optical system illustrates this point.

The Optical Modulation Transfer Function is also frequently referred to as frequency response, sine wave response, or contrast transfer.

If we assume an object consisting of alternating light and dark bands, the brightness (luminance, radiance) of which varies according to a cosine (or sine) function, as indicated by the upper part of Fig. 11.12, the distribution of brightness can be expressed mathematically as

$$G(x) = b_0 + b_1 \cos 2\pi v x \qquad (11.29)$$

where v is the frequency of the brightness variation in cycles per unit length, $(b_0 + b_1)$ is the maximum brightness, $(b_0 - b_1)$ is the minimum brightness and x is the spatial coordinate perpendicular to the bands. The modulation of this pattern is then

$$M_0 = \frac{(b_0 + b_1) - (b_0 - b_1)}{(b_0 + b_1) + (b_0 - b_1)} = \frac{b_1}{b_0} \qquad (11.30)$$

When this line pattern is imaged by an optical system, each point in the object will be imaged as a blur. The energy distribution within this blur will depend on the relative aperture of the system and the aberrations present. Since we are dealing with a linear object, the image of each line element can be described by the line spread function (Section 11.5, Fig. 11.6) indicated in Fig. 11.12 as $A(\delta)$. We now assume (for convenience) that the dimensions x and $(1/v)$ in Eq. 11.29 are the corresponding dimensions in the image. It is apparent that the image energy distribution at a position x is the summation of the product of $G(x)$ and $A(\delta)$, and can be expressed as

$$F(x) = \int A(\delta) \, G(x - \delta) \, d\delta \qquad (11.31)$$

Combining Eqs. 11.29 and 11.31, we get

$$F(x) = b_0 \int A(\delta) \, d\delta + b_1 \int A(\delta) \cos 2\pi v (x - \delta) \, d\delta \qquad (11.32)$$

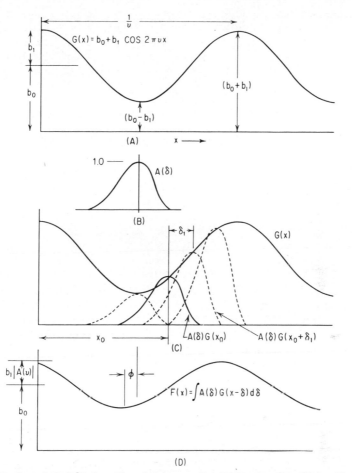

FIG. 11.12. Convolution of the object brightness distribution function $G(x)$ with the line spread function $A(\delta)$. A) The object function, $G(x) = b_0 + b_1 \cos 2\pi vx$, plotted against x. B) The line spread function $A(\delta)$. Note the asymmetry. C) Illustrating the manner in which $G(x)$ is modified by $A(\delta)$. A point (or more accurately, a line element) at x_0 is imaged by the system as $G(x_0)$ times $A(\delta)$. Similarly at $x_0 + \delta_1$, the image of the line element is described by $A(\delta)G(x_0 + \delta_1)$. Thus the image function at a given x has a value equal to the summation of the contributions from all the points whose spread-out images reach x. D) The image function $F(x) = \int A(\delta)G(x - s)\,d\delta$ has been shifted by ϕ and has a modulation $M_i = M_0 |A(v)|$.

After normalizing by dividing by $\int A(\delta)\,d\delta$, Eq. 11.32 can be transformed to

$$F(x) = b_0 + b_1 |A(v)| \cos(2\pi vx - \phi)$$

$$= b_0 + b_1 A_c(v) \cos 2\pi vx + b_1 A_s(v) \cos 2\pi vx$$

(11.33)

where

$$|A(v)| = \left[A_c^2(v) + A_s^2(v)\right]^{\frac{1}{2}} \tag{11.34}$$

and

$$A_c(v) = \frac{\int A(\delta) \cos 2\pi v \delta \, d\delta}{\int A(\delta) \, d\delta} \tag{11.35}$$

$$A_s(v) = \frac{\int A(\delta) \sin 2\pi v \delta \, d\delta}{\int A(\delta) \, d\delta} \tag{11.36}$$

$$\cos \phi = \frac{A_c(v)}{|A(v)|} \tag{11.37}$$

$$\tan \phi = \frac{A_s(v)}{A_c(v)} \tag{11.38}$$

Note that the resulting image energy distribution $F(x)$ is still modulated by a cosine function of the same frequency v, demonstrating that a cosine distribution object is always imaged as a cosine distribution image. If the line spread function $A(\delta)$ is asymmetrical, a phase shift ϕ is introduced.

The modulation in the image is given by

$$M_i = \frac{b_1}{b_0} |A(v)| = M_o |A(v)| \tag{11.39}$$

and $|A(v)|$ is the modulation transfer function.

$$\text{MTF}(v) = |A(v)| = \frac{M_i}{M_o}$$

11.9 Computation of the Modulation Transfer Function

We will illustrate the computation of the Modulation Transfer Function by a grossly simplified and abbreviated example. In actual practice, a much larger number of points must be used in the computation if accurate results are to be obtained.

We assume that a spot diagram has been prepared from ray trace data as shown in Fig. 11.13A. The line spread function is determined by integrating the spot diagram in one direction; in practice, one assumes an increment Δx and counts all the spots between the lines bounding the increment. A normalized plot of N_x against x then represents the line spread function $A(x)$.

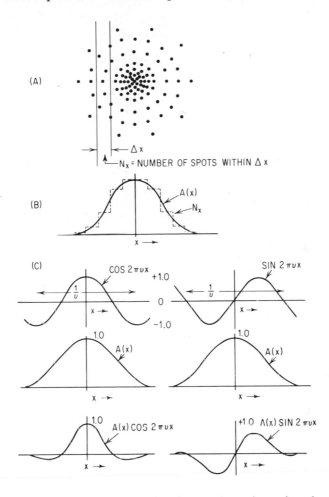

FIG. 11.13. The calculation of the Modulation Transfer factor for a given frequency, v. A. The spot diagram is summed in one direction by counting the number of spots (ray intersections) in each increment, Δx. B. The number of spots is plotted against x to get the line spread function $A(x)$. $A(x)$ is usually normalized to peak at unity. C. $A(x) \cos 2\pi v x$ and $A(x) \sin 2\pi v x$ are generated by point for point multiplication of $A(x)$ by the trigonometric functions. Then $\int A(x) \cos 2\pi v x\,dx$ and $\int A(x) \cos 2\pi v\,dx$ are the areas under their respective curves (remembering that area below the x-axis is negative). Similarly $\int A(x)\,dx$ is the area under the curve of $A(x)$ vs. x. These values are used in Equations 11.34 through 11.39 to get the MTF and phase shift ϕ for the frequency v.

Since real spread functions are rarely (if ever) represented by ordinary analytic functions, we cannot use Eqs. 11.35 and 11.36 in their integral form. A close approximation (which lends itself nicely to electronic computer usage) is given by the equivalent summation equations,

$$A_c(v) = \frac{\Sigma A(x) \cos 2\pi vx \, \Delta x}{\Sigma A(x) \, \Delta x} \tag{11.40}$$

$$A_s(v) = \frac{\Sigma A(x) \sin 2\pi vx \, \Delta x}{\Sigma A(x) \, \Delta x} \tag{11.41}$$

As a numerical example, we will determine the value of the Modulation Transfer Function for a frequency of $v = 0.1$, that is one tenth cycle per unit length. The values of $A(x)$, the line spread function, which we will use are given in Line 2 of Fig. 11.14 for various values of x. Line 4 of the table gives the values of $2\pi vx$ for these same values of x, and Lines 5 and 6 give the values of $\cos 2\pi vx$ and $\sin 2\pi vx$ for each point.

Now Lines 7 and 8 give $A(x) \cos 2\pi vs$ and $A(x) \sin 2\pi vx$, respectively. Since Δx is equal to 1.0 in this example, we can obtain the required summations for Eqs. 11.40 and 11.41 by summing across Lines 2, 7 and 8, giving us

$$\Sigma A(x) \, \Delta x = +5.10$$

$$\Sigma A(x) \cos 2\pi vx \, \Delta x = +2.51236$$

$$\Sigma A(x) \sin 2\pi vx \, \Delta x = 0.0$$

Note that the last value is a foregone conclusion when $A(x)$ is a symmetrical function of x, since the positive and negative values of the sine function on either side of $x = 0$ cause one side to cancel the other when summed. Thus, when $A(x)$ is symmetrical, the labor of the calculation can be reduced by a factor of four, since only one half of the cosine function need be evaluated.

Inserting the above values into Eqs. 11.40 and 11.41, we find that

$$A_c(0.1) = \frac{2.51236}{5.1} = +.4926$$

$$A_s(0.1) = \frac{0.0}{5.1} = 0.0$$

and that by Eqs. 11.34 and 11.38

$$\text{MTF}(0.1) = |A(0.1)| = [.493^2 + 0^2]^{\frac{1}{2}} = 0.493$$

$$\tan \phi = \frac{0}{+2.512} = 0.0$$

(1) x $(\Delta x = 1.0)$	-4.5	-3.5	-2.5	-1.5	-0.5	$+0.5$	$+1.5$	$+2.5$	$+3.5$	$+4.5$
(2) $A(x)$	0.05	0.2	0.5	0.8	1.0	1.0	0.8	0.5	0.2	0.05
									$\Sigma A(x)\,\Delta x = +5.10$	
(3) vx	$-.45$	$-.35$	$-.25$	$-.15$	$-.05$	$+.05$	$+.15$	$+.25$	$+.35$	$+.45$
(4) $2\pi vx$	$-.9\pi$ $(-162°)$	$-.7\pi$ $(-126°)$	$-.5\pi$ $(-90°)$	$-.3\pi$ $(-54°)$	$-.1\pi$ $(-18°)$	$+.1\pi$ $(+18°)$	$+.3\pi$ $(+54°)$	$+.5\pi$ $(+90°)$	$+.7\pi$ $(+126°)$	$+.9\pi$ $(+162°)$
(5) $\cos 2\pi vx$	$-.95106$	$-.58779$	0	$+.58779$	$+.95106$	$+.95106$	$+.58779$	0.0	$-.58779$	$-.95106$
(6) $\sin 2\pi vx$	$-.30902$	$-.80902$	-1.0	$-.80902$	$-.30902$	$+.30902$	$+.80902$	$+1.0$	$+.80902$	$+.30902$
(7) $A(x)\cos 2\pi vx$	$-.04755$	$-.11756$	0.0	$+.47023$	$+.95106$	$+.95106$	$+.47023$	0.0	$-.11756$	$-.04755$
								$\Sigma A(x)\cos 2\pi vx\,\Delta x = +2.51236$		
(8) $A(x)\sin 2\pi vx$	$-.01545$	$-.16180$	-0.5	$-.64722$	$-.30902$	$+.30902$	$+.64722$	$+0.5$	$+.16180$	$+.01545$
								$\Sigma A(x)\sin 2\pi vx\,\Delta x = 0.0$		

FIG. 11.14. Numerical computation of the Modulation Transfer Function for a frequency of $v = 0.1$ cycles per millimeter, from the line spread function $A(x)$ given in Line 2.

Thus, for a frequency $v = 0.1$ cycles per unit length, we find a modulation transfer factor of 49%. This calculation can be repeated for several values of v and a plot of the MTF against frequency, similar in appearance to those of Fig. 11.11, can be prepared from the results. As mentioned above, a much smaller value of Δx must be used if accurate results are to be obtained.

Once the Modulation Transfer Function has been determined (plotted) for a range of frequencies, it is possible to determine an analogous function for the modulation transfer of a square wave pattern, i.e. a bar chart of the type shown in Fig. 11.10. This is done by resolving the square wave into its Fourier components and taking the sine wave response to each component. Thus, for a given frequency v, the square wave modulation transfer $S(v)$ is given by the following equation (in which MTF (v) is written $M(v)$ for clarity).

$$S(v) = \frac{4}{\pi}\left[M(v) - \frac{M(3v)}{3} + \frac{M(5v)}{5} - \frac{M(7v)}{7} + \cdots\right] \qquad (11.40)$$

A rough indication of the practical meaning of resolution can be gained from the following, which lists the resolution required to photograph printed or typewritten copy.

Excellent reproduction (reproduces serifs, etc.) requires 8 resolution line pairs per the height of a lower case letter e.

Legible (easily) reproduction requires 5 line pairs per letter height.

Decipherable (e, c, o partly closed) requires 3 line pairs per letter height.

11.10 Special Modulation Transfer Functions
Diffraction Limited Systems

The preceding sections have discussed MTF in purely geometrical terms; the techniques set forth above are applicable only when the aberrations are large. When they are small, the interactions between the diffraction effects of the system aperture and the aberrations become very complex. If there are no aberrations present, the MTF of a system is related to the size of the diffraction pattern (which is a function of the numerical aperture of the system and the wavelength of the light used) and is given by

$$\text{MTF}(v) = \frac{2}{\pi}(\phi - \cos\phi \sin\phi)(\cos\theta)^k \qquad (11.41)$$

where

$$\phi = \cos^{-1}\left(\frac{\lambda v}{2\,\text{N.A.}}\right)$$

and v is the frequency in cycles per millimeter, λ is the wavelength in millimeters, N.A. is the numerical aperture $(N'\sin U')$, θ is the half field angle and $k = 1$ for radial lines or 3 for tangential lines.*

It is apparent that MTF (v) is equal to zero when ϕ is zero; thus, the "limiting resolution" for an aberration free system is

$$v_0 = \frac{2\,\text{N.A.}}{\lambda} = \frac{1}{\lambda(f/\#)} \tag{11.42}$$

where λ is in millimeters, $(f/\#)$ is the relative aperture of the system, and v_0 is in cycles per millimeter.

A plot of Eq. 11.41 is shown in Fig. 11.15; the frequency scale is in terms of v_0, the limiting frequency given by Eq. 11.42. It should be noted that for *ordinary* systems, this level of performance cannot be exceeded. An MTF curve derived from the *raytrace* data

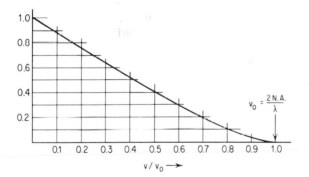

FIG. 11.15. The Modulation Transfer Function of an aberration free system. Note that frequency is given in terms of the limiting resolution frequency v_0. This curve is based on diffraction effects.

---

*Equation 11.41 applies to circular apertures. For apertures of *any* shape, the diffraction MTF is equal to the (normalized) area common to the aperture and the aperture displaced. Equation 11.41 is thus the (normalized) area common to two circles of radius R, as their centers are separated by an amount equal to $2vR/v_0$. For a rectangular aperture the plot of MTF would thus be a straight line. The cut-off frequency, v_0, is computed from Eq. 11.42 in each case using the width (i.e. the $f/\#$ or N.A.) in the direction of the resolution.

of a well corrected lens will sometimes exceed the values of Fig. 11.15; such results are, of course, incorrect and derive from the fact that the light ray concept only partially describes the behavior of electromagnetic radiation.

The effects of small amounts of defocusing on the diffraction limited MTF are shown in Fig. 11.16. Note that curve B corresponds to the depth of focus allowed by one "Rayleigh Limit" as discussed in Sections 11.2 and 11.4. The small effect produced by an O.P.D. of one-quarter wavelength indicates the astuteness of Rayleigh's selection of this amount as one which would not "sensibly" affect the image quality.

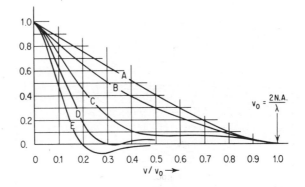

FIG. 11.16. The effect of defocusing on the modulation transfer function of an aberration free system

A) In Focus   ; $OPD = 0.0$
B) Defocus $= \lambda/2N \sin^2 u$; $OPD = \lambda/4$
C) Defocus $= \lambda/N \sin^2 u$; $OPD = \lambda/2$
D) Defocus $= 3\lambda/2N \sin^2 u$; $OPD = 3\lambda/4$
E) Defocus $= 2\lambda/N \sin^2 u$; $OPD = \lambda$

(Based on diffraction-wavefront effects - *not* geometrical calculation)

By way of comparison, Fig. 11.17 shows the MTF plots which would be obtained by geometrical calculations of a perfect system defocused by the same amounts. The agreement between Fig. 11.16, whose curves are derived from wave front analysis, and Fig. 11.17 is poor for small amounts of OPD. However, when the defocusing is sufficient to introduce an OPD of one wavelength or more, the agreement becomes much better. Note that all the curves of Fig. 11.17 are of the same family and that one can be derived from another by a simple ratioing of the frequency scale. These curves are representations of

$$\text{MTF}(v) = \frac{2J_1(\pi Bv)}{\pi Bv} = \frac{J_1(2\pi \delta \text{N.A.} v)}{\pi \delta \text{N.A.} v} \qquad (11.43)$$

where $J_1(\ )$ indicates the first order Bessel Function*, B is the diameter of the blur spot produced by defocusing, δ is the longitudinal defocusing, N.A. is the Numerical Aperture and v is the frequency in cycles per unit length.

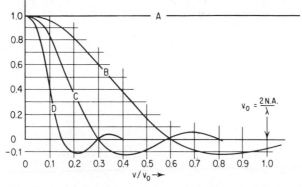

FIG. 11.17. The effect of defocusing on the *geometrically* calculated modulation transfer function of an aberration free system.

$$\text{A) In Focus} \qquad\qquad ; \qquad \text{OPD} = 0.0$$
$$\text{B) Defocus} = \lambda/2N \sin^2 u \; ; \qquad \text{OPD} = \lambda/4$$
$$\text{C) Defocus} = \lambda/N \sin^2 u \; \; ; \qquad \text{OPD} = \lambda/2$$
$$\text{D) Defocus} - 2\lambda/N \sin^2 u \; ; \qquad \text{OPD} = \lambda$$

These geometrically derived plots are in poor agreement with the exact diffraction derived plots of Fig. 11.16 when the defocusing is small. The agreement at OPD = λ(curve *D* above, curve *E* in Fig. 11.16) is fair; the match at OPD = 2λ is still better.

Note that in Figs. 11.16 and 11.17, some of the curves show a negative value for the MTF. This indicates that the phase shift in the image (ϕ in Eq. 11.38) is 180° and that the image is light where it should be dark and vice versa. This is known as spurious resolution (since the lines can be seen, but are not true images of the object) and is a phenomenon which is frequently observed in defocused, well corrected lenses or in lenses whose image of a point is a nearly uniformly illuminated circular blur.

In Fig. 11.18, the effects of third order spherical aberration on MTF are shown. Note once again that the effect of an amount of aberration corresponding to the Rayleigh Limit (OPD = $\lambda/4$) is quite modest. The situation here is quite similar to the defocusing case, in that MTF curves based on geometrical calculations are in

---

*

$$J_n(x) = \sum_{k=0}^{\infty} \frac{(-1)^k x^{n+2k}}{2^{(n+2k)} k! (n+k)!}$$

$$J_1(x) = \frac{x}{2} - \frac{\left(\frac{x}{2}\right)^3}{1^2 2} + \frac{\left(\frac{x}{2}\right)^5}{1^2 2^2 3} - \ldots$$

poor agreement with Fig. 11.18 where the aberration is small, but in quite reasonable agreement where the aberration is to the order of one or two wavelengths of O.P.D.

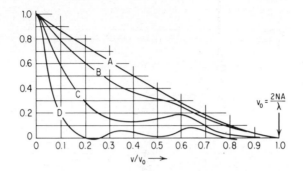

FIG. 11.18. The effect of third order spherical aberration on the modulation transfer function.

$$\text{A) } LA_m = 0.0 \qquad ; \qquad OPD = 0$$
$$\text{B) } LA_m = 4\lambda/N \sin^2 u \ ; \qquad OPD = \lambda/4$$
$$\text{C) } LA_m = 8\lambda/N \sin^2 u \ ; \qquad OPD = \lambda/2$$
$$\text{D) } LA_m = 16\lambda/N \sin^2 u \ ; \qquad OPD = \lambda$$

These curves are based on diffraction-wave front computations for an image plane midway between the marginal and paraxial focii.

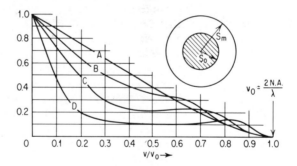

FIG. 11.19. The effect of a central obscuration on the modulation transfer function of an aberration free system.

$$\text{A) } s_0/s_m = 0.0$$
$$\text{B) } s_0/s_m = 0.25$$
$$\text{C) } s_0/s_m = 0.5$$
$$\text{D) } s_0/s_m = 0.75$$

Since the wave front analysis leading to MTF curves of the type shown in Figs. 11.15, 11.16, 11.18, and 11.19 is difficult except in certain specific cases (and those are by no means simple), it has been proposed that for practical cases, the geometrical MTF be multiplied by the wave front MTF for a perfect lens

(Fig. 11.15) on a point for point basis to produce an approximation to the actual result. Thus, if we represent the geometrically computed MTF at some frequency v_i by GMTF (v_i) and the wave front MTF of a perfect lens at the same frequency by DMTF (v_i), then the product

$$GMTF(v_i) \times DMTF(v_i)$$

is a somewhat better representation of the true MTF at v_i than is GMTF (v_i). In general, this product tends to err toward the low side; this tends to make for conservative, or safe, predictions of performance. The reader may test this thesis by multiplying the ordinates of Fig. 11.15 by those of Fig. 11.17 and comparing the results with Fig. 11.16.

Figure 11.19 shows the effect of a central obstruction in the aperture of a diffraction limited system. Note that the introduction of a disc into the aperture* drops the response at low frequencies, but raises it at high frequencies (although it cannot change v_0, the cut off frequency). Thus, a system of this type tends to show reduced contrast on coarse targets and a somewhat higher limit of resolution (when used with a system which requires a modulation of more than zero to detect resolution).

As mentioned previously, Modulation Transfer Functions can be established for image forming systems which are not optical systems. Figure 11.20 shows the MTF curves for a number of photographic emulsions. Since the MTF of a film is computed on the basis of equivalent relative exposures derived from density measurements on films exposed to sinusoidal test patterns, it is

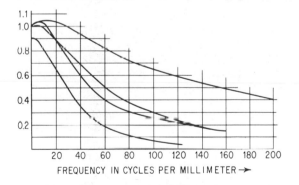

FIG. 11.20. Modulation Transfer Functions of several photographic emulsions.

Apodization is the use of a variable transmission filter or coating at the aperture to modify the diffraction pattern. Coatings which reduce the transmission at the center of the aperture tend to "favor" the response at high frequencies, coatings which reduce transmission at the edge of the aperture tend to favor the lower frequencies.

possible to have an MTF greater than unity. This results from the chemical effects of development of the film on adjacent areas and will be noticed at the low frequency end of the curves in Fig. 11.20.

11.11 Radial Energy Distribution

The data of a spot diagram can be presented in the form of a Radial Energy Distribution plot. If the blur spot is symmetrical, it is apparent that a small circular aperture centered in the image would pass a portion of the total energy and block the rest. A larger aperture would pass a greater portion of the energy and so on. A graph of the encircled fraction of the energy (that is the fraction of the ray intersection points in the spot diagram) plotted against the radius (semi-diameter) of the aperture is called the radial energy distribution curve.

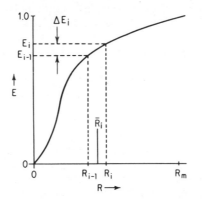

FIG. 11.21. Radial energy distribution plot. The curve indicates the fraction (E) of the total energy in an image pattern which falls within a circle of Radius R. Thus all the energy is encompassed by a circle of radius R_m; E_i of the energy by a circle of radius R_i.

A radial energy distribution curve, such as the example shown in Fig. 11.21, can be used to compute the Modulation Transfer Function for an optical system, by means of the summation equation

$$\text{MTF}\,(v) \;=\; \sum_{i=1}^{i=m} \Delta E_i\, J_0\!\left[2\pi\,v\,\overline{R}_i\right]$$

where v is the frequency in cycles per unit length, ΔE_i is the difference $(E_i - E_{i-1})$ between two values of E, the fractional energy, $\overline{R}_i$ is the average $\frac{1}{2}(R_i + R_{i-1})$ of the corresponding values of the of the radius and J_0 (   ) indicates the zero order Bessel Function.*

Although the radial energy distribution is (strictly speaking) valid only for point images which have rotational symmetry, i.e. for images on the optical axis, it is sometimes used to predict approximate averaged resolution for off axis points. This procedure, while it cannot yield separate radial and tangential values for resolution, does serve to give the designer a rough idea of the state of correction of the system.

References

1. Conrady, "Applied Optics and Optical Design", Oxford (London) 1929. (Reprinted by Dover Publications, N.Y., 1957)

2. Wolf, "Progress in Optics", Vol. I Interscience, N. Y.

3. Perrin, "Methods of Appraising Photographic Systems", Journal of the Society of Motion Picture and Television Engineers Vol. 69, pp 151-156 (Part I), pp 239-248 (Part II) (March, April 1960), Extensive Bibliography

4. "Optical Image Evaluation", N.B.S. Circular 526, Government Printing Office, Washington, 1954

5. Hertzberger, "Modern Geometrical Optics", Interscience, N.Y., 1958

Exercises

1. The longitudinal spherical aberration of a spherical reflector is equal to $y^2/8f$ (to a third order approximation). What is the maximum diameter a 36 inch focal length reflector may have without exceeding an OPD of one quarter wavelength for visual light, $\lambda = 20 \times 10^{-6}$ inches? (Use Eq. 11.16)
   Ans. 4.7 inches

2. The third order sagittal coma of a parabolic reflector is given by $-y^2\theta/4f$, where θ is the half field angle in radians. What field will a 5″ diameter $f/8$ reflector cover without exceeding the Rayleigh limit? ($\lambda - 2 \times 10^{-5}$ inches)
   Ans. ±.0041 radians (0.47° total field)

3. An $f/5$ system is defocused by 0.05 mm. What is the modulation transfer factor for a "sine wave" target with a spatial frequency (at the image) of 120 cycles per mm? Use Fig. 11.16, Eq. 11.2, and assume $\lambda = 0.5$ microns.
   Ans. 0.23

---

* $J_0(x) = 1 - \left(\frac{x}{2}\right)^2 + \frac{\left(\frac{x}{2}\right)^4}{1^2\, 2^2} - \frac{\left(\frac{x}{2}\right)^6}{1^2\, 2^2\, 3^2} + \cdots$

12

The Design of Optical Systems: General

12.1

In the immediately preceding chapters, we have been concerned with the analysis of optical systems, in the sense that the constructional parameters of the system were given and our object was the determination of the resultant performance characteristics. In this chapter we take up the synthesis of optical systems; here the desired performance is given and the constructional parameters are to be determined. A large part of the synthesis process is, of course, concerned with analysis, since optical design is in great measure a systematic application of the cut-and-try process.

There is no "direct" method of optical design for original systems; that is, there is no sure procedure that will lead (without fore-knowledge) from a set of performance specifications to a suitable design. However, when it is known that a certain type of design or configuration is capable of meeting a given performance level, it is a fairly straightforward process for a competent designer to produce a design of the required type. Further, modest improvements to existing designs can almost always be effected by well-established techniques. Thus, it is apparent that a good portion of the ammunition in a designer's arsenal consists of an intimate and detailed knowledge of a wide range of designs, their characteristics, limitations, idiosyncrasies, and potentials. Here is one part of the Art in optical design; basically it consists of the choice of the point at which the designer begins his efforts.

The electronic computer, in the course of little more than a decade, radically modified the techniques used by optical designers. Previously a designer resorted to all manner of ingenious techniques to avoid tracing rays because of the great expenditure of time and effort involved. The computer reduced ray tracing time by several orders of magnitude, and it is now easier to trace rays through a system than it is to speculate, infer, or interpolate from incomplete data. A computer can even be made to carry through the entire design process from start to finish, more or less without human intervention. The results produced by such a process are nonetheless intricately dependent on the starting point elected (as well as the manner in which the computer has been programmed), so that a great deal of art (if perhaps somewhat less personal satisfaction) is present even in the most automatic technique.

The ordinary design process can be broken down into four stages, as follows: First, the choice of the type of design to be executed, that is, the number and types of elements and their general configuration. Second, the determination of the powers, materials, thicknesses and spacings of the elements. These are usually selected to control the chromatic aberrations and the Petzval curvature of the system, as well as the focal length (or magnifying power), working distances, field of view and aperture. (Choices made at this stage may affect the performance of the final system tremendously, and can mean the difference between success and failure in many cases). In the third stage, the shapes of the elements or components are adjusted to correct the basic aberrations to the desired values. The fourth stage is the reduction of the residual aberrations to an acceptable level. If the choices exercised in the first three stages have been fortuitous, the fourth stage may be totally unnecessary. At the other extreme, the end result of the first three stages may be so hopeless that a fresh start from stage one is the only alternative. Stage four conceals a pitfall for the unwary, since at this stage a certain amount of analysis is unavoidable and the designer sometimes must choose between another round of design effort (and the possible improvement of the system) and a complete, full dress modulation transfer computation.

In fully automatic computer design procedures, a portion of stage one and all of stages two, three and four may be accomplished more or less simultaneously (using an approach that might take a human computer a lifetime or two to slog his way through). Computer design techniques are discussed in Section 12.8.

The basic principles of optical design will be illustrated by three detailed examples in the following sections. A simple meniscus (box) camera lens will be used to show the effects of bending and stop shift techniques, as well as the handling of a simplified exercise in satisfying more requirements than there are available degrees of freedom. An achromatic telescope objective will introduce material choice, achromatism and multiple bending techniques. An air spaced (Cooke) triplet anastigmat will illustrate the problem of controlling all the first and third order aberrations in a system with just a sufficient number of degrees of freedom to accomplish this, and will further illustrate the technique of material selection. The design characteristics of several additional types of optical systems are discussed in Chapter 13.

12.2 The Simple Meniscus Camera Lens

There are just two elements to work with in the design of a meniscus camera lens, the lens itself and the aperture stop. If, for the moment, we restrict ourselves to a thin, spherical-surfaced element, the parameters which we may choose or adjust are: the material of the lens, its focal length, its shape (or bending), the position of the stop and the diameter of the stop. With these degrees of freedom we must design a lens which will produce an acceptable image on a given size of film. This implies that all the aberrations of the system must be "sufficiently" small. It is immediately apparent that the spherical aberration will be undercorrected and that the Petzval curvature will be inward curving (and equal to $-h^2 \phi/2n$); these are the immutable characteristics of a simple lens. Thus, the element power and the size of the aperture must be chosen small enough so that the effects of these aberrations are tolerable. The lens material is usually chosen as common crown glass, on the basis of cost, since a box camera lens must be inexpensive. A high index crown does not produce enough improvement in the Petzval curvature to warrant its increased cost; a flint glass would introduce increased chromatic aberrations.

We find ourselves with just two uncommitted degrees of freedom, namely the bending of the lens and the position of the stop. Now in a simple undercorrected system it is axiomatic that for a given (i.e. fixed) shape of the lens (or lenses) the position of the stop for which the coma is zero is also the position for which the astigmatism is the most overcorrected (i.e. most backward curving). Since the Petzval surface will be inward (toward the lens) curving, overcorrected astigmatism is desirable.

Thus the design technique is straightforward: we choose (arbitrarily) a shape for the lens, determine the stop position at which coma is zero and evaluate the aberrations. By repeating this process for several bendings and graphing the aberrations as a function of the shape, we can then choose the best design.

There are several ways in which this can be accomplished. Since this is a simple lens of moderate aperture and field, the third order aberrations are quite representative of the system and one would be quite safe in relying on them. The design could also be handled by trigonometric ray tracing. For this example we will work out the design using the thin lens (G-sum) third order aberration equations and then check the results by ray tracing.

Assuming that the glass has an index of 1.50 and a V-value of 62.5, we will set up the G-sum equations for a focal length of 10, an aperture diameter of 1.0, and an image height of 3 (all in arbitrary units and all subject to scaling and adjustment later). Thus, the element power $\phi = 1/10 = 0.1$, and the total curvature $c = c_1 - c_2 = \phi/(N - 1) = 0.2$. With the object at infinity, $v_1 = 0$. Using the G-values worked out in Example G of Chapter 10, we find that the spherical and coma (stop at the lens) given by Eqs. 10.9M and 10.9N are

$$SC = -2.91666 \, C_1^2 + 1.0 \, C_1 - 0.1125$$

$$CC = -0.0625 C_1 + 0.01125$$

Now the position of the stop can be determined by solving Eq. 10.9G for Q when CC^* is zero.

$$CC^* = 0 = CC + SCQu'_k$$

$$Q = \frac{-CC}{u'_k SC} = \frac{-20 CC}{SC}$$

Eqs. 10.9P, 10.9Q and 10.9S give us

$$AC = -0.45$$

$$PC = -0.30$$

$$LchC = -0.16$$

and by substituting the above into Eqs. 10.9H, 10.9J and 10.9L, we get the following expressions for the third order astigmatism, distortion, and lateral color with the stop as defined by Q above.

$$AC^* = -0.45 + 40\,(CC)\,Q + (SC)\,Q^2$$

$$DC^* = -0.0825\,Q + 3\,(CC)\,Q^2 + 0.05\,(SC)\,Q^3$$

$$TchC^* = .008\,Q$$

We now select several values for C_1 and evaluate the third order aberrations for each. The results are indicated in the tabulation of Fig. 12.1 and the graph of Fig. 12.2. Note that $X_s = PC^* + AC^*$ and $X_t = PC^* + 3\,AC^*$ (to third order accuracy).

A study of Fig. 12.2 can be quite rewarding. First, we note that there are two regions which appear most promising, namely the meniscus shapes at either side of the graph. On the left, the lens is concave to the incident light and (since Q is positive) the stop is in front of the lens. To the right the lens is convex to the incident light and the stop is behind the lens. Both forms have more under-corrected spherical aberration than the less strongly bent shapes, but both have their field curvature "artificially" flattened by over-corrected astigmatism. Note that the form with the least spherical aberration (where $CC = 0$ and the stop is in contact) has the most strongly inward curving field. This is characteristic of any thin optical system with the stop in contact, since by Eqs. 10.9Q and 10.9H

$$\text{Stop in contact } X_T = PC^* + 3\,AC^* = \frac{-h^2 \phi (3N + 1)}{2N}$$

C_1	-0.4	-0.2	0.0	$+0.2$	$+0.4$	$+0.6$	$+0.8$
ΣSC	$-.98$	$-.43$	$-.11$	$-.03$	$-.18$	$-.56$	-1.18
ΣCC	$+.036$	$+.024$	$+.011$	$-.001$	$-.014$	$-.026$	$-.039$
Q	$+.74$	$+1.11$	$+2.00$	$-.86$	-1.53	$-.93$	$-.66$
l_p	-1.23	-1.84	-3.33	$+1.43$	$+2.55$	$+1.56$	$+1.26$
ΣAC^*	$+.087$	$+.077$	0.00	$-.429$	$-.028$	$+.040$	$+.059$
X_s	$-.21$	$-.22$	$-.30$	$-.73$	$-.33$	$-.26$	$-.24$
X_t	$-.04$	$-.07$	$-.30$	-1.59	$-.38$	$-.18$	$-.12$
ΣDC	$-.02$	$-.03$	$-.08$	$+.07$	$+.06$	$+.03$	$+.02$
% Dist.	-0.7%	-1.1%	-2.5%	$+2.3\%$	$+2.1\%$	$+1.0\%$	$+0.7\%$
$\Sigma Tch\,C$	$-.006$	$-.009$	$-.016$	$+.007$	$+.012$	$+.007$	$+.005$

FIG. 12.1. Tabulation of the third order aberrations of a thin lens with the stop at the coma-free position, for various values of C_1.

Selecting the bending $C_1 = -0.2$ for further investigation, we note that $Q = +1.11$ (from Fig. 12.1). Since $Q = y_p/y$ and $y = 0.5$ we find $y_p = 0.555$. The slope of the principal ray in object space which will yield an image height $h = +3$ with a focal length of $+10$ is $u_p = -0.3$. The stop position is thus

$$l_p = \frac{y_p}{u_p} = \frac{0.555}{-0.3} = -1.85$$

or 1.85 units to the left of the lens.

We must of course convert our thin lens to a real lens. A ray with a slope of -0.3 through the upper edge of the stop (diameter = 1.0) will strike the lens at a height of 1.05, and we shall assume a diameter of twice this for the lens. We determine the curvature of the second surface from $C_2 = C_1 - C = -0.2 - 0.2 = -0.4$, and compute the sagittal heights of the surfaces for the diameter of 2.10. Thus for our lens to have an edge thickness of 0.1, it must have a center thickness of C.T. = E.T. + SH_1 − SH_2 = 0.1 − 0.11 + 0.23 = 0.22. We now trace an oblique fan of four equally spaced meridional rays through the system and calculate two values of coma (by Eq. 10.7I), one from the upper three rays and one from the lower three. By linear interpolation between the two overlapping three ray bundles, we find that a bundle with a chief ray axial

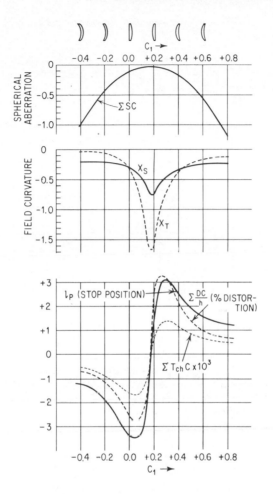

FIG. 12.2. The third order aberrations of a thin lens ($f = 10$, $y = 0.5$, $h = 3$, $N = 1.5$) with the stop at the coma-free position, plotted as a function of the curvature of the first surface (C_1).

intorcopt of $L_{pr} - -1.004$ will have zero coma. This is the stop position for the thick lens.

The results of a ray trace analysis are shown in Fig. 12.3. The aberrations forecast by the thin lens third order computations are shown as circled points, and the agreement with the actual ray trace is quite good. Note that complete T.O.A. plots could be derived from our knowledge of the manner in which the T.O.A. vary with aperture and image height (see the tabulation of Fig. 3.15). For example, knowing that third order spherical varies as Y^2, and that $SC = -.429$ for $Y = 0.5$, we could determine that $SC = -.107$ for $Y = 0.25$ and plot it accordingly.

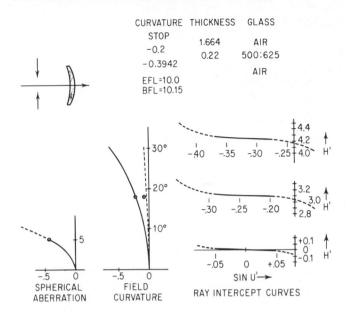

FIG. 12.3. The aberrations of a rear meniscus camera lens. The circled points indicate the aberrations predicted by the thin third order aberration equations (G-sums).

To complete the design we would next scale the entire system to the actual focal length desired. (Note that all the linear dimensions of any system, including the aberrations, may be multiplied by the same constant to effect a change in scale. No additional computation is necessary.) Next an appropriate size for the aperture would be selected, i.e. one which would reduce the aberration blurs to sizes commensurate with the intended application.

At the start of this section we assumed that the lens would be thin and its surfaces spherical. If we increase the thickness of a meniscus lens and maintain its focal length at a constant value by adjusting one of the radii, it is apparent from the thick lens focal length equation (Eq. 2.28) that we must either reduce the power of the convex surface or increase the power of the concave surface to maintain the focal length as the thickness is increased. Either change will have the effect of reducing the inward Petzval curvature of field. This principle (i.e. separation of positive and negative surfaces, elements or components in order to reduce the Petzval sum) is a powerful one and is the basis of all anastigmat designs.

The value of aspheric surfaces is limited in a design as simple as the box camera lens. However, if the lens is moulded from plastic, an aspheric surface is as easy to produce as a spherical one; many simple cameras now have aspheric plastic objectives. The aspheric surface affords the designer additional freedom to modify the system to advantage.

12.3 The Symmetrical Principle

In an optical system which is *completely* symmetrical, coma, distortion and lateral color are identically zero. To have complete symmetry a system must operate at unit magnification and the elements behind the stop must be mirror images of those ahead of the stop. This is a principle of great utility, not only for systems working at unit power, but even for systems working at infinite conjugates. This is due to the fact that, although coma, distortion and lateral color are not completely eliminated under these conditions, they tend to be drastically reduced when the elements of any system are made symmetrical, or approximately so. For this reason most lenses which cover an appreciable field with low distortion and low coma tend to be generally symmetrical in construction.

If we were to apply this principle to the meniscus camera lens, we would simply use two identical menisci equidistant on either side of the stop. The resulting lens would be practically free of coma, distortion and lateral color. The Periscopic lens, shown in Fig. 12.4, makes use of this principle. Symmetry, plus the thick meniscus principle (to flatten the field) achieves a very remarkable

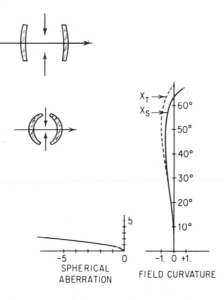

FIG. 12.4. Symmetrical (simple) Meniscus Lenses. The upper sketch shows a periscopic type lens composed of two identical meniscus lenses. The lower sketch shows the Hypergon (U.S. Patent 706,650-1902) whose nearly concentric construction allows it to cover a total field of 135° at $f/30$. The inner and outer radii of the Hypergon differ by only one half per cent, producing a very flat Petzval curvature. Aberrations for a focal length of 100.

astigmatic field coverage of $\pm 67°$ for the Hypergon lens, which is also shown in Fig. 12.4. This is accomplished at the expense of a heavily undercorrected spherical aberration which limits its useful speed to about $f/30$.

12.4 Achromatic Objectives (Thin Lens Theory)

An achromatic doublet is composed of two elements, a positive crown glass element and a negative flint glass element. (Stated more generally, an achromatic doublet consists of a low relative dispersion element of the same sign power as the doublet and a high relative dispersion element of opposite sign.) As degrees of freedom we have the choice of glass types for the elements, the powers of the two elements, and the shapes of the two elements.

Since the lens is to be free of chromatic aberration, we must assign the element powers to the determination of focal length and the control of chromatic aberration. Again we begin by using the thin lens third order aberration equations; assigning the subscripts a and b to the two elements, Eq. 10.9S gives us

$$\Sigma LchC = LchC_a + LchC_b = \frac{-Y_a^2 \phi_a}{V_a u_k'^2} + \frac{-Y_b^2 \phi_b}{V_b u_k'^2}$$

Since the elements are to be cemented together or very nearly in contact, we can substitute $y_a = y_b = y$ and $u_k' = y/f$ to get

$$\Sigma LchC = -f^2 \left[\frac{\phi_a}{V_a} + \frac{\phi_b}{V_b} \right] \tag{12.1}$$

We now set $\Sigma LchC = 0$ (or some other value, if desired) and make a simultaneous solution of Eq. 12.1 with

$$\frac{1}{f} = \phi_a + \phi_b \tag{12.2}$$

to get the necessary powers for the elements.

$$\phi_a = \frac{V_a}{f(V_a - V_b)} \tag{12.3}$$

$$\phi_b = \frac{V_b}{f(V_b - V_a)} = \frac{-\phi_a V_b}{V_a} \tag{12.4}$$

Having determined ϕ_a and ϕ_b, we can now write thin lens equations for the third order spherical and coma in terms of the

shapes of the elements (after tracing a marginal (thin lens) paraxial ray to determine the values for u'_k of the combination and v
(or v') for each element). Since the aperture stop will be at the lens,
$Q = 0.0$ and the coma will be given by Eq. 10.9N. After the appropriate substitutions for h, y, $C_a = \phi_a/(N_a - 1)$, $C_b = \phi_b/(N_b - 1)$ and
the G-factors, we arrive at an equation of the following general
form:

$$\Sigma CC = CC_a + CC_b = K_1 C_1 + K_2 + K_3 C_3 + K_4$$
$$= K_1 C_1 + K_3 C_3 + (K_2 + K_4) \qquad (12.5)$$

where C_1 and C_3 are the curvatures of the first surfaces of the
elements (Fig. 12.5) and K_1 through
K_4 are constants. (Note that by using
the alternate form of Eq. 10.9N for
element a, the equation could be written in C_2 and C_3, the curvatures of the
adjacent inner surfaces). Now for any
desired value of ΣCC, we find that

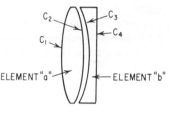

$$C_3 = \frac{\Sigma CC - K_1 C_1 - K_2 - K_4}{K_3}$$

or, combining constants

FIG. 12.5. Achromatic Doublet

$$C_3 = K_5 C_1 + K_6 \qquad (12.6)$$

Thus for any shape of element a, Eq. 12.6 indicates the unique
shape for element b which will give the desired amount of coma.

In similar fashion we can write an expression for the thin lens
third order spherical (using Eq. 10.9M) in the following form:

$$\Sigma SC = SC_a + SC_b =$$
$$K_7 C_1^2 + K_8 C_1 + K_9 + K_{10} C_3^2 + K_{11} C_3 + K_{12} \qquad (12.7)$$

By substituting the value for C_3 from Eq. 12.6 into 12.7, and
combining constants, we get a simple quadratic equation in C_1 of
the form

$$0 = C_1^2 + K_{13} C_1 + K_{14} \qquad (12.8)$$

which can be solved for the value of C_1. When used with the value
of C_3 given by Eq. 12.6, this will yield a doublet with spherical
and coma of the desired amounts. (Note that there may be one, two
or no solutions to Eq. 12.8.)

For a first try, one would use the above procedure with $\Sigma LchC$,
ΣSC, and ΣCC equal to zero (or whatever values are desired). Next,
appropriate thicknesses are inserted, and the system tested by ray
tracing to determine the actual values of spherical, coma (or

O.S.C.) and axial color. If these are not within tolerable limits, the thin lens solution can be repeated using (for the desired $\Sigma L\text{ch}C$, ΣSC, and ΣCC) the negatives of the corresponding values determined by ray tracing. This process converges to a solution very rapidly.

12.5 Achromatic Objectives (Design Forms)

Depending on the choice of glass, the relative aperture, the desired values of the aberrations and also on which solution to the quadratic was selected, the procedure outlined in Section 12.4 will result in an objective with one of the forms sketched in Fig. 12.6. In general the edge contact form and, for lenses of modest (up to 2 or 3 inches) diameter, the cemented form are preferred, primarily because the relationship between the elements (as regards mutual concentricity about the axis and lack of tilt) can be more accurately maintained in fabrication. The crown-in-front forms are more commonly used because the front element is more frequently exposed to the rigors of weather; crown glasses are in general more resistant to weathering than flint glasses.

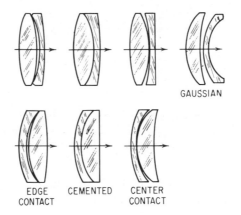

FIG. 12.6. Various forms of achromatic doublets. The upper row are crown-in-front doublets and the lower row are flint-in-front. The curvatures are exaggerated for clarity. The center contact form is usually avoided because it is more difficult to manufacture. The shapes indicated are for lenses corrected for a distant object to the left.

If one followed the procedure of Section 12.4, a design resulting in a cemented doublet (i.e. $C_2 = C_3$) would be a lucky accident. When a cemented interface is necessary, an alternate procedure is followed. The spherical and coma contribution equations are written in C_2 and C_3 (instead of C_1 and C_3) and C_2 is set equal to C_3, resulting in equations in C_2 (or C_3) which may then be solved for either the desired coma or spherical. If these equations are plotted as a function of the shape of the doublet (i.e. vs. C_1 or C_2 or C_4) the resulting graph will look like one of those in Fig. 12.7, in which ΣSC is a parabola and ΣCC is a straight line. In the left plot there is no solution for spherical, in the center plot the solutions for spherical and coma occur at the same bending, and on the right there are two possible solutions for spherical with equal and opposite signed amounts of coma, and often with pronounced meniscus shapes. (These latter solutions are valuable if one desires to utilize the doublets in a symmetrical combination about a central stop; the coma can then be used to reduce or overcorrect the astigmatism.) The exact form obtained is dependent primarily on the types of glass chosen. In general, the spherical aberration parabola can be raised by selecting a new flint glass with a lower index and higher V-value, or by selecting a new crown with a higher index and lower V. Thus the strongly meniscus solutions result from a glass pair with a small difference in V-value. Results approximating those in the middle graph of Fig. 12.7 can be obtained with BSC-2 (517:645) and EDF-1 (649:338).

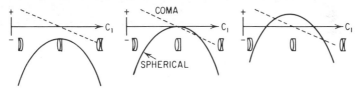

FIG. 12.7. The variation of spherical aberration (solid line) and coma (dashed) as a function of the shape of a cemented achromatic doublet. Depending on the materials used there may be two forms with zero spherical (right), one form (center) or no form (left). The center graph is the preferred type since spherical and coma are both corrected.

Figure 12.8 shows the spherical aberration and the spherochromatism of a typical cemented doublet. As previously noted, the field curvature of a thin system with stop in contact is strongly inward and cannot be modified unless the stop is shifted. Thus, systems of this type are limited to applications which require good imagery over relatively small fields (a few degrees from the axis).

It is occasionally desirable to produce a doublet objective with both the zonal and marginal spherical simultaneously corrected. This can be accomplished by using the air space of a broken contact doublet as an added degree of freedom. The design is begun exactly as in Section 12.4, except that two (or more) thick lens solutions are derived, one with a minimum air space and the other(s) with an increased space. The calculated zonal spherical is then plotted against the size of the air space, and the air space with LA_z equal to zero is selected; this form will usually have no zonal OSC. Speeds of $f/6$ or $f/7$ can be attained with practically no spherical or axial coma over the entire aperture. Good glass choices are a Light Barium Crown combined with either a Dense Flint or an Extra Dense Flint; either crown-in-front or flint-in-front forms are possible. In this type of lens the residual axial aberration consists almost solely of Secondary Spectrum.

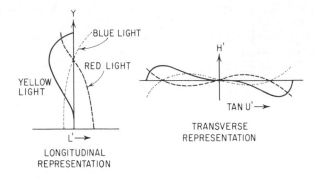

FIG. 12.8. The spherical aberration and spherochromatism of an achromatic doublet. Note that the chromatic is shown corrected at the 0.7 zone. This is good practice for small amounts of aberration, but results in a blue flare or halo when the spherochromatism is large. For this reason the chromatic zone of correction is frequently shifted toward the margin. The residuals as shown (i.e. undercorrected zonal and blue overcorrected more than red) are typical of almost all optical systems.

Spherochromatism, which is the variation of spherical aberration as a function of wavelength, can be corrected by a change in the spacing between elements (or components) which differ in the sign of their contributions to spherical and chromatic aberration. This general principle may be applied to the doublet achromat in a manner paralleling the use of the air space to correct zonal spherical; indeed, the basic principle is the same for both aberrations. Spherochromatism in a thin doublet arises from the fact that the chromatic overcorrection at the contact surfaces is disproportionately larger at the margin than near the axis. This is because of the larger angles of incidence at the margin. Thus if the chromatic is corrected at the axis, it is overcorrected at the margin; this is another way of saying that the spherical is more overcorrected for short wavelengths than long. Now, if the airspace between elements is increased, as indicated in Fig. 12.9, the blue marginal ray, having been refracted more strongly by the crown element, will strike the flint element at a lower height than will the red ray. Thus the refraction of the blue ray at the flint will be lessened relative to the red, and its overcorrection reduced accordingly. A very similar argument can be applied to the reduction of an undercorrected zonal spherical (which is caused by an overcorrected fifth order spherical) by use of an increased air space. *Both principles are applicable to more complex lenses as well.*

One method of effecting a simultaneous elimination of both spherochromatism and zonal spherical is indicated in Fig. 12.9C. The doublet plus singlet configuration introduces still another degree of freedom, namely the balance of positive (crown) power between the two components, which can be used with the air space to bring about the correction. The air spaced triplet shown in 12.9D is also capable of good correction, but is more difficult to manufacture.

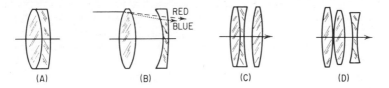

FIG. 12.9. The ordinary spherochromatism of a doublet can be corrected by increasing the the air space (shown highly exaggerated in B). This reduces the height at which the blue ray strikes the flint by a greater amount than for the red ray, thus reducing the overcorrection of the marginal blue ray. Sketches C) and D) show triplet forms which can be used to correct spherochromatism and spherical zonal residuals simultaneously.

The secondary spectrum contribution of a thin lens is given by Eq. 10.9U; combining this with the requirements for achromatism (Eqs. 12.3 and 12.4) we find that the secondary spectrum of a thin achromatic doublet is given by

$$SS = \frac{f(P_b - P_a)}{(V_a - V_b)} = \frac{-f\,\Delta P}{\Delta V} \tag{12.9}$$

For the ordinary glass combinations used in doublets the ratio $\Delta P/\Delta V$ is essentially constant, and the visual secondary spectrum is about 0.0004 to 0.0005 of the focal length. Similarly, the secondary spectrum of an achromatized combination of two separated components can be shown to be

$$SS = \frac{\Delta P}{D\Delta V}\,[f^2 + B(L - 2f)] \tag{12.10}$$

where D is the air space, B the back focus, and $L = B + D$ is the length from front component to the focal point. Again it is apparent that the ratio $\Delta P/\Delta V$ is the governing factor. Note that in this case the secondary color of two spaced positive lenses is less than that of a thin doublet of the same focal length; conversely the secondary color of a telephoto lens (positive front component, negative rear component) is greater than the corresponding thin doublet.

There are a few glasses which will reduce the secondary spectrum, for example, DBC-1 (611:588) used with EK-320 (744:458) or EK-325 (744:456) as the flint element will reduce the visual secondary spectrum to about one third or one tenth of the ordinary value. Note however, that for these glasses $V_a - V_b$ is about 13, and the powers of the individual elements required for achromatism are much higher than with an ordinary pair of glasses (in which the value of $Va - Vb$ is about 30). This increase in element power causes a corresponding increase in the other residual aberrations. There are a few other glasses, notably the Schott K_z FS glasses, which have unusual partial dispersions, but in general these glasses work poorly in the shop and lack chemical stability.

As mentioned in Chapter 7, calcium fluoride (CaF_2—Fluorite) may be combined with an ordinary glass (selected so that $Pa = Pb$) to make an achromat that is essentially free of secondary spectrum. It is also worth noting that there are no glass pairs which will form a useful achromat in the 1.0 to 1.5 micron spectral band; fluorite can be combined with a suitable glass to make an achromat for this region. Silicon and germanium are useful for longer wavelengths, as are BaF_2 and CaF_2.

Hertzberger* has developed a design technique which makes use of a triplet combination of three different glasses to achieve a lens in which four wavelengths are brought to a common focus and which maintains good spherical correction. He has termed this type of system a "Superachromat" to distinguish it from the older designation of "Apochromat" which implied color correction in three wavelengths (and spherical correction at two). Previously taught techniques for producing an apochromat from three different glasses were practically worthless, since the resulting lenses, although corrected for three wavelengths, usually had large residual aberrations (including a tertiary residual chromatic, often almost as large as the ordinary secondary spectrum).

Although a knowledge of the thin lens theory outlined in Section 12.4 is of inestimable value to a designer, the approach set forth there is seldom used when an electronic computer is available. For a manually operated (i.e. non-"automatic") computer, the designer may start with the element powers indicated by Eqs. 12.3 and 12.4, but he will usually make an educated guess at an appropriate starting shape. Then a series of bendings are made; each bending is adjusted so that the chromatic and over-all power are precisely as desired (by adjusting the relative powers of the elements slightly). Each bending is analyzed by ray tracing (or perhaps by use of the third order surface contributions) and a suitable form is rapidly derived (if one exists). This is admittedly an inelegant approach, but the high speed of the computer makes it more efficient in terms of the designer's efforts.

12.6 The Cooke Triplet Anastigmat

The Cooke triplet is composed of two outer positive crown elements and an inner flint element, with relatively large airspaces

---

*See Hertzberger & McClure "The Design of Superachromatic Lenses" Applied Optics, Vol. 2, 553-560, June 1963.

separating the elements. This type of lens is especially interesting because there are just enough available degrees of freedom to allow the designer to correct the primary aberrations. The basic principle used to flatten the field curvature (i.e., the Petzval Sum) is quite simple: the contribution that an element makes to the power of a system is proportional to $y\phi$, and the contribution to the chromatic varies with $y^2\phi$. However, the contribution to the Petzval curvature is a function of ϕ alone, and is independent of y. Now in a thin (compact) system, all the elements have essentially the same value of y and the powers of the elements are determined by the requirements of focal length and chromatic correction; consequently, the Petzval radius of a thin doublet rarely exceeds 1.5 or 2 times the focal length. However, when the negative elements of a system are spaced away from the positive elements (so that the ray height y at the negative elements is reduced), the power of the negative elements must be increased to maintain the focal length and chromatic correction of the system. As a result, their overcorrecting contribution to the Petzval curvature is increased. Thus, by the proper choice of spacing, the Petzval radius can be lengthened to several times the system focal length and the field proportionately "flattened."

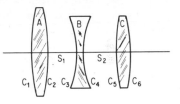

FIG. 12.10. The Cooke Triplet Anastigmat.

From Fig. 12.10, which shows a schematic triplet, we can determine the available degrees of freedom. They are:

1. Three powers (ϕ_a, ϕ_b, ϕ_c)

2. Two spaces (S_1, S_2)

3. Three shapes (C_1, C_2, C_3)

4. Glass choice

5. Thicknesses

Of these, Items 1, 2 and 3 will be of immediate interest; they total eight variables. Item 4, glass choice, is an important tool, but we will reserve its discussion until later. Item 5, element thickness, is only marginally effective; in regard to the primary corrections, its effect overlaps that of the spacings.

With these eight degrees of freedom, the designer wishes to correct (or control) the following primary characteristics and aberrations.

1. Focal length

2. Axial (longitudinal) chromatic aberration

3. Lateral chromatic aberration

4. Petzval curvature

5. Spherical aberration

6. Coma

7. Astigmatism

8. Distortion

Thus, there are just the necessary eight degrees of freedom to control the eight primary corrections.

Power and Spacing Solution: The first four items listed immediately above can be seen (by reference to the thin lens third order aberration equations) to be functions of element power and ray height (which is a function of spacing); they are independent of element shape. Thus, it is necessary that the powers and spaces be chosen to satisfy these four conditions, which may be expressed as follows:

$$\text{Desired} \ \ \Phi \ = \ \frac{1}{y_a} \ \sum y\phi \tag{12.11}$$

$$\text{Desired} \ \ \Sigma LchC \ = \ -\frac{1}{u_k'^2} \ \sum \frac{y^2\phi}{V} \tag{12.12}$$

$$\text{Desired} \ \ \Sigma TchC^* \ = \ -\frac{1}{u_k'} \ \sum \frac{y\,y_p\,\phi}{V} \tag{12.13}$$

$$\text{Desired} \ \ \Sigma PC \ = \ -\frac{h^2}{2} \ \sum \frac{\phi}{N} \tag{12.14}$$

where the summation is over the three elements. These expressions are essentially the same as those of Section 10.9 and the meanings of the symbols are given there.

The four conditions above must be satisfied by the choice of five variables (three powers plus two spacings). There is one more variable than necessary; this "extra" is utilized in a later step to control one of the remaining aberrations (usually spherical or distortion). There are almost as many ways of solving this set of equations as there are designers. Stephens* has worked out the

---

*R. E. Stephens, J. Opt. Soc. Am., V38, p1032 (1948)

algebraic solution for the triplet, and his paper gives explicit equations for the values of the powers and spaces. An iterative approximation technique (which may be easily modified to apply to systems with more than three components) along the following lines is an alternate method.

1. Assume a value for the ratio of the powers of elements c and a. This will be the "extra" degree of freedom mentioned above. ($K = \phi_c/\phi_a = 1.2$ is a typical value.)

2. Choose a value (arbitrary) for ϕ_a. (In the absence of prior experience, $\phi_a = 1.5\ \Phi$ is suitable.) This determines ϕ_c, since from step 1, $\phi_c = K\phi_a$ and also determines ϕ_b, since Eq. 12.14 can be solved for ϕ_b when ϕ_a, ϕ_c, h and ΣPC are known or assumed.

3. Choose a value for S_1 (one-fifth to one-tenth of the focal length is suitable).

4. Solve for the value of S_2 which will satisfy Eq. 12.12 (assume that u'_k is equal to Φy_a.) This is done by tracing a ray through elements a and b to determine y_a, y_b and u'_b. Then find S_2 to yield the value of y_c which satisfies Eq. 12.12.

5. Trace a principal ray (thin lens paraxial) through the desired stop position, which may be conveniently placed at element b to minimize the labor. Again assume u'_k as in 4 and determine $\Sigma TchC^*$.

6. Repeat from step 3 with a new choice for S_1 until $\Sigma TchC^*$ is as desired. (As a second guess for S_1, try the average of S_1 and S_2 from the first try.)

7. Determine the system power Φ. If not as desired, scale the value of ϕ_a used in step 2 and repeat from step 2 until a solution is obtained.

Graphs of the relationships between S_1 and $\Sigma TchC^*$ and between ϕ_a and Φ are useful in steps 6 and 7.

Element Shape Solution: When the element powers and spacings have been determined, there are three unused degrees of freedom, namely the shapes of the three elements (plus the "extra", K, mentioned in step 1 above). These variables must be adjusted so that the spherical, coma, astigmatism and distortion are corrected to their desired values. Referring to the thin lens contribution equations of Section 10.9, the aberrations can be seen to be quadratic functions of the element shapes; thus, a simultaneous algebraic solution cannot be used and some sort of successive approximation procedure is necessary.

Thin lens marginal and principal rays are traced through the three elements. The principal ray is traced so that the aperture stop is at lens b; both y_p and Q for lens b will be zero.

1. Assume an (arbitrary) value for C_1 and calculate AC_a^* for element a by Eq. 10.9H (a value of $C_1 = 2.5\Phi$ is a reasonable first choice).

2. Since the stop is located at element b, AC_b is a constant (Eq. 10.9P). Now solve Eq. 10.9H for the shape of element c, that is, the value of C_5, which will give AC_c^* which will yield the desired ΣAC when combined with AC_a^* and AC_b. Normally there are two solutions for C_5 and the most reasonable one is used.

3. Now CC_a^* and CC_c^* are calculated from Eq. 10.9G. Since the equation for CC_b is linear in C_3 (Eq. 10.9N, since $Q_b = 0$), it can be solved for the unique value of C_3 which will yield the desired ΣCC^*.

4. The value of ΣSC is now determined from Eq. 10.9

5. The procedure is repeated from step 1 with a new value of C_1, and a graph of ΣSC is plotted against C_1. The shape (C_1) for which ΣSC is equal to the desired value is chosen and the corresponding values of C_3 and C_5 are determined so that ΣSC, ΣAC^* and ΣCC^* are simultaneously as desired.

6. If ΣDC^* is within acceptable limits, well and good; if not, a new power and space solution must be made with a different value of $K = \phi_c/\phi_a$. The value of ΣDC^* can be plotted for several values of K as an aid in effecting a solution.

Note that in step 5, there may be two, one, or no solutions for the desired ΣSC. The best triplets seem to result from cases where the plot of ΣSC just barely reaches the desired level.

The next step is the addition of thickness to the design. Center thicknesses for the crown elements are chosen to give workable edge thicknesses; the second surface curvatures $(C_2, C_4$ and $C_6)$ are adjusted to hold the thick element powers exactly to the thin lens powers. Air spaces are chosen so that the principal points of the elements are spaced apart by the thin lens spacings. In this way, the thick lens triplet will have exactly the same focal length as the thin lens version.

The thick lens is now submitted to a trigonometric ray trace analysis and the values of the seven primary aberrations are determined. If (as is likely) the aberrations are not as desired, a new round of design is initiated, with the new "desired" thin lens aberration values adjusted to offset the difference between the ray tracing results and the desired final values. For example, if the original "desired" ΣSC was -0.2 and the ray tracing yielded a marginal spherical, $LA_m = +0.2$, the new "desired" ΣSC would be set at -0.4, assuming that the desired end result was $LA_m = 0.0$.

Initial Choice of Desired Aberration Values: In general, the initial choice for the "desired" third order aberration sums should be small, undercorrected amounts, since the higher order aberrations are usually overcorrecting. Spherical, coma and axial chromatic follow this rule. Since the Cooke triplet is relatively symmetrical, the residuals of distortion and lateral color are small, and initial "desired" values of zero are appropriate. The "desired" Petzval sum should be definitely negative. For high speed lenses, the Petzval radius is frequently as short as two or three times the focal length; moderate aperture systems ($f/3.5$) usually have $\rho = -3f$ to $-4f$; slow systems may have $\rho = -5f$ or longer. One reason for this relationship is that the flatter (less undercorrected) the Petzval surface is made, the higher the element powers; hence the higher the residual aberrations, especially zonal spherical. The value chosen for the desired ΣPC is also an important factor in determining whether or not there is a solution for step 5 in the curvature determination process. The "desired" astigmatism sum is best set slightly positive, about one-third the absolute value of the Petzval sum, so that the inward curvature of the Petzval surface is offset.

Glass Choice: The choice of the glass to be used in the triplet is one of the most important design factors. From field (Petzval) curvature considerations, it is desirable that the positive elements have a high index of refraction and the negative element a low one. As usual, the V-value of the positive elements should be high and that of the negative element low in order to effect chromatic correction. For the positive elements, one of the dense barium crowns is the usual choice, although the light barium crowns on one hand and the rare earth glasses on the other are frequently used. Although triplet designs are possible with ordinary crown glass or even plastics, their performance is relatively poor.

It turns out that the inter-related requirements of Eqs. 12.11 through 12.14 lead to long systems (i.e. S_1 and S_2 are large) when the difference between the V-values of the positive and negative elements is large. A lens with a large vertex length will, at any given diameter, vignette at a smaller angle than will a short lens. Further, the longer the lens, 1) the smaller the spherical zonal and 2) the smaller the field coverage (that is, the higher order astigmatism and coma are greater and limit the angle over which a good image can be obtained when the lens is long). Thus, long systems are appropriate for high speed—small angle systems; short systems for small aperture—wide angle applications. As a very rough rule of thumb, the vertex length of a triplet is frequently equal to the diameter of the front element.

The length of the triplet can be controlled by the choice of the glasses used. For example, if a shorter system is desired, the substitution of a flint with a higher V-value (or a crown with a lower V-value) will produce the necessary change. To get a longer system, use a higher *V*-value crown and/or a lower *V*-value flint. (However, note that a system which is too long will have no solution for the curvatures. The ray height on the negative element may be so low that its overcorrecting contribution to the spherical aberration is insufficient to offset the undercorrection of the positive elements simultaneously with the requirements for coma and astigmatism correction.)

Interestingly enough, *the relationship between vertex length and zonal spherical and field coverage is a general one and applies to most anastigmats.* Thus, if an anastigmat design has too much zonal spherical and more than enough coverage, one can simply choose new glasses to lengthen the system and strike the desired balance between field and aperture, or vice-versa. There are, of course, limits to the effectiveness of this technique.

In general, the higher the index of the crown (positive) elements and the lower the index of the flint, the better the design will be. In other words, with all else equal, a triplet with a greater index difference will have a smaller zonal spherical and/or a wider field coverage.

Figure 10.5 showed a triplet of relatively high aperture and modest field coverage. Figures 12.11 and 12.12 illustrate triplets of increasingly smaller aperture and wider fields of view.

RADIUS	THICKNESS	GLASS
1. +40.1		
	6.0	613:585
2. -537.0		
	10.0	AIR
3. -47.0		
	1.0	621:362
4. +40.0		
	10.8	AIR
5. +234.5		
	6.0	613:585
6. -37.9		

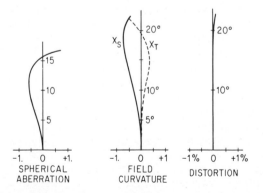

FIG. 12.11. A Cooke triplet anastigmat of moderate aperture and coverage. Compare with Figs. 10.5 and 12.12. English Patent 155,640-1919. Focal length is 100 units. This design is of the type made for use in slide projectors.

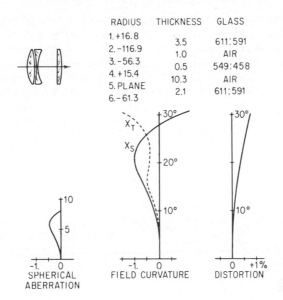

RADIUS	THICKNESS	GLASS
1. +16.8	3.5	611:591
2. -116.9	1.0	AIR
3. -56.3	0.5	549:458
4. +15.4	10.3	AIR
5. PLANE	2.1	611:591
6. -61.3		

SPHERICAL ABERRATION

FIELD CURVATURE

DISTORTION

FIG. 12.12. A Cooke triplet of small aperture and wide coverage. Compare with Figs. 10-5 and 12.11. German Patent 287,089-1913. Focal length is 100 units.

12.7 A Generalized Design Technique

The preceding sections have outlined specific design approaches for three particular types of optical systems. This section will describe a generalized approach to optical design. Because of the varied nature of different types of optical systems, this description will be unnecessarily elaborate for many simple cases and must, because of limitations of space and knowledge, fall short of completeness for elaborate and specialized systems. The reader will recognize generalizations of most of the procedures set forth in the preceding sections.

General Considerations: The first step in the design process is the organization of the requirements to be imposed on the optical system into terms of aperture, focal length, field coverage, resolution (or blur spot size), spectral band width, transmission, mechanical limitations, and the like. In elaborate systems, e.g., telescopes, design may be profitably carried out member by member; if this is to be done, requirements for the individual members are established.

The general configuration of the system is established next. Ordinarily the designer will conduct a survey of known designs (books, technical periodicals, the patent files, and the designer's own experience are the primary sources) to determine whether the performance requirements are within the "state of the art." If so, the designer will select a type of system which is just capable of meeting the requirements (i.e. the most "economical" choice) and will proceed to adjust its parameters to achieve the optimum balance of correction for the particular application at hand. If the performance requirements are beyond the capability of any known design, the designer will select a design form which he feels is "most likely to succeed". He will analyze it thoroughly to determine its shortcomings and then attempt to improve its characteristics. In many instances, there is no directly applicable prior art upon which to base further effort. In such circumstances, a thorough analysis of the first order (Gaussian) requirements is conducted and a system is invented, utilizing the basic design principles exemplified in known designs on a piecemeal basis to accomplish the necessary ends.

In the following paragraphs, we assume that the general design type has been established, either by selection or invention. The next major step in the design process is the correction of the primary aberrations, or at least as many of them as are necessary and feasible. This procedure has been accurately described as the art of solving a number (say n) of second (or higher) order equations in m unknowns; one must ascertain initially that m (the number of *effective* variables) exceeds or equals n (the number of aberrations).

Correction of the Primary Aberrations: The powers and spacings of the elements which are to comprise the system can usually be determined on a highly rational basis. First, the elements must be so arranged as to provide the desired focal length, aperture, field, and so on for the system. Throughout this entire design stage, the value of an accurate scale drawing cannot be overemphasized; such a drawing will prevent attempts to design impossible elements, such as those with negative edge thickness, or with hyper-hemispheric negative surfaces. If the ray paths are roughed in (on a first order basis), one is also less apt to require magnifications and apertures which lead to slope or incidence angles which exceed 90°.

The usual method for correction of the aberrations is bending the elements, that is changing the shape of the elements while maintaining a constant power and position. However, certain aberrations are unaffected (or affected only slightly) by bending. These are axial (longitudinal) chromatic aberration, lateral color, Petzval curvature and, to a certain extent, distortion. The chromatic aberrations and the Petzval curvature must be corrected in the original layout if they are ever to be corrected. The third order contribution equations, especially the thin lens versions, are most useful at this stage, and it is ordinarily a relatively straightforward procedure to adjust the system so that the $\Sigma Lch\,C$, $\Sigma Tch\,C^*$ and ΣPC are equal to values which have been selected as desirable (or at least acceptable).

It then remains to correct the spherical, coma, astigmatism, and distortion to their desired values. A number of alternate procedures are often available at this step. Unless the designer has prior experience with the type of system under construction, or unless the design effort is a minor modification of an existing design, it is probably best at this stage to make a graph of the aberration contributions from each element (or component) as a function of the element shape. Then, from a set of such graphs, a region (or regions) in which a solution is possible can be selected. These graphs can be plotted from data obtained by the use of the thin lens contribution equations, the surface contribution equations, or in some cases by direct ray tracing. The latter two methods are appropriate for electronic computer work; one element at a time is bent and the changes in the final aberrations are plotted. The thin lens expressions have the advantage that the "n equations in m unknowns" are explicitly that, and can be handled analytically.

When a "region of solution" is selected (by whatever means), a method of differential correction is usually applied. The partial differentials of the aberrations against shape, $\partial A/\partial C$ (or $\Delta A/\Delta C$), are computed, along with the values of the aberrations for a trial prescription. The desired amount of change of each aberration (ΔA) is determined from the anlysis of a trial prescription and the necessary number (n) of simultaneous equations of the general form

$$\Delta A_n = \sum_{i=1}^{i=K} \left(\frac{\partial A_n}{\partial C}\right)_i \Delta C_i \qquad (12.15)$$

are set up and solved to yield the required values of ΔC_i. Because of the non-linearity of the equations (i.e. the partials vary as the shape is changed), the first solution is seldom precise. However, the pre-selection of the "region of solution" limits the size of the ΔC's so that the linear simultaneous solution of Eqs. 12.15 is a good approximation; a series of such solutions converges rapidly on the required design form.

It is sometimes advisable to limit the number of parameters used in the technique described above. Because a limited number of aberrations are to be controlled, the problem is simplified if only an equal number of variables are used, *provided that these variables are effective and admit of a solution.* The preliminary graphs of the aberrations (vs. element shapes) and the subsequent selection of a "region of solution" are strongly recommended as insurance against ineffective parameters and insoluble sets of simultaneous equations.

Certain systems lend themselves to an iterative technique which can be a powerful design tool. For example, assume that three aberrations, A, B and C are to be corrected by the adjustment of three parameters, x, y and z. An initial trial prescription is modified by changing one of the parameters, say z, until one of the aberrations, say C, is "corrected". Then parameter y is arbitrarily changed and a new value of z is determined to maintain the correction of C. Parameter y is varied in this manner until the aberrations B and C are simultaneously corrected. Then parameter x is changed; with each change of x, parameters y and z are adjusted as above to hold the aberrations B and C at the desired values. Parameter x is varied in this manner until aberration A is brought to correction simultaneously with B and C. In such a process, graphs of C vs. z, B vs. y, and A vs. x are quite useful.

If the thin lens aberration expressions have been used in any of the preceding steps, it is necessary to add thickness to the elements. This is generally done by adjusting the secondary curvature of each thick element to hold the thick element power equal to the thin lens element power. The spacing between elements is then adjusted so that the spacing of the thick element principal points is equal to the thin lens spacing. This method serves to retain the over-all system power and working distances at the same values as the thin lens systems. Some designers prefer to adjust the secondary curvatures to maintain the Petzval curvature precisely. The exact procedure used to go from thin to thick is not critical; what may be important is that the procedure of introducing thickness be rigorously consistent (in order that the differential trigonometric correction method will be accurate).

Trigonometric Correction: When the third order aberrations have been brought to desired values, it is necessary to trace rays trigonometrically to determine the actual state of correction of the system. It will usually differ by a small amount from that predicted by the third order aberrations. However, a step or two of differential correction as outlined five paragraphs above will usually bring the trigonometric correction home; in most systems, the change in the trigonometrically determined aberrations is quite close to the change predicted by third order aberration calculations.

Reduction of Residual Aberrations: After the primary aberrations have been brought to correction, the design is tested for residual aberrations. The primary aberrations are generally corrected for only a single zone of the aperture or field and can be expected to depart from correction in all other zones, as previously discussed in Chapter 3. Several general principles can be given for the reduction of residuals; their variety and extent make a catalog of specific remedies too extensive for inclusion.

If there were any "left over" parameters that were not used in the correction of the primary aberrations, these may be systematically varied and their effects on the residuals noted and used. In addition to the obvious and continuously variable parameters of bendings, powers and spacings, the choice of glass types is often an effective "left over". Also the possibility that more than one "region of solution" exists should not be overlooked, since this is, in effect, an extra parameter.

An analysis of the source of the third order surface contributions will often pinpoint one or two surfaces or elements which are especially heavy contributors. The elimination or reduction of a single large contribution will often reduce residual aberrations. This can be accomplished by introducing a correcting element near the offender (for example, convert a single element into a compound component, perhaps an achromat); or by splitting the offending element into two elements whose total power equals that of the original; or (infrequently) by shifting the offender to a location where the incidence angles of the rays on its surfaces are reduced. Compounding or splitting an element introduces two new variable parameters; the ratio of the powers of the two elements and the shape of the added element. An additional possibility is that a drastically different shape for the troublesome element may reduce its contribution to an acceptable level.

The specific remedies for spherochromatism, zonal spherical and field coverage set forth in Sections 12.5 and 12.6 have fairly general applicability. Another specific is the introduction of a zero power meniscus element or a concentric meniscus element into the system. Depending on how and where it is used, a meniscus can be effective in modifying zonal spherical, Petzval curvature, or astigmatism.

An aspheric surface can be a powerful design tool for the reduction of residuals or the elimination of primary aberrations (especially distortion, astigmatism, and spherical) which will yield to no other design techniques. One should, if at all possible, temper one's enthusiasm for the easy way out which the aspheric surface represents with the knowledge that several spherical elements may be added to a design for less than the cost of producing a single precise aspheric surface. As a consequence of this fact, aspherics are seldom used except where absolutely necessary (as to produce eyepieces with barrel distortion to offset the pincushion distortion inherent in most electronic image tubes), or where cost is no object (as in one-of-a-kind instruments), or where the required precision of the surface is very low (as in moulded condenser elements).

In general, where residuals are a problem, it is wise to reconsider the initial power and space layout for the entire system. It is sometimes possible to revise the layout in such a way that the powers of the elements or the "work" ($y\phi$ or $y_\mu\phi$) done by the elements can be reduced. This is an extremely rapid and effective way of reducing residuals. An initial choice of too small (i.e., too positive) a value for the Petzval sum will result in elements of high power and large residuals. A change to allow a more inward curving field is the obvious remedy for this situation.

Aberration Balancing: The final stage in the optical design process consists of balancing the aberrations, or "touching up" the design. Here the experienced designer frequently departs from what may seem to be the best state of correction in order to minimize the overall effects of the residual aberrations. In the presence of zonal spherical, spherochromatism, and astigmatism, the interrelationships of the aberrations with each other, and with the position selected for the focal plane, often allow an improvement to be made by selecting a deliberately uncorrected state. We have previously (Section 11.3) seen that the best spherical correction as regards OPD occurs when the marginal spherical is zero and the reference plane is shifted toward the zonal focus; the minimum *geometrical* blur spot size (Section 11.7) requires that the marginal spherical be undercorrected. Thus, if the application of the system is such that resolutions well below the limiting resolution are of prime importance, and if the zonal spherical is large in terms of OPD, then an undercorrected marginal spherical is in order. An overcorrected marginal spherical is rarely desirable; it does permit a higher resolution and reduces focus shift when the system is stopped down, but it reduces the image contrast.

Another reason for preferring a slightly undercorrected spherical is that the oblique spherical aberration ($y^3 h^2$) is almost always overcorrected and the axial undercorrection will counterbalance this tendency. The overcorrected oblique spherical also causes the *effective* field curvature to be more backward curving than indicated by the x_s and x_t curves given by Coddington's equations (Eqs. 10.6). This is especially true for the tangential field curvature. For this reason the astigmatism is seldom made overcorrected enough to cause a backward curving tangential field; ordinarily one desires a correction somewhere between $x_t = 0$ and $x_t = x_t = x_p$. Note that the focus position is usually chosen inside the paraxial focal plane and that the field curvature should be judged with this in mind.

We have previously noted that the Petzval curvature in most anastigmats is preferably left somewhat inward curving in order to minimize element powers and aberration contributions.

The obvious choice of the 0.707 zone of the aperture as the zone at which to correct the longitudinal chromatic is rarely the best choice. In the presence of spherochromatism and an undercorrected zonal spherical, the inward shift of the best focus from the paraxial focus allows the overcorrected spherical of the blue light to produce a halo or blue haze in the image. This can be eliminated, or reduced, by correcting the chromatic at a larger zone of the aperture.

The reader should bear in mind that the preceding comments are intended to apply to normal types of lenses in which (as is usually the case) the higher-order residuals are somewhat larger than desirable.

12.8 Automatic Design by Electronic Computer

The fantastically high computation speed of the electronic computer makes it possible to perform a major portion of the optical design task on an "automatic" basis. Two different philosophies have been used with considerable success. One is essentially a duplication of the process that a designer goes through in correcting the primary aberrations of a system. The computer is presented with an initial prescription and a set of desired values for the aberrations. The machine then computes the partial differentials of the aberrations with respect to each parameter (curvature, spacing, etc.) which is to be adjusted and establishes a set of simultaneous equations (Eqs. 12.15), which it then solves for the necessary changes in the parameters. Since this solution is an approximate one, the computer then applies these changes to the prescription and continues to repeat the process until the aberrations are at the desired values. When there are more variable parameters than system characteristics to be controlled, there is no unique solution to the simultaneous equations; in this case, the computer will add another requirement, namely that the sum of the squares of the parameter changes be a minimum. This allows a solution to be found and has the added advantage that it holds the system close to the original prescription. Since the solution of simultaneous equations may call for excessively large changes to be applied, the computer is usually instructed to scale down the changes if they exceed a certain predetermined value.

This "simultaneous" technique is an extremely useful one. Most moderate sized computers are capable of handling this problem without difficulty and several computer manufacturers offer a program of this type, based on third order aberration contributions. Since the designer is in rather close control of the situation, this technique is, in effect, simply an automation of conventional methods. Thus, the designer should have a fairly good knowledge of the system and the system must have a solution reasonably close to the initial prescription. Otherwise, the computer may take off after a non-existent or extreme solution. This type of approach is very efficient for making modest changes in designs or for "touching-up" a design. It also makes easy work of systems with exceedingly complex inter-relationships of the variables, such as the older meniscus anastigmats of the Dagor or Protar type (which are otherwise quite difficult).

There are many other approaches to automatic design; almost all of them are characterized by the use of a "merit function". The merit function is a single numerical value which indicates to the computer whether any given change has improved the lens or not. Obviously, representing the total performance of a lens system by a single number is a rather tricky business and considerable care must be taken in the choice of the merit function; at times it seems that the "design" of the merit function is more demanding than the design of the lens which the merit function is intended to represent. Most approaches use a merit function of the following sort: A large number of rays are traced from each of several points in the field of view. For each image point, the distance of each ray intersection (with the image plane) from the "ideal" location for that ray is computed and the sum of the squares of these distances is taken. Then the sum of the sums for the several points is the merit function. Since the merit function will be large if the image blur spot is large, it is apparent that a small value of the merit function is desirable.

The construction of the merit function as described above is far from the most desirable scheme of things, and in practice many refinements are used. Since the outer portions of the field are frequently less critical than the center, the individual sums may be weighted to take this into account. A modest amount of computation will indicate that, in the presence of a constant fifth order spherical aberration, the smallest value of the sum of the squares of the ray displacements does *not* represent the best solution from an O.P.D. standpoint. One scheme uses a reduced weighting of large ray displacements in an attempt to take this into account. Other programs use an O.P.D. calculation in the merit function. The choice of the "ideal" intersection point for the rays (for off axis points) is a complex matter; the use of the Gaussian image point is quite misleading if any amount of distortion is present. Similarly the use of the image plane intersection of the principal ray as the ideal point can yield a distorted evaluation in the presence of coma. Frequently the separately computed values of distortion and lateral chromatic aberration are added (suitably weighted) into the merit function, and the computer selects the centroid of the blur spot as the "ideal" point. A "hybrid" program, using *both* the "simultaneous" and "merit function" techniques can take advantage of the best characteristics of each.

In any case, the computer is instructed to vary the constructional parameters of the system in such a way that the merit function is reduced. Since there are usually many more ray intersections to be controlled than there are available parameters, a simultaneous solution is out of the question. Generally, a sort of least squares fit is arrived at, usually by a steepest descent technique which tends to change most rapidly those parameters which are most effective in reducing the merit function. A detailed description of the mathematical philosophies and techniques involved is beyond the scope and size of this volume, and the interested reader is referred to the technical journal references in the bibliography at the rear of the book.

Despite the apparent difficulties in establishing a completely automatic design program of this sort, the successful programs are tremendously powerful. In general, they require a very large and very fast computer. As an extreme example, such a program can take an original prescription consisting simply of a number of plane parallel plates of glass and modify not only the radii, thicknesses and spacings, but also the glass constants (within the range of available glasses) and arrive at a reasonably successful design. One or two outstandingly successful programs have been able to consistently produce practical improvements on finished designs; this ability stems largely from the fact that a computer can "manage" a much greater number of parameters than a human designer can keep track of, and is thus able to "wring out" improvements after the human designer has called it quits.

It should be emphasized that although automatic design programs make it possible for the tyro to produce a design, a considerable knowledge of optics is still important. In even the most automatic programs, it is necessary to give the machine a starting point, to create or weight the merit function and to evaluate the results. It is also frequently necessary to change (or "design") the merit function in the light of interim results. The net effect of all these human choices is to direct the machine to one region or another of the "map" consisting of all possible designs, and a basic understanding of design theory is *absolutely* necessary to utilize the machine to fullest advantage.

References

1. Conrady, "Applied Optics and Optical Design - Vol. 1," Oxford (London) 1929. Volumes 1 and 2 are also published by Dover Publications, New York.
2. Jacobs, "Fundamentals of Optical Engineering," McGraw-Hill, 1943.
3. Hertzberger, "Modern Geometrical Optics," Interscience, 1958.
4. "Dictionary of Applied Physics," Vol. IV, Macmillan and Co., London, 1923.
5. MIL-HDBK-140: Handbook of Optical Design
6. Martin, "Technical Optics," Vols. 1 and 2, Pitman, 1950.
7. "Handbuch der Wissenschaftlichen und Angewandten Photographie," Vol. 1, "Das Photographische Objektiv," (1932), and "Erganzungswerk" (1943), Springer, Vienna. Reprinted by Edwards Brothers, Inc., Ann Arbor, 1944 and 1946. Contains constructional and aberration data for a great number of photographic objectives.
8. Kingslake, "Lenses in Photography," Garden City, 1951.
9. Greenleaf, "Photographic Optics," Macmillan, 1950.
10. Cox, "A System of Optical Design," Focal, 1965; contains construction and aberration data from recent optical design patents.

Exercises

The exercises for this chapter take the form of suggestions for individual design projects; as such, there can be no "right" answers, and none are given. The effort involved in each exercise is considerable, and it is likely that only those interested in obtaining first hand experience in optical design will wish to undertake these exercises. The casual reader will, however, be amply rewarded by mentally reviewing the steps he would follow in attempting the exercises.

1. Design a symmetrical double meniscus objective of the periscopic type. Select a bending (a ratio of 3:2 for the curvatures is appropriate), determine the proper spacing for a flattened field and calculate the thin lens third order aberrations for the combination. Analyze the final design by raytracing and compare the results with the third order calculations. The student may wish to repeat the process for several additional bendings, perhaps including the Hypergon, and to compare the results of each, noting the variations of aperture and coverage.

2. Design an achromatic doublet objective using BSC-2 (517:645) and EDF-1 (649:338). Correct the spherical aberration for an aperture of $f/3.5$. Raytrace marginal and zonal rays in C, D and F light to evaluate the axial image. Compare the coma obtained by raytracing an oblique fan with the O.S.C. calculation.

3. Design an objective lens consisting of two closely spaced doublets of BSC-2 and EDF-1. Vary 1) the distribution of powers between the doublets, and 2) the chromatic correction of each doublet to optimize the correction of zonal spherical and spherochromatic.

4. Design a Cooke triplet anastigmat. For a minimal exercise, duplicate the design of Fig. 10.5, using the same glasses and the same power and space layout as a starting point. For a more ambitious project, design the same lens, but derive the power and space layout without recourse to the data of the Figure.

13

The Design of Optical Systems: Particular

13.1 Telescope Systems

The design of a telescopic system begins with a first order layout of the powers and spacings of the objective, erectors, field lenses, prisms, and eyepiece, as required to produce the desired magnification, field of view, aperture (pupil), and image orientation. Then the individual components are designed so that the telescope, as an entire system, is corrected. Usually the eyepiece is designed first; the design is carried out as if the eyepiece were imaging an infinitely distant object through an aperture stop located at the system exit pupil. That is, the rays are traced in the reverse direction from the direction in which the light travels in the actual instrument. Usually a principal ray is traced from the objective (or the aperture stop) through the eyepiece to locate the exit pupil, then an oblique bundle can be traced in the reversed direction (from the eye) to evaluate the off-axis imagery. Almost all optical design is done in this manner, by tracing rays from long conjugate to short, largely for convenience, because the focus variations (due to aberrations and small power changes) are smaller and more readily managed at the short conjugate.

The erectors, if there are any, are usually designed next; their design is frequently included in the eyepiece design by considering the erector and eyepiece as a single unit. (Alternatively the erector may be considered as a part of the objective; the choice is usually determined by the location of the reticle.) Usually the objective is designed last and its spherical and chromatic aberrations are adjusted to compensate for any undercorrection of the eyepiece. Note that prisms must be included in the design process if they are "inside" the system, since they contribute aberrations which must be offset by the objective and eyepiece. Prisms are introduced into the calculation as plane parallel plates of appropriate thickness.

An eyepiece is a rather unusual system, in that it must cover a fairly wide field of view through a relatively small aperture (the exit pupil) which is *outside* the system. The external aperture stop and wide field force the designer to use extreme care with regard to coma, distortion, lateral color, and curvature of field; the first three mentioned can become especially difficult, since even approximate symmetry about the stop (which is used in many lens systems to reduce these aberrations) is not possible. On the other hand, the small relative aperture of an eyepiece tends to hold spherical and axial chromatic aberrations to reasonable values. Typically an eyepiece is fairly well corrected for coma for one zone of the field (a fifth order coma of the $y^2 h^3$ type is common in wide angle eyepieces) and the field is usually artificially flattened by overcorrected astigmatism which offsets the undercorrected Petzval curvature. Lateral color may or may not be well corrected; frequently some undercorrection exists to offset the effect of prisms. There is almost always some pin cushion distortion apparent (note that when an eyepiece is traced from long to short conjugate, the sign of the distortion is reversed). An eyepiece can be considered "reasonably" corrected for distortion if it has 3% to 5%; 8% to 12% distortion is not uncommon in eyepieces covering total fields of 60° or 70°. About the only way to eliminate this distortion is by the use of aspheric surfaces, a not very attractive solution. One should remember that, in many applications, the function of the outer portions of the field of view is to orient the user and to locate objects which are then brought to the center of the field for more detailed examination. Thus, eyepiece correction off axis need not be as good as that of a camera lens, for example.

Because the eyepiece is subject to a final evaluation by a visual process, it is sometimes difficult to predict, from ray tracing results alone, just what the visual impression will be. For this reason, it is frequently useful to begin an eyepiece design on the lens bench, by mocking up an eyepiece out of available elements. A series of mockups will yields a good grasp of the more promising orientations and arrangements of the elements. The designer can then use these as starting points for his design effort with reasonable assurance that the visual "feel" of the finished design will be acceptable.

The Huygenian Eyepiece: The Huygenian eyepiece (Fig. 13.1A) consists of two plano convex elements, an eyelens and a field lens, with the plane surface of each toward the eye. The focal plane is between the elements. For a given set of powers of the elements, the spacing can be adjusted to eliminate lateral color. The required spacing is approximately equal to the average of the focal lengths of the elements. The only remaining degree of freedom is the ratio of powers between the elements. This is used to eliminate coma (and thus artificially flatten the field, as discussed in Section 12.2). Since the image plane is between the lenses and is viewed by the eyelens alone, it is not well corrected and is unsuitable for use with a reticle.

The Ramsden Eyepiece: The Ramsden eyepiece (Fig. 13.1B) also consists of two plano convex elements, but the plane surface of the field lens faces away from the eye. The spacing is made about 30% shorter than the Huygenian to allow an external focal plane, and for this reason lateral color is not fully corrected. Coma is corrected as in the Huygenian by varying the ratio of field lens power to eyelens power. The Ramsden eyepiece can be used with a reticle.

The Kellner Eyepiece: The Kellner (Fig. 13.1C) is simply a Ramsden eyepiece with an achromatized eyelens to reduce the lateral color. It is frequently used in low cost binoculars.

The relative characteristics of the three simple eyepieces described above are summarized in the table of Fig. 13.2. They are almost invariably made in plano-convex form and little is gained by departing from this form. Since these eyepieces are chiefly noted for their low cost, the usual material for the single elements is common crown; indeed they are frequently made from selected window glass by grinding and polishing only the convex surface. In the Kellner eye-

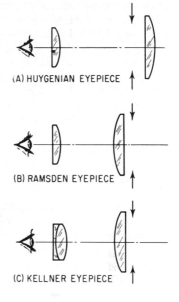

(A) HUYGENIAN EYEPIECE

(B) RAMSDEN EYEPIECE

(C) KELLNER EYEPIECE

FIG. 13.1. Three Basic Eyepiece Forms.

lens, the index difference across the cemented face is critical; usually a barium crown is used to keep the overcorrection of the astigmatism from becoming too large when a wide field of view is desired. Departure from the plano-convex form, in favor of a biconvex shape, is not uncommon in the Kellner eyepiece. The half

	Huygenian	Ramsden	Kellner
Spherical	1.	0.2	0.2
Chromatic (Axial)	1.	0.5	0.2
Lateral Color (C. D. M.)	0.0	0.01	0.003
Distortion	1.	0.5	0.2
Field Curvature	1.	$\sim$0.7	$\sim$0.7
Eye Relief	1.	1.5 to 3.	1.5 to 3.
Coma	0.0	0.0	0.0
M. P. Tolerance*	1.	5.	5.
Ratio, High Power‡	2.3	1.4	0.8
Ratio, Low Power‡	1.3	1.0	0.7

FIG. 13.2. The relative characteristics of three simple eyepieces. *The M.P. tolerance is the relative ability to retain the desired state of correction when used at magnifications other than that for which the eyepiece was originally designed. ‡Ratio of the focal length of the field lens to the focal length of the eyelens; high and low power refer to the power of the telescope, with which the eyepiece is to be used.

field covered by these eyepieces is to the order of ±15°, more or less, depending on the performance required.

The Orthoscopic Eyepiece: The orthoscopic eyepiece (Fig. 13.3A) consists of a single element eyelens (usually plano convex) and a cemented triplet (usually symmetrical). The eyelens is frequently of light barium crown or light flint glass and the triplet is composed of borosilicate crown and dense flint glass. This is a better eyepiece than the preceding simple types and is used for half fields of ±20° to ±25°. The Petzval curvature is about 20% less than that of the Ramsden or Kellner, although higher order astigmatism causes a strongly backward curving tangential field at angles of more than 18° or 20° from the axis. (This high order astigmatism is the characteristic which limits the field coverage of most eyepieces; some control can often be achieved by lowering the index difference across cemented surfaces.) The eye relief is long, to the order of 80% of focal length. Distortion correction is quite good.

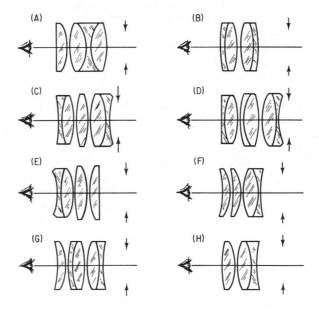

FIG. 13.3. Eyepiece Designs. A) Orthoscopic. B) Symmetrical. C) Erfle. D) Erfle. E) Berthele.

The Symmetrical, or Plossl, Eyepiece: This eyepiece is composed of two achromatic doublets (usually identical) with their crown elements facing each other (Fig. 13.3B). It is usually executed in borosilicate crown and extra dense flint (649:338) glass, although it can be improved by raising the index of both elements. It shares the long eye relief (0.8F) and field characteristics of the

Orthoscopic, but is in general a somewhat superior eyepiece, except that its distortion is typically 30% to 50% greater than the Orthoscopic. It is widely used in military instruments and as a general purpose eyepiece of moderate (to ±25°) field. A similar eyepiece with both flints facing the eye is occasionally used.

The Erfle Eyepiece: This eyepiece (Fig. 13.3C) is probably the most widely used wide field (±30°) eyepiece. The eye relief is long (0.8F) but working distance is quite short. The Petzval sum is about 40% less than the orthoscopic or symmetrical types, and distortion is about the same as the orthoscopic (for the same angular field). The type shown in Fig. 13.3C usually has under-corrected lateral color which can be reduced by use of an achromatic center lens as in Fig. 13.3D. Glasses used are usually dense barium crown and extra dense flint.

Magnifiers: Magnifiers and viewer lenses are basically the same as eyepieces, with one notable exception: there is no fixed exit pupil. This means that the eye is free to take any position and therefore the magnifier must be insensitive to pupil shift. For this reason, magnifiers tend to be symmetrical in configuration. Two plano convex lenses with convex surfaces facing or a symmetrical (Plössl) construction are common for better grade magnifiers. Where cost is important and a single element must be used, the following arrangements are good. If the eye is always close to the magnifier, use a plano convex form with plano surface toward the eye. If the eye is always far from the magnifier, use a plano convex form with the convex surface toward the eye. If the eye position is variable, as in a general purpose magnifier, an equiconvex form is probably the best choice.

Note that the eyepieces of instruments which use an electronic image tube, such as the Sniperscope, fall into the category of magnifiers, since they are used to view a diffuse image on the phosphor surface of the image tube. As such they must be designed so that they perform well with the eye in a wide range of locations.

Erectors: Erector systems come in all sizes and shapes. Occasionally a single element may serve as an erector, or two simple elements in the general form of a Huygenian eyepiece may be used, as in the Terrestrial eyepiece shown in Fig. 13.4A. This form of eyepiece is widely used in surveying instruments, occasionally with an achromatic eyelens or second erector lens. A popular erector for gun scopes is illustrated in Fig. 13.4B and consists of a single element plus a low power, overcorrecting doublet. Photographic objective systems are occasionally used as erectors; symmetrical forms of the Cooke triplet, the Dogmar or the Double Gauss being the most popular. Probably the most widely used erector consists of two achromats, crown elements facing, with a modest spacing between them.

As previously mentioned, erectors are usually designed in conjunction with either the eyepiece or objective of a telescopic

system. Considerable care should be taken in the first order layout of any telescope to be certain that the work load placed on the
erector is not impossibly large. The introduction of suitable field
lenses is often necessary to reduce the height of the principal ray
at the erector, although this does produce an undesirable increase
in the Petzval curvature.

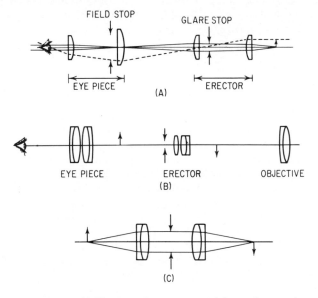

FIG. 13.4. Erector Systems. A) The four element terrestrial erecting eyepiece. B) Typical
Gunsight optical system. C) Symmetrical Doublet Erector.

Objective systems: For most telescopic systems, the objective
will be an ordinary achromatic doublet, or one of the variations
described in Section 12.5. A photographic type objective may be
used where a wide field is desired, Cooke triplets and Tessars
being the most commonly used. A Petzval objective is useful when
high relative apertures are necessary; the construction of a
Petzval objective (Section 13.3) is such that its rear lens acts as
a sort of field lens, and this characteristic is occasionally useful.
For high power telescopes where it is desirable to keep the system
as short as possible, a telephoto type of construction is valuable.
The front component is an achromatic doublet and the rear is a
negative lens, either simple or achromatic. The focal length is
usually 20% to 50% longer than the over-all length of the objective.
Either the Petzval or Telephoto type of objective can be used as
an internal focusing objective (Fig. 13.5), where focusing is accomplished by shifting the rear (inside) component, making a more
easily sealed instrument. Surveying instruments and theodolites
conventionally use the telephoto form with the focusing lens located

about two-thirds of the way from the front component to the focal plane so that the stadia "constant" will remain constant as the instrument is focused. Alignment telescopes use a positive focusing lens of high power placed near the focal plane at infinity focus; thus, a modest shift of the focusing lens toward the front component allows the system to be focused at extremely short distances, or even on the objective itself. Note that a system which works over a wide range of magnifications (as this type of focusing lens does) should be designed so that the change of aberration contribution is small as the magnification is varied.

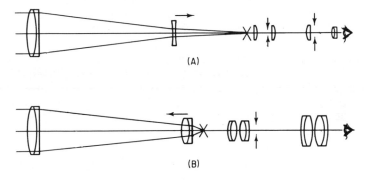

FIG. 13.5. Telescopic Systems. A) Typical surveying telescope with negative focusing lens and terrestrial eyepiece. Note that the objective is telephoto, in that its effective focal length is longer than the objective. B) Alignment telescope. The strong positive focusing lens, when shifted forward, allows the instrument to focus at extremely short distances.

13.2 Microscope Objectives

Microscope objectives may be divided into three major classes; those designed to work with the object under a cover glass, those designed to work with no cover glass, and immersion objectives, which are designed to contact a liquid in which the object is immersed. All types are designed by ray tracing from the long conjugate to the short; the effects of the cover glass (when used) must be taken into account by including it in the ray trace analysis. Standard cover glass thickness is 0.18 mm.

Microscope objectives are designed to work at specific conjugates, and their correction will suffer if they are used at other distances. For "cover glass" objectives and immersion objectives, the standard distance from object plane to image plane is 180 mm. For metallurgical types (no cover glass), the standard distance is 240 mm. The chief effect of changing the tube length or cover glass thickness from their nominal values is to overcorrect or undercorrect the spherical aberration; an objective which has been improperly adjusted at the factory may be reclaimed by using a nonstandard tube length or cover glass, if the defect is not too serious.

Note that ordinary microscope objectives are designed to yield an essentially perfect image, and aberrations (on axis at least) should be reduced to well below the Rayleigh limit if at all possible. Micro-objectives for projection or photography may be corrected with more emphasis on the outer portions of the field, depending on the exact application for which they are intended.

Low Power Objectives: These are usually ordinary achromatic doublets, or occasionally three element systems. The 32 mm NA 0.10 or 0.12 is the most common and produces a magnification of about 4×. A 48 mm NA 0.08 is also occasionally encountered. These may be designed in exactly the same manner as the achromatic objective discussed in Sections 12.4 and 12.5, except that the "object" will be located at 150 mm (more or less) instead of at infinity.

Medium Power Objectives: These are usually composed of two widely spaced achromatic doublets. The most common objective is the 10×, 16 mm, which is available in several forms. The ordinary achromatic 10× objective has an N.A. of 0.25 and is probably the most widely used of all objectives. The divisible or separable (Lister) objective is designed so that it can be used as a 16 mm or, by removing (ot swinging to one side) the front doublet, as a 32 mm objective. This is accomplished at the sacrifice of astigmatism correction, since both components must be free from spherical and coma and thus no correction of field curvature is possible. An apochromatic 16 mm objective is also available with an N.A. of 0.3; fluorite (CaF_2) is used in place of crown glass to reduce the secondary spectrum.

The power layout for this type of objective is usually arranged so that the product $y\phi$ for each doublet is the same; in this way the "work" (bending of the marginal ray) is evenly divided. Conventionally the second doublet is placed midway between the first doublet and the image formed by the first doublet. (Note that the preceding refers to ray tracing sequence—in use the "second" doublet is near the object to be magnified and the "first" doublet is nearer the actual image.) This relatively large spacing allows the cemented surface of the second doublet to overcorrect the astigmatism and flatten the field (assuming the stop to be at the first doublet). This layout leads to an arrangement with the space about equal to the focal length of the objective, the focal length of the first doublet approximately twice that of the objective, and that of the second doublet about equal to that of the objective. Note that this arrangement is similar to that of a high speed Petzval type projection lens.

Ordinarily three sets of shapes for the two components can be found for which spherical and coma are corrected. One form will be that of the divisible objective, with the spherical contribution zero for each doublet; this is usually the form with the poorest field curvature.

High Power Objectives: If the surface contribution equation for the spherical aberration of a single surface is solved for zero spherical, three solutions are found. One case occurs when the object and image are at the surface, and is of little interest. A second is of more value; when object and image both lie at the center of curvature, there is no spherical aberration introduced (and the axial rays are not deviated). The third case, usually called the aplanatic case, allows the convergence of a cone of rays to be increased (or decreased) without the introduction of spherical aberration and occurs when the following relationships are satisfied.

$$L = R\left(\frac{N' + N}{N}\right) \tag{13.1}$$

$$L' = R\left(\frac{N' + N}{N'}\right) - \frac{N}{N'} L \tag{13.2}$$

$$U = I' \tag{13.3}$$

$$U' = I \tag{13.4}$$

$$\frac{N'}{N} = \frac{\sin U'}{\sin U} \tag{13.5}$$

Note that if any of the above are satisfied, all are satisfied, and that, since no spherical is introduced, if $L = l$, then $L' = l'$. It is also worth noting that coma is zero for all three cases and that astigmatism is zero for the first and third cases and overcorrecting between.

This principle is used in the "aplanatic front" of an oil immersion microscope. The object is immersed in an oil whose index of refraction matches that of the first lens. R_1 (as shown in Fig. 13.7) is chosen to satisfy Eq. 13.1; this results in a hyper-hemispheric form for the first element. R_2 is chosen so that the image formed by R_1 is at its center of curvature; R_3 is chosen to satisfy Eq. 13.1. Note that $\sin U$ is reduced by a factor of N at each element, and that the "aplanatic front" reduces the numerical aperture of the cone of rays from a large value (as high as $NA = N \sin U = 1.4$) to a value which a more conventional "back" system can handle.

The Amici objective (Fig. 13.6C) consists of a hyper-hemispheric front element combined with a Lister type back combination. Since the Amici is usually a dry objective, the radius of the hyper-hemisphere is frequently chosen somewhat flatter than that called for by the aplanatic case to offset the spherical introduced by the (dry) plano surface. The space between the hyper-hemisphere and the adjacent doublet is kept small to reduce the lateral color

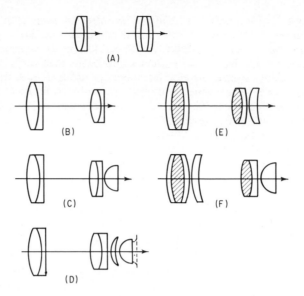

FIG. 13.6. Microscope Objectives. A) Low power-achromatic doublet or triplet. B) Lister objective 10× N.A. 0.25. C) Amici objective 20× N.A. 0.5 to 40× N.A. 0.8. D) Immersion objective. E) Apochromatic 10× N.A. 0.3. Shading indicates Fluorite (CaF$_2$). F) Apochromatic 50× N.A. 0.95.

introduced by the front element. The standard 4 mm 40×, *NA* 0.65 to 0.85 objectives are usually Amici objectives. The working distance (object to front surface) is quite small in the Amici, to the order of a half millimeter. Since there is a direct relationship between zonal spherical and working distance in this type of objective, the higher N.A. versions tend to have very short working distances.

The Oil-Immersion objective utilizes the full "aplanatic front" and may be combined with a Lister type back, as shown in Fig. 13.6D, or a more complex arrangement. Both the Amici and Immersion types are frequently designed with Fluorite (CaF$_2$) crowns to reduce or eliminate secondary spectrum.

Note that although the aplanatic front is a classic text book case, departures from the exact aplanatic form are common. For

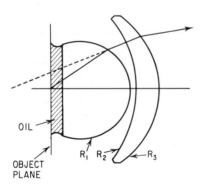

FIG. 13.7. The Aplanatic Front. The object is immersed in a fluid whose index matches that of the hyperhemispheric first element. R_1 is an aplanatic surface. The image formed by R_1 is at the center of curvature of R_2. R_3 is an aplanatic surface of the same type as R_1.

example, it is possible to find a meniscus lens of higher power than the aplanatic case which will introduce overcorrected spherical. This not only reduces the "work" that the back elements must accomplish, but also reduces the correction load as regards spherical aberration (but not chromatic). Aplanatic-front objectives have a residual lateral color, resulting from the separation of the undercorrected "front" and the overcorrecting "back". Special "compensating" eyepieces with opposite amounts of lateral color are used to correct this situation.

Reflecting Objectives: Objectives for use in the ultraviolet or infrared spectral regions are frequently made in reflecting form, because of the difficulty of finding suitable refracting materials for these spectral regions. The central obscuration required by such a construction will modify the diffraction pattern of the image, reducing the contrast of coarse targets and improving the contrast for fine details, as indicated in Chapter 11.

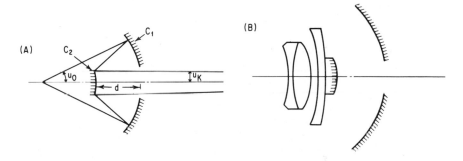

FIG. 13.8. Reflecting microscope objectives. A) Concentric 30× N.A. 0.5. B) Ultraviolet 50× N.A. 0.7. Fused quartz and calcium fluoride are used for the refracting elements.

The basic construction for such an objective is shown in Fig. 13.8A and consists of two nearly concentric mirrors. The design of such a system may be started as follows:

$$y_1 = 1.0 \tag{13.6}$$

$$y_2 = 1 - d(2c_1 - u_0) \tag{13.7}$$

Then to get a desired value for u_k

$$c_2 = \frac{u_k + 2c_1 - u_0}{2[1 - d(2c_1 - u_0)]} \tag{13.8}$$

The third order spherical aberration for the system can be

obtained from the surface contribution equations of Chapter 10. Substituting the above values, we find

$$\Sigma SC = \frac{c_1(c_1 - u_0)^2}{u_k^2} + \frac{[1 - d(2c_1 - u_0)][u_k - (2c_1 - u_0)][(2c_1 - u_0)^2 - u_k^2]}{8u_k^2}$$

(13.9)

The third order spherical will be equal to zero when

$$d = \frac{u_k(1-m)[4c_1^2 - 2u_k c_1(m-1) - u_k^2(m+1)^2]}{(2c_1 - mu_k)[-8c_1^3 + 4u_k c_1^2(3m+1) - 2u_k^2 c_1(3m-1)(m+1) + u_k^3(m-1)(m+1)^2]}$$

(13.10)

where

$$m = u_0/u_k$$

(13.11)

Thus, for a given magnification (m), one selects c_1 and u_k (note that u_k and m determine u_0, which indirectly determines the working distance, since $y_1 = 1.0$) and solves for d by Eq. 13.10 and for c_2 by Eq. 13.8. Although the family of solutions is theoretically large, only those of the general type shown in Fig. 13.8A are practical; one is also constrained to keep the obscuration due to surface c_2 to a small value. (Note that correction is possible when $u_k = 0$, but not when $u_0 = 0$, for the mirror orientations shown in the figure.)

The resulting system not only has zero third order spherical, but even the higher orders tend to be exceedingly small; by proper choice of parameters a delightfully simple but nonetheless useful objective can be obtained. The two mirror system is limited to about 30× at NA = 0.5. For higher magnifications and numerical apertures, it is necessary to introduce additional refracting elements to maintain correction, as indicated in the sketch of the 50×, N.A. 0.7 ultraviolet objective in Fig. 13.8B.

13.3 Photographic Objectives

In this section, we will outline the basic design principles of the photographic objective, and for this purpose we will classify objectives according to their relationship to, or derivation from, a few major categories: a) Meniscus types, b) Cooke Triplet types, c) Petzval types, d) Telephoto types. These categories are quite arbitrary and are chosen for their value as illustrations of design features, rather than any historic or generic implications.

Meniscus Anastigmats: In this category, we include those objectives which derive their field correction primarily from the use of a thick meniscus. As mentioned in Sections 12.1 and 12.2, a thick meniscus element has a greatly reduced inward Petzval curvature in comparison with a biconvex element of the same power; indeed, the Petzval sum can be overcorrected if the thickness is made great enough. The simplest example of this type of lens is the Goerz Hypergon (Fig. 12.4) which consists of two symmetrical menisci. Because the convex and concave radii are nearly equal, the Petzval sum is very small, and the fact that the surfaces are nearly concentric about the stop enables the lens to cover an extremely wide (135°) field, although at a very low aperture ($f/30$).

To obtain an increased aperture, it is necessary to correct the spherical and chromatic aberrations. This can be accomplished by the addition of negative flint elements, as in the Topogon lens, Fig. 13.9. Note that the construction of this lens is also very nearly concentric about the stop; lenses of this type cover total fields of 75° to 90° at speeds of $F/6.3$ to $F/11$.

FIG. 13.9. The Topogon lens (U.S. Patent 2,031,792-1936) covers 90° to 100° at a speed of $f/8$.

Attempts to design a system consisting of symmetrical cemented meniscus doublets in the last half of the 19th century were only partially successful. If the spherical aberration was corrected by means of a dispersing (i.e. with negative power) cemented surface, the higher order overcorrected astigmatism necessary to artificially flatten the field tended to become quite large at wide angles. If a high index crown and low index flint were used, the resulting collective cemented surface was incapable of correcting the spherical. In 1890, Rudolph (Zeiss) designed the *Protar*, Fig. 13.10, which used a low power "old" achromat (i.e. low index crown, high index flint) front component and a "new" achromat (high index crown and low index flint) rear component. The dispersive cemented surface of the front component was used to correct the spherical while the collective cemented surface of the rear kept the astigmatism in control. Note that the components are thick menisci, which allows correction of the Petzval sum, while the general symmetry helps to control the coma and distortion. Lenses of the Protar type cover total fields of 60° to 90° at speeds of $f/8$ to $f/18$.

FIC. 13.10. The Zeiss Protar (U.S. Patent 895,045-1908).

A few years later, Rudolph and von Hoegh (Goerz), working independently, combined the two components of the Protar into a single cemented component, which contained both the required dispersing and collective cemented surfaces. The Goerz *Dagor* is shown in Fig. 13.11, and is composed of a symmetrical pair of cemented triplets. Each half of such a lens can be designed to be

FIG. 13.11. The Goerz Dagor (U.S. Patent 528,155, 1894). The glasses used are 613: 563, 568:560 and 515: 547, from the left. The construction is symmetrical about the stop.

corrected independently so that photographers were able to remove the front element to get two different focal lengths. A great variety of designs based on this principle were produced around the turn of the century, using three, four and even five cemented elements in each component, although very little was gained from the added elements. Protars and Dagors are still used for wide angle photography because of the fine definition obtained over a wide field, especially when used at a reduced aperture.

The additional degree of freedom gained by breaking the contact of the inside crowns of the Dagor construction proved to be of more value than additional elements. Lenses of this type (Fig. 13.12) are probably the best of the wide angle meniscus systems and cover fields up to 70° total at speeds of $f/5.6$ (or faster for smaller fields). The Meyer *Plasmat*, the Ross *W. A. Express*, and the Zeiss *Orthometer* are of this construction and recently excellent 1:1 copy lenses (symmetrical) have been designed for electro-photography applications.

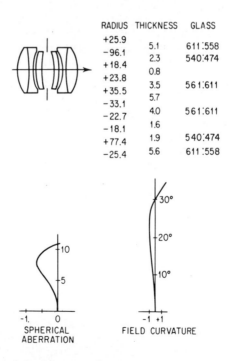

RADIUS	THICKNESS	GLASS
+25.9	5.1	611:558
−96.1	2.3	540:474
+18.4	0.8	
+23.8	3.5	561:611
+35.5	5.7	
−33.1	4.0	561:611
−22.7	1.6	
−18.1	1.9	540:474
+77.4	5.6	611:558
−25.4		

SPHERICAL ABERRATION

FIELD CURVATURE

FIG. 13.12. The Zeiss Orthometar (U.S. Patent 1,792,917). Constructional data and aberration curves for a focal length of 100.

The design of the thick meniscus anastigmats is a complex undertaking because of the close inter-relationship of all the variables. In general the exterior shape and thickness are chosen to control the Petzval sum, and the distance from the stop can be used to adjust the astigmatism. However, the adjustment of element powers to correct chromatic inevitably upsets the balance, as does the bending of the entire meniscus to correct spherical. What is necessary is one simultaneous solution for the relative powers, thicknesses, bendings and spacings; an approach of the type des-scribed in Section 12.7 for the simultaneous solution of the third order aberrations is ideally suited to this problem. The efforts of designers in this direction over the past 75 years have been well spent, and it is exceedingly difficult to improve on the best rep-resentative designs in this category unless one utilizes the newer types of optical glass (e.g. the rare earth glasses).

The Double Gauss (Biotar) type and the Sonnar type of objectives both make use of the thick meniscus principle, although they differ from the preceding meniscus types in that they are used at larger apertures and smaller fields. The Biotar objective in its basic form consists of two negative meniscus inner doublets and two single positive elements as shown in Fig. 13.13. This is an ex-ceedingly powerful design form, and many high performance lenses are modifications or elaborations of this type. If the vertex length is made short and the elements are strongly curved about the central stop, fairly wide fields may be covered.

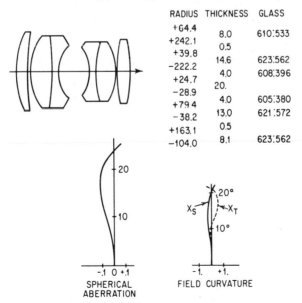

RADIUS	THICKNESS	GLASS
+64.4	8.0	610:533
+242.1	0.5	
+39.8	14.6	623:562
−222.2	4.0	608:396
+24.7	20.	
−28.9	4.0	605:380
+79.4	13.0	621:572
−38.2	0.5	
+163.1	8.1	623:562
−104.0		

FIG. 13.13. The Double Gauss (Biotar) Objective (U.S. Patent 2,117,252-1938). Construc-tional data and aberration curves for a focal length of 100.

FIG. 13.14. The Son-
nar Type Objective.

Conversely, a long system with flatter curves will cover a narrow field at high aperture. One possible design approach is as follows:

1. Select an appropriate vertex length, based on considerations of aperture and field coverage. Prior art is useful in this regard. Usually this length is almost filled with glass, in that the first and last airspaces are minimal and the edge clearance between the central flints is small. Baker, in U.S. Patent 2,532,751, suggests a rule of thumb for the total thickness of the two meniscus doublets plus the central airspace: for narrow fields (less than ±10°), a value of 0.6 to 0.7 times the focal length; for moderate fields (between ±10° and 20°), $0.5F$ to $0.6F$; for fields larger than ±20°, a value of $0.4F$ to $0.5F$.

2. Select glass types. Crowns are usually high index DBC or rare earth. Flints are usually lower in index by several hundredths. The difference in V-value can be used to shape the cemented surfaces; surface 4 is usually made concave to the stop and surface 6 convex to the stop.

3. Make a rough layout of thickness and curvature. Prior art is a useful guide. Use R_5 and R_6 to adjust the Petzval sum and vary R_4 and R_7 to correct the axial and lateral color as desired.

4. Use the third order surface contributions to effect a solution for the desired ΣSC, ΣCC^*, ΣAC^* and ΣDC^*. This can be handled by plotting the contribution of each component against its shape, locating a region of solution and applying a differential correction technique.

5. A trigonometric check and differential correction complete the primary phase of the design.

6. Note that there are many unused degrees of freedom remaining. The distribution of power from front to back elements and the distribution of power between inside and outside crowns may be systematically varied within rather broad limits. The glass and thickness choices are subject to revision as well. Each of these will have an effect on residuals and higher order aberrations.

7. The following comments may be helpful:
   a. Oblique spherical (a fifth order aberration which causes spherical to vary with obliquity, i.e. as $y^3 h^2$) is usually troublesome, causing an off-axis overcorrection which reduces image contrast. This comes from the large angles of

incidence at surface 5 for the upper rim ray and at surface 6 for the lower. This can be reduced (at the expense of other corrections) by increasing the central airspace or by curving the system strongly about the stop to allow a more concentric passage of the rays through these surfaces, or by reducing the thickness of the doublets which will tend to force a more curved configuration on them and increase the zonal spherical. Making the cemented surfaces more collective also tends to reduce the oblique spherical.

b. The longitudinal position of surface 7 can be used to control sphero-chromatism. A shift to the right will reduce the spherical overcorrection of blue light relative to red light.

c. If the index difference across the cemented surfaces is small, the adjustments of R_4 and R_7 for chromatic correction will have a correspondingly small effect on the monochromatic aberrations.

d. The thickness of the cemented doublets (especially the front) has a strong effect on spherical aberration. Increasing the thickness leads to undercorrection, and vice versa. This sensitivity is a common characteristic of thick meniscus systems which, although it makes fabrication difficult, is useful as a design tool.

Common elaborations of the Biotar format include compounding the outer elements into doublets or triplets or converting the meniscus doublets into triplets. Frequently the outer elements are split (after shifting some power from the inner crowns). Some recent designs have advantageously broken the contact at the cemented surface, especially in the front meniscus.

Airspaced Anastigmats: These are systems which utilize a large separation between positive and negative components to correct the Petzval sum. Although it is historically incorrect in several instances, from a design standpoint it is useful to view these lenses as derivatives from the Cooke Triplet, Fig. 13.15 (see also Section 12.6).

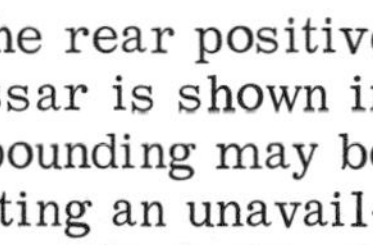

FIG. 13.15. The Cooke Triplet.

The Tessar may be regarded as a triplet with the rear positive element compounded; the classic form of the Tessar is shown in Fig. 13.16. The additional freedom gained by compounding may be regarded as simply a means of artificially generating an unavailable glass type by combining two available glasses; alternatively one may utilize the refractive characteristics of the cemented interface to control the course of the upper rim ray, which is affected strongly by this surface. The Tessar formulation, either as shown, or with the doublet reversed, or even with the front element compounded, is utilized when a performance a bit beyond that of the Cooke triplet is required.

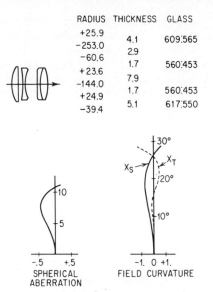

RADIUS	THICKNESS	GLASS
+25.9	4.1	609:565
−253.0	2.9	
−60.6	1.7	560:453
+23.6	7.9	
−144.0	1.7	560:453
+24.9	5.1	617:550
−39.4		

FIG. 13.16. The Tessar Objective (U.S. Patent 1,588,073-1922). Construction and aberration data for a focal length of 100.

A further example of the compounding of the elements of the basic triplet is the Pentac (or Heliar) type, Fig. 13.17, which is simply an extension of the Tessar principle. In the Hektor (Fig. 13.18), all three elements are compounded and the speed can be raised to $f/1.9$ with fields to the order of ±20°. Many "compounded triplets" make use of what is sometimes called a "Merté" surface; the cemented surface of the negative component of the Hektor is an example of such a surface. This is a strongly curved (usually cemented) collective surface so arranged that the angle of incidence increases rapidly toward the margin of the lens. Such a surface contributes a modest amount of undercorrecting spherical to the rays near the axis, since the index break across the surface is not large. As the angle of incidence rises (and it may approach 45°), the spherical aberration contribution rises even more rapidly, and the undercorrecting effect dominates the marginal zone. The result is a spherical aberration curve which contains not only third and fifth order aberration, but a sizeable amount of negative seventh order as well. The spherical aberration shown in Fig. 13.18 is a rather extreme example of this technique. This is an approach which obviously must be used with discretion, since large amounts of high order aberration are delicately balanced. Such a surface is best located near the stop to minimize the disparity of its effects on the

FIG. 13.17. The Pentac-Heliar Anastigmat.

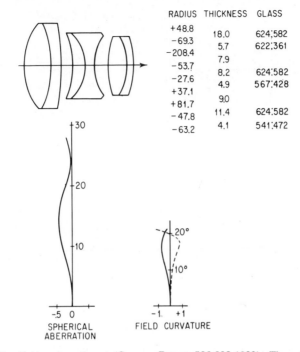

RADIUS	THICKNESS	GLASS
+48.8	18.0	624;582
−69.3	5.7	622;361
−208.4	7.9	
−53.7	8.2	624;582
−27.6	4.9	567;428
+37.1	9.0	
+81.7	11.4	624;582
−47.8	4.1	541;472
−63.2		

FIG. 13.18. The Hektor Anastigmat (German Patent 526,308-1930). The spherical aberration curve shows a large seventh order component which originates at the strongly curved fifth surface. (Focal length, 100)

rim rays; otherwise, the off axis ray intercept curves tend toward a very unpleasant asymmetry.

Another basic technique for the reduction of residuals involves splitting the individual elements into two (or more) elements. A single crown element has about five times as much undercorrected spherical as a "minimum spherical" two element lens of equivalent power and aperture. Thus, a split allows the contributions of the other elements of the system to be reduced, resulting in a corresponding decrease in higher order aberrations. Ordinarily the crown elements of a triplet are split when a larger aperture is desired; Figs. 13.19 and 13.20 are examples of this technique. Since it requires a fairly long system to make this technique effective, the coverage of such systems is usually modest. However, by *compounding* the elements, excellent combinations of aperture and field have been obtained from these forms. Splitting the front crown is usually more profitable than splitting the rear, since the astigmatism at the edge of the field is better controlled in the split front types. Although less frequently encountered, element splitting is also effective to a limited degree in extending field coverage.

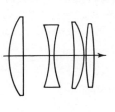

RADIUS	THICKNESS	GLASS
+56.0	9.0	610:588
−1500.0	17.0	
−55.0	2.5	673:320
+80.0	12.0	
−373.0	7.0	610:588
−48.0	1.5	
+194.0	5.0	610:588
−194.0		

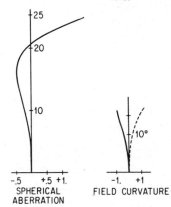

FIG. 13.19. Split Rear Crown Triplet (U.S. Patent 1,540,752-1924). Focal length, 100.

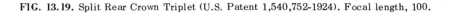

RADIUS	THICKNESS	GLASS
+51.0	8.8	564:607
−441.0	0.0	
+35.3	7.8	564:607
+47.6	8.4	
−254.8	2.0	648:339
+28.3	29.4	
+107.8	4.9	564:607
−60.3		

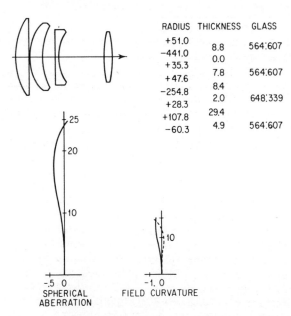

FIG. 13.20. Split Front Crown Triplet (English Patent 237,212-1925). Focal length, 100.

Split flint triplets (Fig. 13.21) may also be regarded as thick meniscus systems with an air lens separating the crown and flint of each half; indeed this was their historical derivation. This form is not especially notable for reduced spherical zonal as are the split crown types, but some of the finest general purpose photographic objectives (e.g. the $f/4.5$ Dogmar and Aviar lenses) have been of this construction. The general symmetry of this design lends itself to a wider angular coverage than does the split crown type, although, as in most "triplet derived" forms, the limit of coverage is sharply defined and image quality tends to fall off rapidly beyond the stigmatic node. (This last comment is less true of systems where the crown-flint spacing is small, since these types are closer to the meniscus lenses than to triplets.) Many excellent process and enlarging lenses are based on this format.

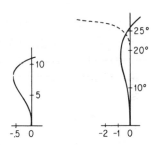

RADIUS	THICKNESS	GLASS
+27.7	4.2	614:563
-103.1	1.8	
-53.9	1.6	569:426
+37.7	5.4	
-63.3	1.6	548:461
+35.1	1.8	
+53.2	3.6	614:563
-35.7		

FIG. 13.21. The Dogmar Anastigmat (U.S. Patent 1,108,307-1914). Focal length, 100.

Lenses for close conjugate work, such as enlarger lenses, are usually airspaced anastigmats. They differ from camera objectives primarily in that they are designed for low magnification ratios, rather than for infinite object distances. Most camera objectives maintain their correction down to object distances to the order of 25 times their focal length, and many do well at even shorter distances. Enlargers, however, are frequently used at magnifications approaching unity, and enlarging lenses are usually designed at conjugate ratios of four or five. A lens which is approximately symmetrical (such as the split flint triplet) makes a good enlarger lens since it is a bit less sensitive to object-image distance changes. Compounded triplets of approximately symmetrical construction are also used, and the Tessar formula is widely used because of its wide field of coverage and relatively simple construction.

Petzval Lenses: The original Petzval Portrait lens (Fig. 13.22) was a relatively close coupled system consisting of two achromatic doublets, the rear doublet with broken contact, with a sizeable airspace between. It covered a modest field at a speed of about $f/3$. The modern version, often referred to as the Petzval Projection lens because of its widespread use as a

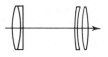

FIG. 13.22. The Petzval Portrait Lens.

motion picture projection objective, utilizes a larger airspace (almost equal to its focal length) and covers half field of ±5° to ±10° at speeds up to $f/1.6$. This type of system (Fig. 12.23) is noted for the excellence of its correction on axis, and also for its strongly inward curving field. The field is artificially flattened by overcorrected astigmatism which is introduced at the cemented surface of the rear doublet. A typical formulation has a thin lens spacing about equal to the focal length, a front doublet with twice the focal length of the system and a rear doublet with a focal length equal to that of the system. Thus, the (thin lens) back focus is about half the focal length and the front vertex to focal plane distance is about 1.5 times the focal length. If the airspace is appreciably shortened, it is necessary to break contact at the rear doublet to maintain the overcorrected astigmatism.

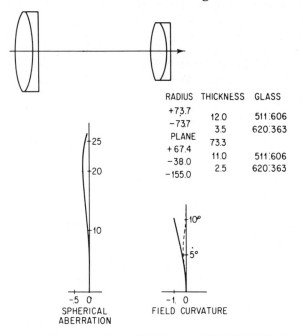

RADIUS	THICKNESS	GLASS
+73.7		
	12.0	511:606
−73.7		
	3.5	620:363
PLANE		
	73.3	
+67.4		
	11.0	511:606
−38.0		
	2.5	620:363
−155.0		

FIG. 13.23. The Petzval Projection Lens (U.S. Patent 1,843,519-1932). Focal length, 100.

The inward curving Petzval surface can be corrected by the use of a negative "field flattener" element near the focal plane, Fig. 13.24. In this location the power contribution $(y\phi)$ of the element is low, but the Petzval field is nicely flattened, and a lens of beautiful definition over a small field can be obtained. The drawback to this is the location of the element near the image plane where dust and dirt can become quite noticeable.

The glasses used in the Petzval lens are usually an ordinary crown (C-1 or BSC-2) and common dense flint (DF-2). Occasionally higher index glass is used and one or both doublets are of the broken contact type.

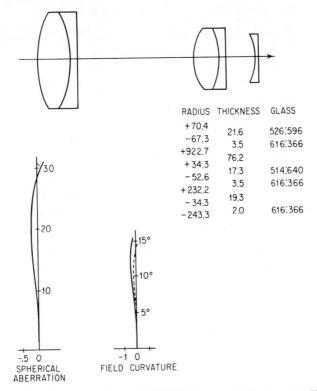

RADIUS	THICKNESS	GLASS
+70.4		
	21.6	526:596
−67.3		
	3.5	616:366
+922.7		
	76.2	
+34.3		
	17.3	514:640
−52.6		
	3.5	616:366
+232.2		
	19.3	
−34.3		
	2.0	616:366
−243.3		

FIG. 13.24. Petzval Projection Lens with Field Flattener (U.S. Patent 2,076,190-1937). Focal length, 100.

An interesting variation on the field flattener Petzval is shown in Fig. 13.25, in which the rear negative element does double duty, serving both as the rear flint and as the field flattener as well. The airspace in the front doublet is necessary to completely correct the aberrations. This lens has a tendency toward fifth order coma of the y^4h type which is introduced by the airspaced front doublet. This aberration is frequently encountered when a strong negative "air lens" is used in this manner to correct spherical.

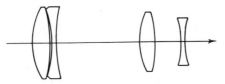

FIG. 13.25. $f/1.6$ Petzval Lens with field flattening effect achieved by large airspace between rear crown and flint.

The already small spherical zonal of the Petzval lens can be reduced still further by splitting the rear doublet into two doublets as indicated in Fig. 13.26 or by the introduction of a meniscus element into the central airspace, Fig. 13.27. One Petzval modification achieved a speed of $f/1.0$ (with an almost spherical field)

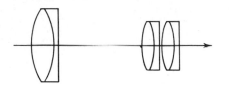 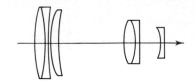

FIG. 13.26. $f/1.3$ Petzval Lens with two rear doublets to reduce spherical zonal. (U.S. Patent 2,158,202-1939).

FIG. 13.27. Field Flattener Petzval Lens with front crown split into two elements to reduce spherical zonal. (U.S. Patent 2,541,484-1951).

by splitting off a sizeable part of the power of each crown element into separate plano-convex elements. Other modifications have made use of strongly meniscus correctors to reduce the zone, or of thick concentric meniscus elements to improve the field. Two recent designs which are used as $2''$ $f/1.4$ projection lenses for 16 mm motion pictures are shown in Fig. 13.28.

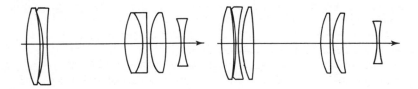

FIG. 13.28. High Performance $2''$ $f/1.4$ 16 mm motion picture projection lenses. (Left is U.S. Patent 2,989,895-1961. Right is U.S. Patent 3,255,664-1966.)

Telephoto Lenses: Telephoto lenses are arbitrarily defined as lenses whose length from front vertex to film plane is less than the focal length. This is achieved by a positive front component separated from a negative rear component as indicated in Fig. 13.29. Several forms of telephoto lenses are also shown; distortion correction is usually achieved by splitting the rear component. A common difficulty of the telephoto and reverse telephoto lenses is a strong inclination toward an overcorrected Petzval sum.

Reverse Telephoto Lenses: By reversing the basic power arrangement of the telephoto, a back focal length which is longer than the effective focal length may be achieved. This (Fig. 13.30) is a useful form when prisms or mirrors are necessary between the lens and the image plane; it also allows the use of a short focal length projection lens with a condenser designed for longer

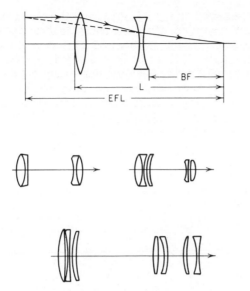

FIG. 13.29. Telephoto Lenses: A focal length which is greater than the physical length of the lens is achieved by a positive front member widely separated from a negative rear member.

lenses, since the pupil position is well away from the image plane. The construction is usually a strong negative achromat in front, combined with a modification of a standard objective. Biotars, triplets and Petzvals have all been used for the rear member. It is usually necessary to split the negative achromat and bend it toward the rear member to achieve good correction. In extreme forms ("sky-lenses") coverage can exceed ±90° with a very strongly meniscus negative lens. Obviously in order to image 180° or more on a finite sized flat film, a large amount of distortion is unavoidable.

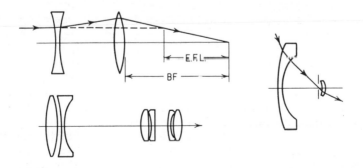

FIG. 13.30. The reverse Telephoto is characterized by a long-back focus which is useful for short focal length lenses. In extreme forms (right hand sketch) the coverage can be made to exceed 180°.

Afocal Attachments: These usually take the form of Galilean or reversed-Galilean telescopes as indicated in Fig. 13.31. The focal length of the "prime" lens is multiplied by the magnification of the telescopic attachment. The field of view limits the power of the telephoto types to about 1.5×, but the wide angle type of attachment is useful to about 0.5× or lower. Such systems are, of course, designed to use an external stop (that of the prime lens) and frequently require a bit of "stopping-down" to achieve satisfactory imagery, especially in the simpler constructions.

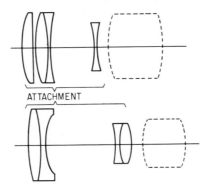

FIG. 13.31. The focal length of a prime lens can be modified by the use of an afocal attachment, which is basically a Galilean Telescope. The upper sketch shows a "telephoto" attachment which increases the focal length. The lower system is a "wide-angle" which reduces the focal length.

13.4 Condenser Systems

The condenser in a projection system is quite analogous to the field lens in a telescope or radiometer. The function of the condenser is illustrated in Fig. 13.32. The upper sketch shows a projection system without a condenser. It is apparent that for the axial object point *A* only about half the lens area can be used, for point *B* only a small fraction of the lens is utilized, and that no light from the lamp passing through point *C* can pass through the projection lens. The result is that the illumination at the projected image is not as high as it might be, and drops off rapidly away from the axis. This can be alleviated somewhat by moving the lamp closer to the film, and, in a very few cases, this solution is satisfactory, if inefficient. However, the filament is usually not uniform enough to allow it to be projected directly without producing objectionable non-uniformity of illumination at the image.

The projection condenser shown in the lower sketch of Fig. 13.32 images the lamp filament directly into the aperture of the projection lens. If the image size is equal to (or greater than) the

aperture size, the illumination is optimized, and if the condenser
has a sufficient diameter, the illumination over the full image
field is as uniform as possible. The requirements for an ideal
condenser may be expressed as follows: the image of the filament
must completely fill the projection lens aperture through a small
pinhole placed anywhere in the field (i.e. at the film plane).

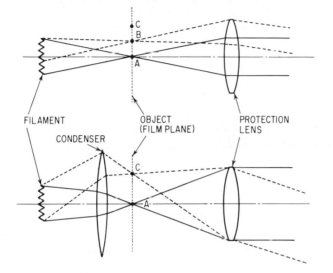

FIG. 13.32. The projection condenser produces an image of the source in the pupil of the
projection lens. Note that the minimum condenser diameter for optimum illumination at the
image of point C is determined by a line through C and the opposite rim of the pupil.

The chief aberrations of concern in condenser systems are
usually spherical and chromatic aberrations; coma, field curvature
and distortion are of secondary importance in ordinary systems. Figure 13.33 is an exaggerated sketch of a condenser

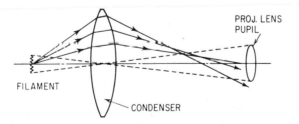

FIG. 13.33. Spherical aberration in a condenser can cause the rays through the margin of
the condenser to completely miss the aperture of the projection lens.

afflicted with spherical aberration. Note that the filament image
formed by the marginal zone of the condenser completely misses

the projection lens aperture, resulting in a marked fall off in illumination at the edge of the field. This situation could be alleviated by reducing the condenser power so that the marginal focus was at the aperture; however, in difficult cases this will result in a dark zonal ring in the field because the zonal rays will miss the aperture. The effects of chromatic aberration are similar, except that one end of the spectrum (red or blue) may miss the aperture and cause an unevenly colored field of view.

Except in unusual cases (e.g. some microscope condensers) chromatic effects can be held to a tolerable level without achromatizing. Spherical aberration is controlled by splitting the condenser into two or three elements of approximately equal power and bending each element toward the "minimum spherical" shape, as indicated in Fig. 13.34A and B. An aspheric surface can be moulded on one of the elements to reduce the spherical, as in Fig. 13.34C. The aspheric is usually a simple paraboloid, and a moulded surface can be sufficiently precise to meet the requirements of a condenser system.

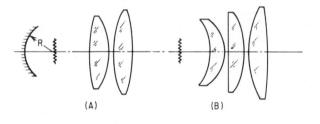

(A)   (B)

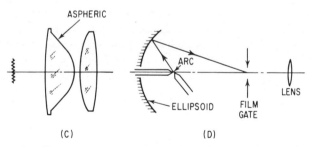

(C)   (D)

FIG. 13.34. Condensing Systems: A) Two element design with reflector concentric to source. B) Three elements, shaped to minimize spherical. C) Aspheric surfaces can be used to reduce spherical. D) The crater of a carbon arc can be imaged directly at the film gate by an ellipsoidal mirror.

When the light source is uniformly bright, it can be imaged directly on the film gate. In arc motion picture projectors, an ellipsoidal mirror is used for this purpose, as shown in Fig. 13.34D. Note that for full illumination the mirror must be large enough to accept the ray from the bottom of the projection lens aperture through the top of the film gate, just as in the regular condenser.

The ellipsoidal mirror is used since it has no spherical aberration when the arc is at one focus and the image (film gate) is at the other. Note that an ellipsoid does have a substantial coma however, and thus off-axis imagery through the margin of the mirror may depart considerably from that predicted by first order optics.

Most condensing systems can be improved by the addition of a spherical reflector behind the light source, as indicated in Fig. 13.34A. If the source is at the center of curvature, the mirror images the source back on itself, effectively increasing the average brightness of the source. With a lamp filament of relatively open construction, such as a V shape, or two parallel coils, the increase in illumination may approach the reflectivity of the reflector, i.e. 85 to 90%. The gain is much less in a tightly packed source, but even a biplane filament will gain 5 or 10% from a properly aligned reflector.

13.5 Reflecting Systems

The increasing use of optical systems in the non-visual regions of the spectrum, i.e., the ultraviolet and infrared regions, has resulted in a corresponding increase in the use of reflecting optics. This is due primarily to the difficulty in procuring completely satisfactory refractive materials for these regions, and secondarily, to the fact that many of the applications permit the use of relatively unsophisticated mirror systems.

The material difficulty is of two kinds. Many applications require the use of a broad spectral band, and a refractive material must transmit well over the full band to be of value. Secondly, chromatic aberration can be difficult to correct over a wide spectral band, and the residual secondary spectrum is often intolerable. A review of Chapter 7 will demonstrate quite clearly the advantages of a reflector in this regard; an ordinary aluminized mirror actually has much better reflectance in the infrared than in the visible and (with special attention) aluminum mirrors suitable for the ultraviolet can be fabricated. Pure reflecting systems are, of course, completely free of chromatic aberration over any desired band width.

The Spherical Mirror: The simplest reflecting objective is the spherical mirror. For distant objects the spherical mirror has undercorrected spherical aberration, but the aberration is only one-eighth of that of an equivalent glass lens at "minimum bending." The sphere is an especially interesting system when the aperture stop is located at the center of curvature, as shown in Fig. 13.35, because the system is then concentric and any line through the center of the stop may be considered to be the optical axis. The image quality is thus practically uniform for any angle of obliquity and the only aberration present is spherical aberration. Coma and astigmatism are zero, and the image surface is a sphere of radius

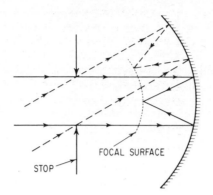

FIG. 13.35. A spherical reflector with the stop at its center of curvature forms its image on a concentric spherical focal surface. The image is free of coma and astigmatism when the stop is at this position.

approximately equal to the focal length, centered about the center of curvature. We can approximate the spherical aberration by use of the third order surface contribution equations. Setting $N = -N' = 1.0$ in Eq. 10.8G, we find that

$$SC = \frac{y^2}{4R} \qquad \left[SC = \frac{(m - 1)^2}{4R} y^2 \right] \qquad (13.12)$$

where y is the semi-aperture, R is the radius, and m is the magnification. The first expression applies for an infinite object distance and the bracketed expression applies to finite conjugates. Using Eq. 11.26 to determine the minimum diameter of the blur spot, B, we find that

$$B = \frac{y^3}{4R^2} \qquad \left[B = \frac{(m - 1)^2 y^3}{(m + 1) 4R^2} \right] \qquad (13.13)$$

This expression can be converted into the angular blur (in radians) by dividing by the image distance l' (or focal length) to get

$$\beta = \frac{y^3}{2R^3} \qquad \left[\beta = \frac{(m - 1)^2 y^3}{(m + 1)^2 2R^3} \right] \qquad (13.14)$$

By substituting $f = R/2$ and $(f/\#) = f/2y = R/4y$ = relative aperture, we obtain the following convenient expression for the angular blur size of a spherical mirror as a function of its speed (for infinite object distance)

$$\beta = \frac{1}{128\,(f/\#)^3} = \frac{.00781}{(f/\#)^3} \text{ radians} \tag{13.15}$$

Although this is exact for the third order spherical only, the expression is quite reliable up to speeds of $f/2$. At $f/1$ the exact value of β is .0091, at $f/0.75$ it is about .024, and at $f/0.5$ it is about 0.13 radians.

When the stop is **not** at the center of curvature, coma and astigmatism are present, and (for an infinite object distance) the third order contributions are

$$CC^* = \frac{y^2(l_p - R)u_p}{2R^2} \tag{13.16}$$

$$AC^* = \frac{(l_p - R)^2 u_p^2}{4R} \tag{13.17}$$

$$PC = \frac{u_p^2 R}{4} \tag{13.18}$$

where u_p is the half field angle and l_p is the mirror to stop distance. Note that when l_p is equal to R, CC^* (the sagittal coma) and AC^* (one half the separation of the S and T fields) are zero. For the case of the stop located at the mirror, we find the minimum angular blur sizes to be

$$\text{Coma}_s: \quad \beta = \frac{u_p}{16\,(f/\#)^2} \text{ radians} \tag{13.19}$$

$$\text{Astigmatism}: \quad \beta = \frac{u_p^2}{2\,(f/\#)} \text{ radians} \tag{13.20}$$

Equations 13.15, 13.19 and 13.20 provide a very convenient way of estimating the image size for a spherical mirror, when combined with the knowledge that 1) coma and astigmatism are zero with the stop at the center of curvature and 2) coma varies linearly (per Eq. 13.16) and astigmatism varies quadratically (per Eq. 13.17) with the distance of the stop from the center of curvature. The sum of the spherical, coma, and astigmatism blur angles gives a fair estimate of the effective size of a point image for a spherical mirror.

The Paraboloidal Reflector: Reflecting surfaces generated by rotation of the conic sections (circle, parabola, hyperbola and ellipse) share a valuable optical property, namely that a point object located at one focus is imaged at the other focus without spherical aberration. The paraboloid of revolution, Fig. 13.36, described by the equation

$$x = \frac{y^2}{4f} \tag{13.21}$$

has one focus at f and the other at infinity, and is thus capable of forming perfect (diffraction limited) images of distant *axial* objects.

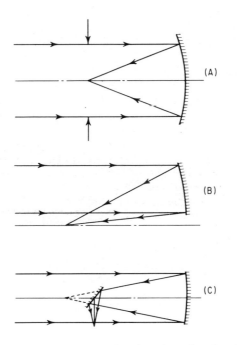

FIG. 13.36. A) A parabolic reflector is free of astigmatism when the stop is at the focus. B) The Herschel mount for a paraboloid uses an off-axis aperture to keep the focus out of the entering beam. C) The Newtonian mount utilizes a 45° plano reflector to direct the focus to an accessible point outside the main tube of the telescope.

However, the paraboloid is not completely free of aberrations; it has both coma and astigmatism. Since it has no spherical aberration, the position of the stop does not change the amount of coma, which is given by Eq. 13.19. The amount of astigmatism *is* modified by the stop position. With the stop at the mirror the astigmatism is given by Eq. 13.20; when the stop is at the focal plane, the astigmatism is zero and the image is located on a spherical surface of radius f.

The Ellipsoid and Hyperboloid: The imaging properties of these conic sections are made use of in the Gregorian and Cassegrain telescopic systems, as indicated in Figs. 13.37 and 13.38, respectively.

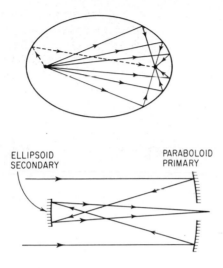

FIG. 13.37. Upper: A point object at one focus of an elliptical reflector is imaged at the other focus without spherical aberration. Lower: The classical Gregorian telescope uses a parabolic primary mirror and an elliptical secondary so that the image is free of spherical.

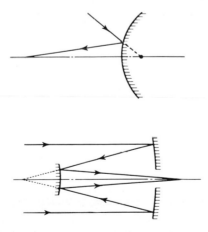

FIG. 13.38. Upper: A ray directed toward one focus of a hyperbola is reflected through the other focus. Lower: The classical Cassegrain objective uses a parabolic primary mirror with a hyperboloid secondary. When the primary image is at the focus of the secondary mirror, the final image has no spherical aberration. If the osculating radii of the surfaces are equal, the Petzval field is flat.

The primary mirror in each of these is a paraboloid which produces an aberration-free axial image at its focus. The secondary mirror is located so that its first focus coincides with the focus of the paraboloid. Thus the final image is located at the second focus of the secondary and is completely free of spherical aberration. The paraboloid, ellipsoid, and hyperboloid all suffer from coma (compare the magnification produced by the dotted vs the solid lines in Fig. 13.37) and astigmatism, so that the image is aberration-free only on the axis.

It should be apparent that either the Gregorian or Cassegrain objective systems could be made up with almost any arbitrary (within reason) surface of rotation for the primary; some surface then could be found for the secondary which would produce a spherical-free image. This is, in effect, an extra degree of freedom which can be used by the designer to improve the off-axis imagery of these systems. The third order surface contribution equations for aspheric surfaces (Eqs. 10.8) are very useful in this regard.

Note that the conics appear to violate the principles of image illumination laid down in Chapter 8. For example, a paraboloid can readily be constructed with a diameter more than twice its focal length; a paraboloid with a speed of say $f/0.25$ is quite feasible and will indeed be free of spherical aberration on the axis, whereas in preceding chapters, we may have led the reader to believe that a speed of $f/0.5$ was the largest aperture attainable.

This apparent paradox can be resolved by an examination of Fig. 13.39 which shows an $f/0.25$ parabola. Note that the focal length is equal to f *only* for the axial zone and that for marginal zones the focal length is much larger; for marginal zones the focal length of a parabola is given by

$$F = f + x = f + \frac{y^2}{4f} \qquad (13.22)$$

The parabola is thus far from an aplanatic (spherical and coma free) system. For an $f/0.25$ paraboloid the marginal zone focal length is twice that of the paraxial zone and the magnification is correspondingly larger. Thus, if the object has a finite size, the image formed by the marginal zones of this mirror will be twice as large as those from the axial zone; this is, of course, nothing but ordinary coma (the "variation of magnification with aperture").

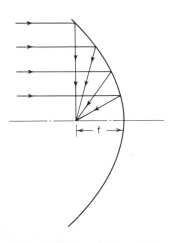

FIG. 13.39. Illustrating the extreme variation of focal length with ray height in an $f/0.25$ parabolic reflector.

The parabola is thus aberration-free only *exactly* on the axis.

The apparent contradiction of our image illumination principles is thus resolved since we had assumed aplanatic systems in their derivations. From another viewpoint, we can remember that although the parabola forms a perfect image of an infinitesimal (geometrical) point, such a point (being infinitesimal) cannot emit a real amount of energy; the moment one increases the object size to any real dimension, the parabola has a real field, the image becomes comatic, and the energy in the image is spread out over a finite blur spot. This reduces the image illumination to that indicated as the maximum in Chapter 8.

The Cassegrain objective system is used (usually in a modified form) in a great variety of applications because of its compactness and the fact that the second reflection places the image behind the primary mirror where it is readily accessible. It suffers from a very serious drawback when an appreciable field of view is required, in that an extreme amount of baffling is necessary to prevent stray radiation from flooding the image area. Figure 13.40 indicates this difficulty and the type of baffles frequently used to overcome this problem. An exterior "sun-shade," which is an extension of the main exterior tube of the scope, is frequently used in addition to the internal baffles.

BAFFLES

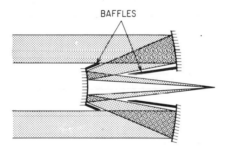

FIG. 13.40. Complex conical baffles are necessary in a Cassegrain objective to prevent stray radiation from flooding the image plane.

Because of their uniaxial character, aspheric surfaces are much more difficult to fabricate than ordinary spherical surfaces. A strong paraboloid may cost an order of magnitude more than the equivalent sphere; ellipsoids and hyperboloids are a bit more difficult, and non-conic asherics are still more difficult. Thus one might well think twice (or three times) before specifying an aspheric. Frequently a spherical system can be found which will do nearly as well at a fraction of the cost. This is also true in refracting systems of moderate size where several ordinary spherical elements can be purchased for the cost of a single aspheric. For very large, one of a kind, systems, however, aspherics are frequently a sound choice. This is because the large systems (e.g., astronomical objectives) are, in the final analysis, hand made, and the aspheric surface adds only a little to the optician's task.

Conic Section through the Origin:

$$y^2 - 2rx + px^2 = 0$$

$$x = \frac{r \pm \sqrt{r^2 - py^2}}{p}$$

$$x = \frac{y^2}{2r} + \frac{1}{2^2\,2!}\frac{py^4}{r^3} + \frac{1 \cdot 3}{2^3\,3!}\frac{p^2y^6}{r^5} + \frac{1 \cdot 3 \cdot 5}{2^4\,4!}\frac{p^3y^8}{r^7} + \frac{1 \cdot 3 \cdot 5 \cdot 7}{2^5\,5!}\frac{p^4y^{10}}{r^9} + \cdots$$

Ellipse	$p > 1$
Circle	$p = 1$
Ellipse	$1 > p > 0$
Parabola	$p = 0$
Hyperbola	$p < 0$

distance to foci:

$$\frac{r}{p}(1 \pm \sqrt{1 - p})$$

magnification:

$$-\left[\frac{1 + \sqrt{1 - p}}{1 - \sqrt{1 - p}}\right]$$

intersects axis at:

$$x = 0 \,, \quad \frac{2r}{p}$$

distance between conic and a circle (i.e., departure from a sphere):

$$\Delta x = \frac{(p - 1)}{2^2\,2!}\frac{y^4}{r^3} + \frac{1 \cdot 3\,(p^2 - 1)}{2^3\,3!}\frac{y^6}{r^5} + \frac{1 \cdot 3 \cdot 5\,(p^3 - 1)}{2^4\,4!}\frac{y^8}{r^7} + \cdots$$

angle between the normal to the conic and the x-axis:

$$\phi = \tan^{-1}\left[\frac{-y}{(r - px)}\right]$$

$$\sin\phi = \frac{-y}{\left[y^2 + (r - px)^2\right]^{\frac{1}{2}}}$$

radius of curvature:

meridional:
$$R_t = \frac{R_s^{\,3}}{r^2} = \frac{\left[y^2 + (r - px)^2\right]^{3/2}}{r^2}$$

sagittal (distance to axis along normal):

$$R_s = \left[y^2 + (r - px)^2\right]^{\frac{1}{2}}$$

The Schmidt System: The Schmidt objective (Fig. 13.41) can be viewed as an attempt to combine the wide uniform image field of the stop-at-the-center sphere with the "perfect" imagery of the paraboloid. In the Schmidt, the reflector is a sphere and the spherical aberration is corrected by a thin refracting aspheric plate at the center of curvature. Thus the concentric character of the sphere is preserved in great measure, while the spherical aberration is completely eliminated (at least for one wavelength).

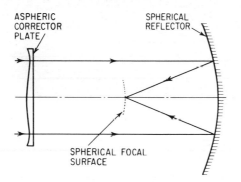

FIG. 13.41. The Schmidt System consists of a spherical reflector with an aspheric corrector plate at its center of curvature. The aspheric surface in the $f/1.$ system shown here is greatly exaggerated.

The aberrations remaining are chromatic variation of spherical and certain higher-order forms of astigmatism or oblique spherical which result from the fact that the off-axis ray bundles do not strike the corrector at the same angle as do the on-axis bundles. The action of a given zone of the corrector is analogous to that of a thin refracting prism. For the axial bundle the prism is near minimum deviation; as the angle of incidence changes, the deviation of the "prism" is increased, introducing a sort of overcorrected oblique spherical. Also, since the action is different in the tangential plane than in the sagittal plane, astigmatism results. The meridional angular blur of a Schmidt system is well approximated by the expression

$$\beta = \frac{u_p^2}{24(f/\#)^3} \text{ radians} \tag{13.23}$$

There are obviously an infinite number of aspheric surfaces which may be used on the corrector plate. If the focus is maintained at the paraxial focus of the mirror, the *paraxial* power of the corrector is zero and it takes the form of a weak concave surface. The best forms have the shape indicated in Fig. 13.41, with a convex paraxial region and the minimum thickness at the 0.707 or 0.866 zone, depending on whether it is desired to minimize chromatic aberration or to optimize the off-axis correction. The performance of the Schmidt can be improved slightly by a) incompletely correcting the axial spherical to compensate for the off-axis overcorrection, b) "bending" the corrector slightly, c) reducing the spacing, d) using a slightly aspheric primary to reduce the load on, and thus the overcorrection introduced by, the corrector. Further improvements have been made by using more than one corrector and by using an achromatized corrector.

The aspheric corrector of the Schmidt is usually easier to fabricate than is the aspheric surface of the paraboloid reflector. This is because the index difference across the (glass) corrector surface is about 0.5 compared to the effective index difference of 2.0 at the reflecting surface of the paraboloid, making it only one-fourth as sensitive to fabrication errors.

The Mangin Mirror: The Mangin mirror is perhaps the simplest of the catadioptric (that is, combined reflecting and refracting) systems. It consists of a second-surface spherical mirror with the power of the first surface chosen to correct the spherical aberration of the reflecting surface. Figure 13.42 shows a Mangin mirror. The design of a Mangin is straightforward. One radius is chosen arbitrarily (a value about 1.6 times the focal length is suitable for the reflector surface) and the other radius is varied systematically until the spherical aberration is corrected. The correction is exact for only one zone, however, and an undercorrected zonal residual remains. The size of the angular blur spot resulting from the zonal spherical can be approximated (for apertures smaller than $f/1.0$) by the empirical expression

$$\beta = \frac{10^{-3}}{4(f/\#)^4} \text{ radians} \tag{13.24}$$

Note that this is the minimum diameter blur and that the "hard core" blur diameter is smaller, as discussed in Chapter 11. At larger apertures, the angular blur predicted by Eq. 13.24 is too small; for example, at f/0.7 the blur is about 0.002 radians, almost twice as large as that given by Eq. 13.24.

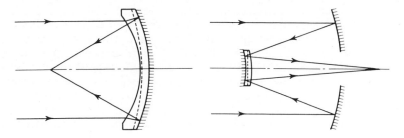

FIG. 13.42. In the Mangin Mirror (left) the spherical aberration of the second surface reflector is corrected by the refracting first surface. In the right hand sketch, the spherical is corrected by a Mangin type secondary. The dotted lines indicate the manner in which color correction can be achieved.

Since the Mangin is roughly equivalent to an (achromatic) reflector plus a pair of simple negative lenses, the system has a large overcorrected chromatic aberration. This can be readily corrected by making an achromatic doublet out of the refracting element. For the simple Mangin, the chromatic angular blur is approximated by

$$\beta = \frac{1}{6V(f/\#)} \text{ radians} \tag{13.25}$$

where V is the Abbe V-value of the material used.

The coma blur of the Mangin primary mirror is approximately one-half of that given by Eq. 13.19. Since the (marginal) spherical is corrected, little change in the coma results from a shift of the stop position.

The Mangin principle may be applied to the secondary mirror of a system as well as to the primary. The right hand sketch of Fig. 13.42 shows a Cassegrain type of system in which the secondary is an achromatic Mangin mirror. Such a system is relatively economical and light in weight, since all surfaces are spheres and only the small secondary needs to be made of high quality optical material. The power of a thin second surface reflecting element is given by

$$\phi = 2C_1(N - 1) - 2C_2N$$

The Bouwers-Maksutov System: The Bouwers (or Maksutov) system may be considered a logical extension of the Mangin mirror principle in which the correcting lens is separated from the mirror to allow two additional degrees of freedom, producing a great improvement in the image quality of the system.

The most popular version of this device is the Bouwers concentric system, shown in Fig. 13.43. In this system, all surfaces are made concentric to the aperture stop, which (as we have noted in the case of the simple spherical mirror) results in a system with uniform image quality over the entire field of view. This is an exceedingly simple system to design, since there are only three degrees of freedom, namely, the three curvatures. One chooses R_1 to set the scale of the lens (a value of R_1 equal to about 85% of the intended focal length is appropriate), R_2 to provide an appropriate thickness for the corrector and then determines the value of R_3 for which the marginal spherical is zero. Because of the concentric construction, coma and astigmatism are zero, and the image is located on a spherical surface which is also concentric to the stop and whose radius equals the focal length of the system. Thus only a few rays need be traced to completely determine the correction of the system.

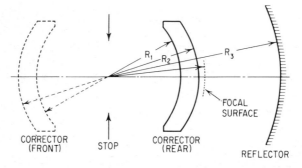

FIG. 13.43. In the Bouwers concentric catadioptric system, all the surfaces are concentric about the aperture. The "front" and "rear" versions of the corrector are identical and produce identical correction. The "rear" system is more compact, but the "front" system can be better corrected, since it can utilize a greater corrector thickness without interference with the focal surface. Occasionally, correctors in both locations are utilized simultaneously.

One of the interesting features of this system is that the concentric corrector element may be inserted anywhere in the system (as long as it remains concentric) and it will produce exactly the same image correction. Two equivalent positions for the corrector are shown in Fig. 13.43. A third position is in the convergent beam, between the mirror and the image.

If we accept the curved focal plane, the only aberrations of the Bouwers concentric system are residual zonal spherical aberration and longitudinal (axial) chromatic aberration. In general, as the corrector thickness is increased, the zonal is reduced and the chromatic is increased.

The concentric system described above is used for most applications requiring a wide field of view. When the field requirements permit, the zonal spherical or the chromatic may be reduced by departing from the concentric mode of construction, although this is, of course, accomplished at the expense of the coma and astigmatism correction.

Another means of effecting chromatic correction is shown in Fig. 13.44A, in which the corrector meniscus is made achromatic. Note that concentricity is destroyed by this technique, although if the crown and flint elements are made of materials with the same index but different V-values (e.g., DBC-2, 617:549, and DF-2, 617:366), the concentricity can be preserved for the wavelength at which the indices match, and only the chromatic correction will vary with obliquity.

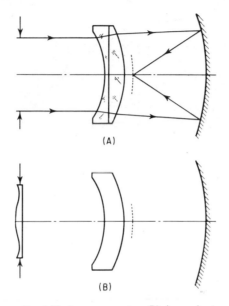

(A)

(B)

FIG. 13.44. A) An achromatized Meniscus corrector. B) An aspheric corrector plate at the stop removes the residual zonal spherical aberration of the concentric system.

A very powerful system results if the concentric Bouwers system is combined with a Schmidt type aspheric corrector plate, as shown in Fig. 13.44B. Since the aspheric plate need only correct the small zonal residual of the concentric system, its effects are relatively weak and the variation of effects with obliquity are correspondingly small. The Baker-Nunn satellite tracking cameras are based on this principle, although their construction is more elaborate, using double meniscus correctors and three (achromatized) aspheric correctors at the stop.

The basic Bouwers-Maksutov meniscus corrector principle has been utilized in a multitude of forms. A few of the possible Cassegrain embodiments of the principle are shown in Fig. 13.45. The reader can probably devise an equal number in a few minutes. An arrangement similar to that shown in Fig. 13.45C is frequently used in homing missile-guidance systems. The corrector makes a reasonably aerodynamic window, or dome, and although the system

is not concentric, the primary and secondary can be gimballed as
a unit about the center of the dome so that the "axial" correction
is maintained as the direction of sight is varied.

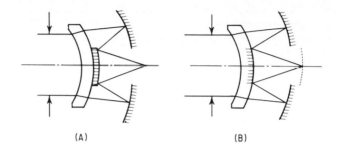

(A)   (B)

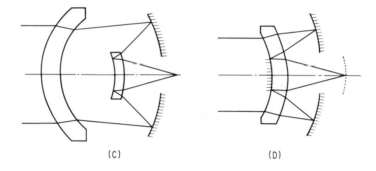

(C)   (D)

FIG. 13.45. Four of the many possible Cassegrain versions of the Meniscus corrector
catadioptric system.

There are a tremendous number of variations of the catadioptric
principle. Refractive correctors in almost every conceivable form
have been combined with mirrors. Positive field lenses have been
used to flatten the overcorrected Petzval surface of the basic
concave reflector, coma correcting field elements have been used
with paraboloids, and multiple non-meniscus correctors have been
used with spheres, to name just a few of the variations on the
device. The basic strength of this general system is, of course, the
relatively small aberration inherent in a spherical reflector; the
corrector's task is to remove the faults without losing the virtues.

13.6 The Rapid Estimation of Blur Sizes
for Simple Optical Systems

It is frequently useful to be able to estimate the size of the
aberration blur produced by an optical system, without going to the

trouble of making a raytrace analysis. In preliminary engineering work, or the preparation of technical proposals, where time is limited, the following material (which is based largely on third-order aberration analysis or empirical studies) can be of great value.

The aberrations are expressed in terms of the angular size β (in radians) of the blur spot which they produce; β may be converted to B, the linear diameter of the blur, by multiplying by the system focal length. In this section the object will be assumed to be at infinity.

Where the blur size for more than one aberration is given, the sum of all the aberration blurs will yield a conservative (i.e., large) estimate of the total blur.

Where the blur is due to chromatic aberration, the blur angle given encompasses the total energy. Occasionally it is of value to know that 75 to 90% of the energy is contained in a blur one-half as large, and 40 to 60% of the energy is contained in a blur one-quarter as large, as that given by the equations. In the visible, the chromatic blur is usually reduced by a factor of about 40 by achromatizing the system.

The blurs given for spherical aberration are the minimum diameter blur sizes; these values are the most useful for work with detectors. For visual or photographic work, a "hard core" focus, as discussed in Chapter 11, is preferable, and the blurs given here should be modified accordingly.

Note that with the exception of Eq. 13.26 and 13.27, all the blurs are based on geometrical considerations. It is, therefore, wise to evaluate Eq. 13.26 or 13.27 first to be certain that the geometrical blurs are not smaller than the diffraction pattern, before basing further effort on the geometrical results.

More complete discussions of the individual systems may be found in the preceding section.

Diffraction Limited Systems: The diameter of the first dark ring of the Airy pattern is given by

$$\beta = \frac{2.44\,\lambda}{D} \text{ radians} \qquad (13.26)$$

$$B = 2.44\,\lambda(f/\#) \qquad (13.27)$$

where λ is the wavelength, D is the clear aperture of the system, $(f/\#) - f/D$ is the relative aperture, and f is the focal length. The "effective" diameter of the blur (for modulation transfer purposes) is one-half the above.

Spherical Mirror:

$$\text{Spherical Aberration}: \quad \beta = \frac{.0078}{(f/\#)^3} \text{ radians} \qquad (13.28)$$

$$\text{Sagittal Coma:} \quad \beta = \frac{(l_p - R)U_p}{16R(f/\#)^2} \text{ radians} \qquad (13.29)$$

$$\text{Astigmatism:} \quad \beta = \frac{(l_p - R)^2 U_p^2}{2R^2(f/\#)} \text{ radians} \qquad (13.30)$$

where l_p is the mirror to stop distance, R is the mirror radius, $(l_p - R)$ is the center to stop distance, and U_p is the half field angle in radians. The focal plane of a spherical mirror is on a spherical surface concentric to the mirror, when the stop is at the center of curvature.

Paraboloidal Mirror:

$$\text{Spherical Aberration:} \quad \beta = 0 \qquad (13.31)$$

$$\text{Sagittal Coma:} \quad \beta = \frac{U_p}{16(f/\#)^2} \text{ radians} \qquad (13.32)$$

$$\text{Astigmatism:} \quad \beta = \frac{(l_p + f)^2 U_p^2}{2f^2(f/\#)} \text{ radians} \qquad (13.33)$$

where the symbols have been defined above.

Schmidt System:

$$\text{Spherical Aberration:} \quad \beta = 0 \qquad (13.34)$$

$$\text{Higher Order Aberrations:} \quad \beta = \frac{U_p^2}{24(f/\#)^3} \text{ radians} \qquad (13.35)$$

Mangin Mirror: (stop at the mirror)

$$\text{Zonal Spherical:} \quad \beta = \frac{10^{-3}}{4(f/\#)^4} \text{ radians} \qquad (13.36)$$

$$\text{Chromatic Aberration:} \quad \beta = \frac{1}{6V(f/\#)} \text{ radians} \qquad (13.37)$$

$$\text{Sagittal Coma:} \quad \beta = \frac{U_p}{32(f/\#)^2} \text{ radians} \qquad (13.38)$$

$$\text{Astigmatism:} \quad \beta = \frac{U_p^2}{2(f/\#)} \text{ radians} \qquad (13.39)$$

Simple Thin Lens (Minimum Spherical Shape):

$$\text{Spherical Aberration}: \quad \beta = \frac{K}{(f/\#)^3} \text{ radians} \qquad (13.40)$$

$$
\begin{aligned}
K &= .067 \quad \text{for } N = 1.5 \\
&= .027 \qquad\qquad = 2.0 \\
&= .0129 \qquad\quad\; = 3.0 \\
&= .0087 \qquad\quad\; = 4.0
\end{aligned}
$$

$$\text{Chromatic Aberration}: \quad \beta = \frac{1}{2V(f/\#)} \text{ radians} \qquad (13.41)$$

$$\text{Sagittal Coma}: \quad \beta = \frac{U_p}{16(N+2)(f/\#)^2} \text{ radians} \qquad (13.42)$$

$$\text{Astigmatism}: \quad \beta = \frac{U_p^2}{2(f/\#)} \text{ radians} \qquad (13.43)$$

where N is the index of refraction, V is the reciprocal relative dispersion, and the stop is at the lens.

Concentric Bouwers: The expressions for monochromatic aberrations are empirical and are derived from the performance graphs and tables given by Bouwers and by Lauroesch and Wing (see references).

Rear Concentric (solid line in Fig. 13.43): The maximum corrector thickness of this form must be limited to keep the image from falling inside the corrector. With the thickest possible corrector:

$$\text{Zonal Spherical}: \quad \beta \approx \frac{4 \times 10^{-4}}{(f/\#)^{5.5}} \text{ radians} \qquad (13.44)$$

General Concentric:

$$\text{Zonal Spherical}: \quad \beta \approx \frac{10^{-4}}{\left(\dfrac{t}{f} + .06\right)(f/\#)^5} \text{ radians} \qquad (13.45)$$

$$\text{Chromatic Aberration}: \quad \beta \approx \frac{tf\,\Delta N}{2N^2 R_1 R_2 (f/\#)} \text{ radians} \qquad (13.46A)$$

$$\text{or very approximately}: \quad \beta \approx 0.6\,\frac{t}{f}\,\frac{\Delta N}{N^2(f/\#)} \text{ radians} \qquad (13.46B)$$

Corrected Concentric:

Higher Order Aberrations : $\quad \beta \approx \dfrac{9.75\,(U_p + 7.2U_p^3)}{(f/\#)^{6.5}}$ radians   (13.47)

where t is the corrector plate thickness, f is the system focal length, ΔN is the dispersion of the corrector material, N is the index of the corrector, and R_1 and R_2 are the radii of the corrector. These expressions apply for corrector index values in the 1.5 to 1.6 range and for relative apertures to the order of $f/1.0$ or $f/2.0$. For speeds faster than $f/1.0$, the monochromatic blur angles are larger than above (e.g., about 20% larger at $f/0.7$). The use of a high index corrector $(N > 2)$ will reduce the monochromatic blur somewhat at high speeds.

The charts of Figs. 13.46, 13.47, 13.48, 13.49, 13.50, and 13.51 are designed to give a rapid, albeit incomplete, estimation of the performance of the systems discussed in this section.

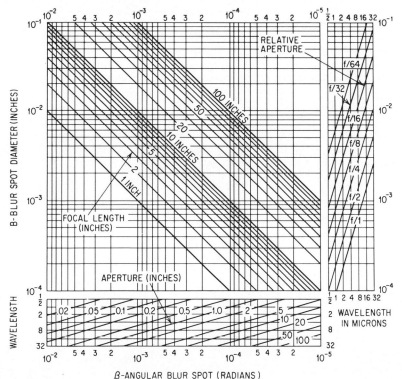

FIG. 13.46. Blur Spot Size Chart for diffraction limited systems. Diameter is that of the first dark ring of the Airy disc.

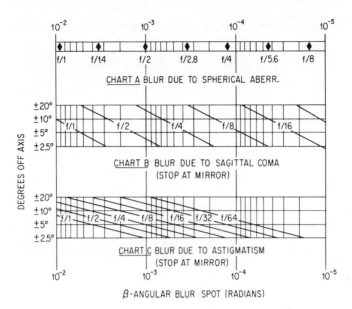

FIG. 13.47. Blur spot size charts for spherical reflector. Charts B and C also apply to a paraboloidal reflector; Fig. 13.46 may be used for a paraboloid on axis.

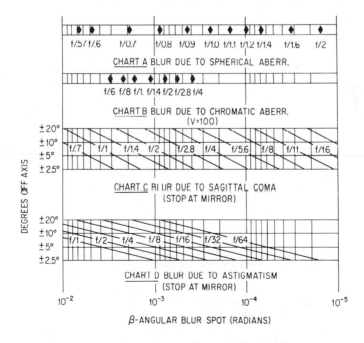

FIG. 13.48. Blur spot size charts for Mangin Mirrors.

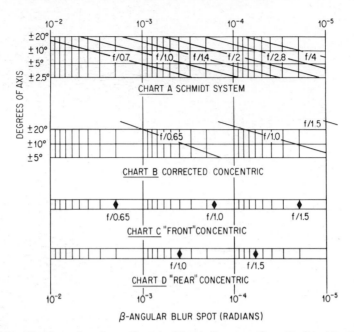

FIG. 13.49. Blur Spot size chart. Chart A: Schmidt Systems. Chart B, C and D: Concentric Bouwers Systems.

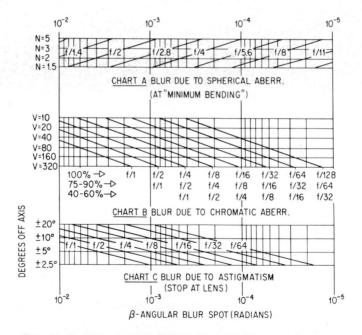

FIG. 13.50. Blur spot size charts for a single refracting element.

Figure 13.46 is used by locating the intersection of a wavelength line with the appropriate diagonal aperture line. The linear blur spot diameter B may be converted to the angular blur spot diameter β by locating the abscissa corresponding to the intersection of the B diameter ordinate with the appropriate diagonal focal length line.

Figures 13.47, 13.48, and 13.50 assume that the aperture stop is in contact with the lens or mirror. The blur size for the angle-dependent aberrations is found by locating the intersection of a horizontal field angle line with the diagonal f-number line.

A very rough idea of the geometrical modulation transfer factor of the system can be obtained by using Fig. 13.51. The total angular blur spot for the system is determined (by summing the individual aberration blurs) and is then multiplied by the desired spatial frequency in cycles per radian. The modulation transfer factor may then be read directly from the figure.

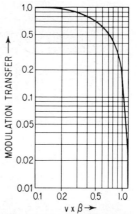

FIG. 13.51. The Modulation Transfer characteristic of a system with an angular blur, β, (in radians) for a sinusoidal object with a spatial frequency of v cycles per radian. This is a plot of M.T.F. $= 2J_1(\pi\beta v)/\pi\beta v$ and assumes that the image blur is a uniformly illuminated disc.

Note that this is very approximate and is not reliable when the total blur is of the same order of magnitude as the Airy disc (see the discussion in Chapter 11).

References

See the references for Chapter 12.

1. Bouwers, "Achievements in Optics," Elsevier, 1950.
2. Journal of the Optical Society of America, Maksutov, Vol. 34, p. 270 (1944); Lauroesch and Wing, V, 49, p. 410 (1959).

Exercises

See the note preceding the exercises for Chapter 12.

1. Design a symmetrical eyepiece, using BSC-2 and EDF-1 glass, for a 10× telescope.

2. Design a separable 10× NA 0.25 microscope objective by determining the zero spherical form for each doublet. Analyze the field curvature of the combination.

3. Design a 20× (8 mm) two-mirror reflecting microscope objective and determine an appropriate combination of aperture and field over which the aberrations will not exceed the Rayleigh limit.

4. Design a split-front-crown triplet (Fig. 13.20), using DBC-1 and EDF-2 glasses, for a speed of $f/2.8$.

5. For an aperture of $f/2$ and a half field of 0.1 radians, determine the relative angular blur spot size of the various systems listed in Section 13.6. Where stop position is critical, consider a) the best position, and b) the most compact arrangement. Assume a V of 100 for the refracting materials.

6. Design an $f/1$ Bouwers concentric system. Achromatize the design (use DBC-2 and DF-2 glass) and analyze the off-axis chromatic and chromatic variation of the aberrations.

14

Optics in Practice

This chapter will briefly survey the factors involved in reducing an optical system to practice. A short description of the optical manufacturing process will be followed by a discussion of the specification and tolerancing of optics for the shop. The mounting of optical elements will be considered next, and the chapter will be concluded with a section on optical laboratory measurement techniques.

14.1 Optical Manufacture

Materials: The starting point for quantity production of optics is most frequently a rough moulded glass blank or pressing. This is made by heating a weighed chunk of glass to a plastic state and pressing it to the desired shape in a metal mould. The blank is made larger than the finished element to allow for the material which will be removed in processing; the amount removed must (at a minimum) be sufficient to clean up the outer layers of the blank which are of low quality and may contain flaws or the powdery sand used in moulding. Typically a lens blank will be 3-mm thicker than the finished lens and 2-mm larger in diameter. A prism blank will be large enough to allow removal of about 2 mm on each surface. These allowances vary somewhat with the size of the piece. When the blank is of an expensive material, such as silicon or one of the more exotic glasses, the blanking allowances are held to the absolute minimum to conserve material.

Although most blanks are single, a cluster form is frequently economical for small elements. A cluster may consist of five or ten blanks connected by a thin web which is ground off to free the individual blanks. If moulded blanks are unobtainable, either because of the small quantity involved or the type of material, a rough blank may be prepared by chipping or sawing a suitable shape from stock material.

Rough blanks can be checked fairly satisfactorily for the presence of strain (which results from poor annealing of the glass) by the use of a polariscope. An accurate check of the index requires that a plano surface be polished on a sample piece; however, if a batch of blanks is known to have been made from a single melt or run of glass, only one or two of the blanks need be checked, because the index within a melt is quite consistent. Since the final annealing process raises the index, the presence of strain is frequently accompanied by a low index value.

Rough Shaping: The preliminary shaping of an element is often accomplished by using diamond charged grinding wheels. In the case of spherical surfaces, the process involved is generating. The blank is rotated in a vacuum chuck and is ground by a rotating annular diamond wheel whose axis is at an angle to the chuck axis, as indicated in Fig. 14.1. The geometry of this arrangement is such that a sphere is generated; the radius is determined by the angle between the axes and by the effective diameter of the diamond tool. The thickness is, of course, governed by how far the work is advanced into the tool. Flat work can be roughed out in a similar manner, with the two axes parallel. Rectangular shapes can be formed by milling, again using diamond tools.

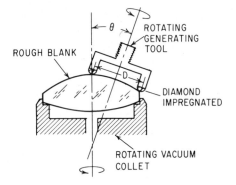

FIG. 14.1. Schematic Diagram of the Generating Process. The annular diamond tool and and the glass blank are both rotated. Since their axes intersect at an angle (θ), the surface of the blank is generated to a sphere of radius $R = D/2 \sin\theta$.

Blocking: It is customary to process optical elements in multiples by fastening or blocking a suitable number on a common support. There are two primary reasons for this: The obvious reason is economy, in that several elements are processed simultaneously; the less apparent reason is that a better surface results when the processing is averaged over the larger area represented by a number of pieces.

The elements are fastened to the blocking tool with pitch, although various compounds of waxes and rosins are also used for special purposes. Pitch has the useful property of adhering tenaciously to almost anything which is hot and not sticking to cold surfaces. The pitch bond is readily broken by chilling the pitch to a brittle state and shocking it with a brisk but light tap. Typically the elements are fastened to the blocker by pitch buttons which are moulded to the back of the elements (suitably warmed); the buttons are then stuck to the heated blocker, as indicated in Fig. 14.2. (The surfaces of the elements are maintained in alignment by placing the buttoned elements into a lay-in tool of the proper radius and then pressing the heated blocker into contact with the pitch buttons.)

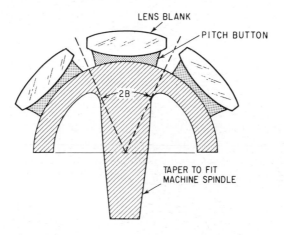

FIG. 14.2. Section of a blocking tool with blanks fastened in place with buttons of blocking pitch. The maximum number of lenses that can be blocked on a tool is determined by the angle B (see Eq. 14.2).

The cost of processing an element is obviously closely related to the number of elements which can be blocked on a tool. There is no simple way to determine this number exactly; however, the following expressions (which are "limiting case" expressions, modified to fit the actual values) are accurate to within about one element per tool.

For plano surfaces:

$$\text{No. per tool} = \frac{3}{4}\left(\frac{D_t}{d}\right)^2 - \frac{1}{2} \tag{14.1}$$

rounded downward to the nearest integer, where D_t is the diameter of the blocking tool and d is the effective diameter of the piece (and should include an allowance for clearance between the elements).

For spherical surfaces:

$$\text{No. per tool} = \frac{1.5}{(\sin B)^2} - \frac{1}{2} \tag{14.2}$$

rounded downward to the nearest integer, where B is the half angle subtended by the lens diameter (plus spacing allowance) from the center of curvature of the surface, as indicated in Fig. 14.2.

Where there are only a few lenses per tool, the following tabulation is convenient:

No. per tool	Maximum D_t/d	Maximum Sin B
2	0.500	0.707
3	0.462	0.655
4	0.412	0.577
5	0.372	0.507
6		0.500
7	0.332	0.447
8	0.301	0.398
9	0.276	0.383

Grinding: The surface of the element is further refined by a series of grinding operations, performed with loose abrasive in a water slurry and cast iron grinding tools. If the elements have not been generated, the grinding process begins with a coarse, fast-cutting emery. Otherwise, it begins with a medium grade and proceeds to a very fine grade which imparts a smooth velvety surface to the glass.

The grinding (and polishing) of a spherical surface is accomplished to a high degree of precision with relatively crude equipment, by taking advantage of a unique property of a spherical surface, namely, that a concave sphere and a convex sphere of the same radius will contact each other intimately regardless of their relative orientations. Thus, if two mating surfaces which are approximately spherical are contacted (with abrasive between them) and randomly moved with respect to each other, the general tendency is for both surfaces to wear away at their high spots and to approach a true spherical surface as they wear. (For a detailed analytical treatment of the subject of relative wear in optical processing, the reader is strongly urged to consult the reference by Deve, listed at the end of this chapter.)

Usually the convex piece (either blocker or grinding tool) is mounted in a power driven spindle and the concave piece is placed on top as shown in Fig. 14.3. The upper tool is constrained only by a ball pin-and-socket arrangement and is free to rotate as driven by its sliding contact with the lower piece; it tends to assume the same angular rate of rotation as the lower piece. The pin is oscillated back and forth so that the relationship between the two tools is continuously varied. By adjusting the offset and amplitude of the motion of the pin, the optician can modify the pattern of wear on the glass and thus effect minute corrections to the value and uniformity of the radius generated by the process.

Each successively finer grade of emery is used until the grinding pits left by the preceding grade are ground out.

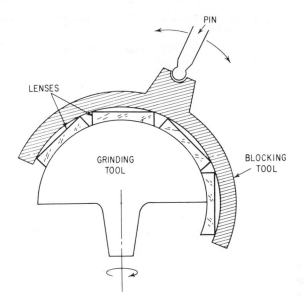

FIG. 14.3. In grinding (or polishing) a semi-random scrubbing action is set up by the rotation of the lower (male) tool about its axis and the back-and-forth oscillation of the upper (female) tool. Note that the upper tool is free to rotate about the ball end of the driving pin and takes on a rotation induced by the lower tool.

Polishing: The mechanics of the polishing process are quite analogous to the grinding process. However, the polishing tool is lined with a layer of pitch and the polishing compound is a slurry of water and rouge (iron oxide) or cerium oxide. The polishing pitch will cold flow and thus takes on the shape of the work in a very short time.

The polishing process is a peculiar one in that it appears to be a combination of different effects. A ground glass surface is a sort of microscopic mountain range, and part of the polishing process consists of planing off the mountain tops by the particles of polishing compound. At the same time, however, an extremely thin layer of glass at the surface seems to melt and flow. The author has seen surfaces (which were perfect under a microscope examination) develop scratches when heated after polishing—apparently a case of scratches which had "flowed" shut in polishing opening up under subsequent heating.

Polishing is continued until the surface is free of any grinding pits or scratches. The accuracy of the radius is checked by the use of a test plate (or test glass). This is a very precisely made master gage which has been polished to an exact radius and which is a true sphere to within a fraction of a wavelength. The test plate is placed in contact with the work, and the difference in shape is determined by the appearance of the interference fringes (Newton's rings)

formed between the two. The relative curvatures of the two surfaces can be determined by noting whether the gage contacts the
work at the edge or the center. If the number of rings are counted,
the difference between the two radii can be determined from the
formula

$$\Delta R \approx N\lambda \left(\frac{2R}{d}\right)^2 \tag{14.3}$$

where ΔR is the radius difference, N is the number of fringes, λ is
the wavelength of the illumination, R is the radius of the test plate,
and d is the diameter over which the measurement is made. One
fringe indicates a change of one-half wavelength in the spacing between the two surfaces. A non-circular fringe pattern is an indication of an aspheric surface.

Corrections are made either by adjustment of the stroke of the
polishing machine or by scraping away portions of the polishing
tool so that the wear is concentrated on the portion of the work which
is too high.

Centering: After both surfaces of an element are polished, the
lens is centered. This is done by grinding the rim of the lens so
that the mechanical axis (defined by the ground edge of the lens)
coincides with the optical axis, which is the line between the centers
of the two surfaces. In *visual* centering the element is fastened
(with pitch) to an accurately trued tubular tool on a rotating spindle.
When the lens is pressed on the tool, the surface against the tool is
automatically aligned with the tool and hence with the axis of rotation. While the pitch is still soft, the operator slides the lens
laterally until the outer surface also runs true. If the lens is rotated slowly, any decentration of either surface is detectable as a
movement of the reflected image (of a nearby target) formed by
that surface, as indicated in Fig. 14.4. For high precision work,
the images may be viewed with a telescope or microscope to increase the operator's sensitivity to the image motion. The periphery of the lens is then ground to the desired diameter with a
diamond charged wheel. Bevels or protective chamfers are usually
ground at this time.

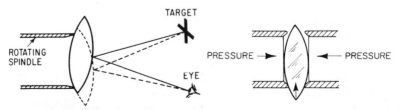

FIG. 14.4. Left: In visual centering the lens is shifted laterally until no motion of the
image of a target reflected from the lens surface can be detected as the lens is rotated.
Right: In mechanical centering the lens is pressed between hollow cylinders. It slides
laterally until its axis coincides with the common axis of the two tools.

For economical production of moderately precise optics, a mechanical centering process is used. In this method, the lens element is gripped between two accurately trued tubular tools. The pressure of the tools causes the lens to slip sideways until the distance between the tools is at a minimum, thus centering the lens. The lens is then rotated against a diamond wheel to grind the diameter to size.

The manufacture of the lens is completed by low reflection coating the surfaces as required and by cementing, if the element is part of a compound component; these processes are outlined in Chapter 7.

Modifications of the standard processing techniques are sometimes required for unusual materials. Brittle materials (e.g., calcium fluoride) must be treated gently, especially in grinding. A finer, softer abrasive is required; sometimes soap is added to the abrasive and soft brass grinding tools are used in place of cast iron. At the other extreme, sapphire (Al_2O_3) cannot be processed with ordinary materials because of its extreme hardness, and diamond power is used for both grinding and polishing.

Non-Spherical Surfaces: Aspherics, cylinders, and toroids do not share the universality of the spherical surface and their manufacture is difficult. While a sphere is readily generated by a random grinding and polishing (because any line through the center is an axis), optical aspherics have only one axis of symmetry. Thus the simple principle of random scrubbing which generates a sphere must be replaced by other means. An ordinary spherical optical surface is a true sphere to within a few millionths of an inch. For aspherics this precision can only be obtained by a combination of exacting measurement and skilled hand correction.

Cylindrical surfaces of moderate radius can be generated by working the piece between centers (i.e., on a lathe). However, any irregularity in the process tends to produce grooves or rings in the surface. This can be counteracted by increasing the rate of working *along* the axis relative to the rate of rotation about the axis. It is difficult to avoid a small amount of taper (i.e., a conical surface) in working cylinders. Large radius cylinders are difficult to swing between centers and are usually handled with a rocking mechanism which constrains the axes of work and tool to parallelism so as to avoid a saddle surface.

Aspherics of rotation, such as paraboloids, ellipsoids, and the like, can be made in modest production quantities if the precision required of the surface is of a relatively low order, as for example in an eyepiece. The usual technique is to use a cam guided grinding rig (with a diamond wheel) to generate the surface as precisely as possible. The problem is then to fine grind and polish the surface without destroying its basic shape. The difficulty is that any random motion which works the surface uniformly tends to change the surface contour toward a spherical form. Extremely flexible tools which can follow the surface contour are required; however, their very flexibility tends to defeat their purpose, which is to smooth or average out small local irregularities left in the surface by the generating process. Pneumatic (i.e., air-filled, elastic) tools have proved quite successful for this purpose.

Where precise aspherics are required, hand correction is practically a necessity. The surface is ground and polished as accurately as possible and is then measured. The measurement technique must be precise enough to detect and quantify the errors. For high-quality work, this means that the measurements must indicate surface distortions of a fraction of a wavelength. The Foucault knife edge test and the Ronchi grating tests are widely used for this purpose; these tests can usually be applied directly to the aspheric surface, although there are many aspheric applications (e.g., the Schmidt corrector plate) where the test must be applied to the complete system to determine the errors in the aspheric.

When the surface errors have been measured and located, the surface is corrected by polishing away the areas which are too high. This can be accomplished with a full-size polisher by scraping away those areas of the polisher which correspond to the low areas of the surface. In making a paraboloid of low aperture, such as used in an astronomical telescope, the surface is close enough to a sphere that the correction can often be effected by modifying the stroke of the polisher. However, for large work and for difficult aspherics, it is usually better to use small tools and to wear down the high zones by a direct attack. A certain amount of delicacy and finesse is required for this approach; if the process is continued for a minute or so longer than required, the result is a depressed ring which then requires that the entire balance of the surface be worn down to match this new low point.

14.2 Optical Specifications and Tolerances

Many otherwise fully competent optical workers come to grief when it is necessary for them to send their work to the shop for fabrication. The two most common difficulties are under-specification, in the sense of incompletely describing what is required, and over-specification, wherein tolerances are established which are much more severe than necessary.

Optical manufacture is an unusual process. If enough time and money are available, almost any degree of precision can be attained. Thus, specifications must be determined on a dual basis: 1) the limits which are determined by the performance requirements of the system, and 2) the expenditure of time and money which is justified by the application. Note well that optical tolerances which represent an equal amount of difficulty to maintain may vary widely in magnitude. For example, it is not difficult to control the sphericity of a surface to one-tenth of a micron; the comparable (in terms of difficulty) tolerance for thickness is about one hundred microns (0.1 mm), three orders of magnitude larger. For this reason it is rare to find "box" tolerances in optical work; each dimension, or at least each class of dimension, is individually toleranced.

Every essential characteristic of an optical part should be spelled out in a clear and unambiguous way. Optical shops are accustomed to this, and if a specification is incomplete, either time must be wasted in questioning the specification to determine what the requirements are, or the shop must arbitrarily establish a tolerance. Either procedure is undesirable.

The following paragraphs are an attempt to provide a general guide to the specification of optics. The discussion will include the basis for the establishment of tolerances, the conventional methods of specifying desired characteristics, and an indication of what tolerances a typical shop may be expected to deliver.

The intelligent choice of specifications and tolerances for optical fabrications is an extremely profitable endeavor. The guiding philosophy in establishing tolerances should be to allow as large a tolerance as the requirement for satisfactory performance of the optical system will permit. Designs should be established with the aim of minimizing the effect produced by production variations of dimensions. Frequently, simple changes in mounting arrangements can be made which will materially reduce fabrication costs without detriment to the performance of a system. One should also be certain that the tightly specified dimensions of a system are the truly critical dimensions, so that time and money are not wasted in adhering to meaningless demands for accuracy.

Surface Quality: The two major characteristics of an optical surface are its quality and its accuracy. Accuracy refers to the dimensional characteristics of a surface, that is, the value and uniformity of the radius. Quality refers to the finish of the surface, and includes such defects as pits, scratches, incomplete or "grey" polish, stains, and the like. Quality is usually extended to similar defects within the element, such as bubbles or inclusions. In general (with the exception of incomplete polish which is almost never acceptable) these factors are "beauty defects" and may be treated as such. The percentage of light absorbed or scattered by such defects is usually a completely negligible percentage of the total radiation passing through the system. However, if the surface is in or near a focal plane, then the size of the defect must be considered relative to the size of the detail it may obscure in the image. Also if a system is *especially* sensitive to stray radiation, such defects may assume a functional importance. In this case, one may evaluate the effect of a defect by comparing its area with that of the system clear aperture at the surface in question.

The standards of military specification MIL-O-13830 are widely utilized in industry. The surface quality is specified by a number such as 80-50, in which the first two digits relate to the *apparent* width of a tolerable scratch in microns and the second two indicate the diameter of a permissible dig, pit, or bubble in hundredths of a millimeter. Thus, a surface specification of 80-50 would permit a scratch of 80 μ *apparent* width and a pit of 0.5-mm diameter. The total length of all scratches and the number of pits is also limited by the specification. In practice, the size of a defect is judged by a visual comparison with a set of graded standard defects. Digs and pits can, of course, be readily measured with a microscope; unfortunately the apparent width of a scratch is not directly related to its physical size, and this portion of the specification is not as well founded as one might desire.

Surface qualities of 80-50 or coarser (i.e., larger) are relatively easily fabricated. Qualities of 60-40 and 40-30 command a small premium in cost. Surfaces with quality specifications of 40-20, 20-10, 10-5 or similar combinations require extremely carful processing and are considerably more expensive to fabricate. Such specifications are usually reserved for field lenses or reticle blanks.

Surface Accuracy: Surface accuracy is usually specified in terms of the wavelength of light from a sodium lamp (0.0005893 mm). It is determined by an interferometric comparison of the surface with a test plate gage, by counting the number of (Newton's) rings or "fringes" and examining the regularity of the rings. As previously mentioned, the space between the surface of the work and the test plate changes one-half wavelength for each fringe. The accuracy of the fit between work and gage is described in terms of the number of fringes seen when the gage is placed in contact with the work.

Test plates are made flat or spherical to an accuracy of a small fraction of a fringe. Spherical test plates, however, have radii which are known to an accuracy only as good as the optical-mechanical means which are used to measure them. Thus the radius of a test plate is frequently known only to an accuracy of about one part in a thousand or one part in ten thousand. Further, test plates are expensive (several hundred dollars per set) and are available as "stock tooling" only in discreet steps. Thus it frequently pays to inquire what radii the optical shop has as standard tooling.

The usual shop specification for surface accuracy is thus with respect to a *specific* test plate, and it takes the form of specifying that the piece must fit the gage within a certain number of rings and must be spherical (or flat in the case of plane surfaces) within a number of rings. A fit of from five to ten rings, with a sphericity (or "regularity") of from one-half to one ring is not a difficult tolerance. Fits of from one to three rings with correspondingly better regularity can be achieved in large-scale production at a very modest increase in cost. Note that an irregularity of a small fraction of a ring is difficult to detect when the fit is poor. Thus, little is gained by specifying a ten ring fit and a quarter ring spher-icity, since the fit must be considerably better than ten rings to be certain that the irregularity is less than one-quarter ring. The change in radius due to a poor fit is frequently negligible in effect. For example, the radius difference between two (approximately) 50-mm radii at a 30-mm diameter which corresponds to five rings is (by Eq. 14.3) only a bit more than 3μ.

If possible, one should avoid specifying accurate surfaces on pieces whose thickness-to-diameter ratio is low. Such elements tend to spring and warp in processing and extreme precautions are necessary to hold an accurate surface figure. A rule of thumb is to make the axial thickness at least one-tenth of the diameter for neg-ative elements; where there is a good edge thickness, one-twentieth or one-thirtieth of the diameter is frequently acceptable. For ex-tremely precise work, especially on plane surfaces, the optician will prefer a thickness of one-fifth to one-third of the diameter.

The effects of errors in radius values (that is, departures from the nominal design radii) are usually not too severe. In fact, it is the practice of many purchasers of optics (including the United States Army Arsenals) *not* to indicate a tolerance on the specified radii, but to specify final performance in terms of focal length and resolution. It is usually possible for a well-tooled optical shop to select judiciously (from its tooling list) nearby radii which produce a result equivalent to the nominal design. If tolerances are specified on radius values, one should bear in mind the fact that most effects produced by a radius variation are not proportional to ΔR, but to $\Delta R/R^2$ or to ΔC. To take a simple example, we can differentiate the thin lens focal length equation

$$\phi = \frac{1}{f} = (N - 1)(C_1 - C_2) = (N - 1)\left(\frac{1}{R_1} - \frac{1}{R_1}\right)$$

with respect to the first surface to get the following:

$$d\phi = (N - 1)\,dC_1$$

$$df = f^2(N - 1)\,dC_1 = f^2(N - 1)\frac{dR_1}{R_1^2}$$

In a more complex system, the change in focal length resulting from a change in the ith curvature is approximated by:

$$df \approx \left(\frac{y_i}{y_1}\right)f^2(N_i' - N_i)\,dc_i$$

$$df \approx \left(\frac{y_i}{y_1}\right)f^2(N_i' - N_i)\frac{dR_i}{R_i^2}$$

Thus, if a uniform tolerance is to be established for all radii in a system, the uniform tolerance should be on a curvature, not on radius. Radius tolerances should be proportional to the square of the radius.

The preceding is, of course, based on focal length considerations only. With regard to aberrations, it is difficult to generalize, since one radius of a system may be very effective in changing a given aberration while another may be totally ineffective. The relative sensitivity is determined by the heights of the axial and principal rays at the surface, the index break across the surface and the angles of incidence at the surface. A good approximation to the effect that any tolerance has on the aberrations of a system can be determined by use of the third-order surface contribution equations.

The effect of surface irregularity is more readily determined. Consider the case where the Newton's rings are not circular; this is an indication of axial astigmatism, since the power in one meridian is stronger than in the other. Here it is convenient to call on the Rayleigh quarter wave criterion. The OPD produced by a "bump" of height H on a surface is equal to $H(N' - N)$, or, expressing it in terms of interference rings (remembering that each fringe represents one-half wavelength),

$$OPD = \frac{1}{2}(\#FR)(N' - N) \text{ wavelengths}$$

where $(\#FR)$ is the number of fringes.

Thus, to stay within the Rayleigh criterion, the total OPD, summed over the whole system, should not exceed one-fourth wavelength; this is expressed by the following inequality:

$$\Sigma(\#FR)(N' - N) < 0.5$$

Thus, a single element of index 1.5 could have one-half fringe of astigmatism (or any other surface irregularity) on each surface before the Rayleigh criterion was exceeded (assuming that the nominal correction was perfect and that the irregularities were additive).

Thickness: The effects of thickness and spacing variation on the performance of a system are readily analyzed, either by raytracing or by a third-order aberration analysis. The importance of thickness variations differs greatly from system to system. In the negative doublets of a Biotar (Double Gauss) objective, the thickness is extremely critical (as regards spherical aberration); for this reason the crown and flint elements are usually selected so that their *combined* thickness is very close to the design nominal. At the other extreme, the thickness variation of a planoconvex eyepiece element may be almost totally ignored, since it ordinarily has little or no effect on anything.

In general, thicknesses and spacings may be expected to be critical where the slope of the marginal ray is large. Anastigmats in
general, and meniscus anastigmats in particular, are prone to this
sensitivity. High-speed lenses, large NA microscope objectives and
the like are usually sensitive.

Unfortunately the thickness of an optical element is not as readily controlled as some of the other characteristics. In production
procedures where many elements are processed on the same block,
the maintenance of a uniform nominal thickness requires precise
blocking and tooling. The grinding operation, while precise enough
in terms of radius, is difficult to control in terms of its extent.
For close thickness control, the generating operation must be accurate and each subsequent grinding stage must be exactly timed
so that the proper finish, radius and thickness are arrived at simultaneously.

A reasonable thickness tolerance for precise work is ±0.1 mm
(±0.004 in.). This can cause a shop some difficulty on certain lens
shapes and on larger lenses; where a relaxation is possible, a tolerance of ±0.15 or ±0.2 mm is more economical. It is (barely)
possible to hold ±0.05 mm in large-scale production by taking extreme care throughout the fabrication procedure. The rejection
rate at this tolerance can become disastrous if the smallest mis-
chance occurs. Of course it is possible, by hand working and
selection, to produce pieces to any desired tolerance level; the
author has seen ±0.01 mm held in moderate production quantities
(although at rather immoderate cost).

Centering: The tolerances in centering are 1) on the diameter
of the piece, and 2) on the accuracy of the centering of the optical
axis with the mechanical axis. If the piece is to be centered (i.e.,
as a separate operation) the diameter can be held to a tolerance of
plus nothing, minus 0.03 mm by ordinary techniques, and this is the
standard tolerance in most shops. A small economy is effected by
a more liberal tolerance. Tighter tolerances are possible, but are
not often necessary.

The concentricity of an element is most conveniently specified
by its deviation. This is the angle by which an element deviates a
ray of light directed toward the mechanical center of the lens. The
deviation angle is an especially useful measure of decentration
since the deviation of a group of elements is simply the (vector)
sum of the deviations of the individual elements. Figure 14.5 is an
exaggerated sketch of a decentered element. The optical and mechanical axes are shown separated by an amount Δ. Since a ray parallel to the optical axis must pass through the focal point, the angular
deviation (δ) of the ray in radians is given by the decentration divided
by the focal length.

$$\delta = \frac{\Delta}{f}$$
(14.4)

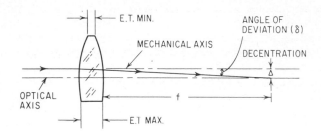

FIG. 14.5. Showing the relationships between the optical and mechanical axes, and the decentration and angle of deviation in a decentered lens.

Note that a decentered element may be regarded as a centered element plus a thin wedge of glass. The angle of the wedge (W) is given by the difference between the maximum and minimum edge thicknesses divided by the diameter of the element

$$W = \frac{E_{\max} - E_{\min}}{d} \qquad (14.5)$$

Since the deviation of a thin prism is given by $D = (N - 1)A$, we can similarly relate the wedge angle of an element to its deviation by

$$\delta = (N - 1)W \qquad (14.6)$$

If an element is centered on a high-production mechanical (clamping) centering machine, the limit on the accuracy of the concentricity obtained is determined by the residual difference in edge thickness which the cylindrical clamping tools cannot "squeeze out." On most machines, this is to the order of 0.0005 in. when residual tooling and spindle errors are taken into account. Thus the residual wedge angle for a lens with a diameter d is given by

$$W = \frac{.0005''}{d}$$

and the resulting deviation is

$$\delta = \frac{.0005'' (N - 1)}{d}$$

Thus, for ordinary lenses $(N = 1.5$ to $1.6)$ a reasonable estimate of the deviation is given by

$$\delta = \frac{1}{d} \text{ minutes} \qquad (14.7)$$

where d is in inches, and the centering is done mechanically.

If the centering is accomplished visually (as indicated in the left-hand sketch of Fig. 14.4), then the ability of the eye to detect motion is the limiting factor. If we assume that the eye can detect an angular motion of 6 or 7×10^{-5} rad, then the deviation will be approximately

$$\delta = (N - 1)\left(\frac{1}{R} \pm .06\right) \pm \text{(contact \& spindle errors)} \qquad (14.8)$$

where δ is in minutes and R is in inches.*

The term $(N - 1)/R$ is from the visually undetected "wobble" of the outer radius and the $0.06 (N - 1)$ term is due to the tilt in the tool which the eye could not detect in the truing of the tool (this is tested by pressing a flat glass plate against the rotating tool and observing any motion in the reflected image). The eye can, of course, be aided by means of a telescope or microscope which will further reduce the amount of decentration which can be detected by a factor equal to the magnification.

Occasionally lenses are not put through a specific centering operation. When this is the case, the concentricity of the finished lens is determined by the wedge angle which is left in by the grinding operations. If the blocking tooling is carefully worked out, it is possible to produce elements with a wedge (that is, the difference between the edge thickness on opposite edges) to the order of 0.1 or 0.2 mm. Centering is often omitted on box camera lenses, condensers, magnifiers, or almost any single element of a simple optical system. Simple elements made from rounded circles of window glass are often left uncentered.

Prism Dimensions and Angles: The linear dimensions of prisms can be held to tolerances approximating those of an ordinary machined part, although the fabrication requirements of a prism are more difficult because of the finish and accuracy requirements of an optical surface. Thus tolerances of 0.1 or 0.2 mm are reasonable and tighter tolerances are possible.

Prism angles can be held to within 5 or 10 min of nominal by the use of reasonably good blocking forms. Indeed it is possible, although exceedingly difficult, to make angles accurate to a few percent of these tolerances if one takes exquisite pains with the design, fabrication, correction, and use of the blocking tools. Usually angles which must be held to tolerances of a few seconds (such as roof angles) are hand corrected. Such angles are checked with an autocollimator, either by comparison with a standard or by using the internal reflections to make the piece a retrodirector. Angles of 90° and 45° can be self-checked in this way since their internal reflections form constant deviation systems of 180° deviation.

Prism size tolerances are usually based on the necessity to limit the displacement errors (lateral or longitudinal) which they

---

*Equation 14.8 assumes that the image reflected from the outer radius is viewed at 10 in. This is obviously impossible if R is a convex surface with a radius longer than 20 in., and is impractical if R is a long concave radius. Thus for $|R| > 20$ in., one should substitute 0.05 for the $1/R$ term.

produce. Angular tolerances are usually established to control angular deviation errors. One can usually find one or two angles in a prism system which are more critical than the others; these can be tightly controlled and the other angles allowed to vary. For example, with respect to the deviation of a pentaprism, an angular error in the 45° angle between the reflecting faces is six times as critical as an error in the 90° angle between the faces, and the other two angles have no effect on the deviation. On occasion, prism tolerances are based on aberration effects. Since a prism is equivalent to a plane parallel plate and introduces overcorrected spherical and chromatic, an increase in prism thickness in a nominally corrected system will overcorrect these aberrations. Some prism angle errors are equivalent to the introduction of thin-wedge prisms into the system. The angular spectral dispersion of a thin wedge is $(N - 1)W/V$ (where W is the wedge angle and V is the Abbe V-value of the glass) and the resultant axial lateral color may limit the allowable angular tolerances.

Materials: The characteristics of the refractive materials used in optical work which are of primary concern are index, dispersion, and transmission. For ordinary optical glass procured from a reputable source, visual transmission is almost never a problem. Only occasionally, where a thick piece of dense glass is used in a critical application, must transmission limits be specified. Similarly, the dispersion, or V-value, is seldom a problem, except in special cases. For apochromatic systems where the partial dispersion ratio is exceedingly critical, very special precautions are required.

The index of refraction is usually of prime concern in optical glass. As indicated in Chapter 7, the standard index tolerance is ±0.001 or ±0.0015 depending on the glass type. The glass supplier can hold the index more closely than this by selection or by extra care in the processing; either increases the cost. In practice the glass supplier will ordinarily use up only a fraction of this tolerance, since the index within a single melt or batch of glass is remarkably consistent. Thus, in a single lot of glass the index may vary only a few digits in the fourth place. However, bear in mind that this variation *may* be centered about a value which is 0.001 or 0.0015 from the nominal index. It is sometimes economical to accept the standard tolerance and to adjust a design parameter (e.g., a spacing) to compensate for the variation of a lot of glass in cases where the index is critical.

Transmission and spectral characteristics are often poorly specified. Ambiguity can usually be avoided by specifying spectral reflection (or transmission) *graphically*, that is, by indicating the area of the reflection (or transmission) vs wavelength plot within which the characteristics of the part must lie. One should also indicate whether or not the spectral characteristics outside the specified region are of importance. For example, in a band pass filter, it is important to indicate how far into the long and short wavelength regions the blocking action of the filter must extend.

Figure 14.6 is a table of typical tolerances and may be used as a guide. Bear in mind, however, that the values given are *typical* and that there are many special cases that this sort of tabulation cannot cover.

	Diameter	Thickness	Radius	Linear Dimension	Angles
Low cost	± 0.2 mm	± 0.5 mm	Gage	± 0.5 mm	Degrees
Commercial	± 0.07 mm	± .25 mm	10 Fr.	± 0.25 mm	± 15'
Precision	± .02 mm	± 0.1 mm	5 Fr.	± 0.1 mm	± 5'-10'
Extra Precise	± .01 mm	± 0.05 mm	As req'd.	As req'd.	Seconds

FIG. 14.6. Tabulation of typical optical fabrication tolerances

Additive tolerances: In analyzing an optical system to determine the tolerances to be applied to specific dimensions, one can readily calculate the partials of the system characteristics with respect to the dimensions under consideration. Thus, one obtains the value of the partial derivative of the focal length (for example) with respect to each thickness, spacing, curvature, and index; likewise for the other characteristics, which may include back focus, magnification, field coverage, as well as the aberrations. Then each dimensional tolerance, multiplied by the appropriate derivative, indicates the contribution of that tolerance to the variation of the characteristic. Now if it were necessary to be *absolutely* certain that (for example) the focal length did not vary more than a certain amount, one would be forced to established the parameter tolerances so that the sum of the derivative-tolerance products did not exceed the allowable variance. Although this is occasionally necessary, one can frequently allow much larger tolerances by taking advantage of the laws of probability.

As a simple example, let us consider a stack of disks, each 0.1 in. thick. We will assume that each disk is made to a tolerance of ±0.005 in. and that the probability of the thickness of the disk being any given value between 0.095 and 0.105 in. is the same as the probability of its being any other value in this range. This situation is represented by the rectangular frequency distribution curve of Fig. 14.7A. Thus, for example, there is one chance in ten that any given disk will have a thickness between 0.095 and 0.096 in. Now if we stack two disks, we know that it is *possible* for their combined thicknesses to range from 0.190 to 0.210 in. However, the *probability* of the combination having either of these extreme thickness values is quite low. Since the probability of either of the disks having a thickness between 0.095 and 0.096 is one in ten, the probability of *both* falling in this range is one in one hundred. Thus, the

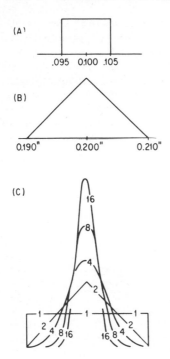

(A)

.095 0.100 .105

(B)

0.190"   0.200"   0.210"

(C)

16

8

4

2

1   1   1
2   2
4 8 16   16 8 4

FIG. 14.7. Showing the manner in which additive tolerances combine in assembly. Plot A shows a uniform probability in a dimension of a single piece. When two such pieces are combined, the resulting frequency distribution is shown in B. Normalized curves for assemblies of 1, 2, 4, 8 and 16 pieces are shown in C.

probability of a pair of disks having a thickness between 0.190 and 0.192 is one in one hundred; similarly for a combined thickness of 0.208 to 0.210 in. The probability of a combined thickness of 0.190 to 0.191 (or 0.209 to 0.210) is much less; one in four hundred.

The frequency distribution curve representing this situation is shown in Fig. 14.7B as a triangular distribution. Figure 14.7C shows frequency distribution curves for one, two, four, eight, and sixteen element assemblies. These curves have been normalized so that the area under each is the same and the extreme variations have been equalized. The important point here is that the probability of an assembly taking on an extreme value is tremendously reduced when the number of elements making up the assembly is increased. For example, in a stack of 16 disks with a nominal total thickness of 1.6 in. and a possible variation in thickness of ±0.080 in., the probability of a random stack having a thickness less than 1.568 in. or more than 1.632 in.(i.e., ±0.032 in.) is less than one in one hundred.

The importance of this in setting tolerances is immediately apparent. In the stacked disks example, if the range of thicknesses represented by 1.568 to 1.632 in. for 16 disks were the greatest variation that could be tolerated, we could be absolutely sure of meeting this requirement *only* by tolerancing each individual disk at ±0.002 in. However, if we were willing to accept a rejection rate of 1% in large scale production, we could set the thickness tolerance at ±0.005 in. If the cost of the pieces made to the tighter tolerance exceeded the cost of the pieces made to the looser tolerance by 1% (plus one-sixteen hundredth of the assembly, processing and final inspection costs), the looser tolerance would result in a less costly product.

In a frequency distribution curve, such as those shown in Fig. 14.7, the area under the curve between two abscissa values represents the (relative) number of pieces which will fall between the two abscissa values. Thus the probability of a characteristic falling between two values is the area under the curve between the two abscissas divided by the total area under the curve.

The "peaking up" characteristic of multiple assemblies can be represented by the two plots shown in Fig. 14.8. The graph on the left shows the percentage of assemblies which fall within a given central fraction of the total tolerance range as a function of the fraction of the range. The number of elements per assembly is indicated on each curve. These curves were derived from Fig. 14.7C. The graph on the right in Fig. 14.8 is simply another way of presenting the same data. If one were interested in an assembly of 10 elements, the intersection of the abscissa corresponding to 10 and the appropriate curve would indicate that all but 0.2% (using the 99.8% curve) of the assemblies would fall within 0.55 of the total tolerance range represented by the sum of all ten tolerances, and that over one-half of the pieces (using the 50% curve) would fall within 0.15 of the total possible range.

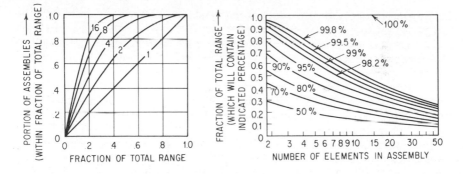

FIG. 14.8. Probability distributions of additive tolerances in multiple assemblies. See text for details.

The preceding discussion has been based upon the assumptions that 1) each individual piece had a rectangular frequency distribution and 2) each tolerance was equal in effect. This is rarely true in practice. The frequency distribution will, of course, depend on the techniques and controls used in fabricating the part, and the tolerance sizes may represent the partial derivative tolerance products from such diverse sources as index, thickness, spacing, and curvature. Note, however, that in Fig. 14.7C the progression of curves may be started at any point. If, for example, the production methods produce a triangular distribution (such as that shown for an assembly of two elements), then the curve marked "four elements" will be the frequency distribution for two elements (of triangular distribution) and so on.

Note also that as more and more elements are included in the assembly, the curve becomes a closer and closer approximation to the "normal distribution curve" which is so useful in statistical analysis (except that the tolerance-type curves do not go to infinity as do "normal" curves). One useful property of the "normal" curve for an additive assembly is that its "peakedness" is proportional to the square root of the number of elements in assembly. Thus if 90% of the individual pieces are expected to fall within some given range, then for an assembly of 16 elements, 90% would be expected to fall within $\sqrt{1/16}$ or one-quarter of the total range. A brief examination will indicate that even the rectangular distribution assumed for Fig. 14.7 and 14.8 tends to follow this rule when there are more than a few elements in the assembly.

A rule of thumb frequently used to establish tolerances may be represented as follows:

$$T \approx \sqrt{\sum_{i=1}^{n} t_i^2} \tag{14.9}$$

This states that the expected total variation T in an assembly is given by the square root of the sum of the squares of the individual tolerances t_i. A study of Fig. 14.8 will indicate that for uniform tolerances with a rectangular distribution, about 90% of the assemblies can be expected to vary from nominal by less than T. For pieces with a triangular frequency distribution, about 98% of the assemblies would be included. Thus Eq. 14.9 provides a fairly good measure of tolerance build-up and is a convenient method of establishing tolerances.

While this section may seem to be a far cry from optical engineering, consider that a simple Cooke triplet has the following dimensions which affect its focal length and aberrations: six curvatures, three thicknesses, two spacings, three indices, and three V-values. These total fourteen for monochromatic characteristics and seventeen for chromatic aberrations. Such a system is eminently qualified for statistical treatment.

While the above may tend to induce a desirable relaxation in tolerances, one or two words of caution are in order. As previously mentioned, the index of refraction distribution within a melt or lot of glass may or may not be centered about the nominal value. When it is centered about a non-nominal value, the preceding analysis is valid only with respect to the "central" value, not the nominal value. Further, in some optical shops, there is a tendency to make lens elements to the high side of the thickness tolerance; this allows scratched surfaces to be reprocessed and will, of course, upset the theoretical probabilities. Another tendency is for polishers to try for a "hollow" test glass fit, that is, one in which there is a convex air lens between the test plate and the work. This is done because a block of lenses which is polished "over" is difficult to bring back.

Thus, the situation is seen to be a complex one, but nonetheless one in which a little careful thought in relaxing tolerances to the greatest allowable extent can pay handsome dividends. For those who wish to avoid the labor of a detailed analysis, the use of Eq. 14.9, or even the assumption that the tolerance build-up will not exceed one-half or one-third of the possible maximum variation, are fairly safe procedures in assemblies of more than a few elements. Above all, when cost is important, one should try to establish tolerances which are readily held by normal shop practices.

14.3 Optical Mounting Techniques

General: In optical systems, just as in precise mechanical devices, it is well to observe the basic principles of kinematics. A body in space has six degrees of freedom (or ways in which it may move). These are translation along the three rectangular coordinate axes and rotation about these three axes. A body is fully constrained when each of these possible movements is *singly* prevented from occuring. If a motion is inhibited by more than one mechanism, then the body is overconstrained and one of two conditions occurs; either all but one of the (multiple) constraints are ineffective or the body is deformed by the multiple constraint.

The laboratory mount indicated in Fig. 14.9 is a classical example of a kinematic mount. Here it is desired to uniquely locate the upper piece with respect to the lower plate. At A the ball-ended rod fits into a conical depression in the plate. This (in combination with gravity or a spring-like pressure at D) constrains the piece from any lateral translations. The V-groove at B eliminates two rotations, that about a vertical axis at A and that about the axis A-C. The contact between the ball end and the plate at C eliminates the final rotation (about axis A-B). Note that there are no extra constraints and that there are no critical tolerances. The distances AB, BC, and CA can vary widely without introducing any binding effects. There is one unique position which will be taken by the piece; the piece may be removed and replaced and will always assume exactly the same position.

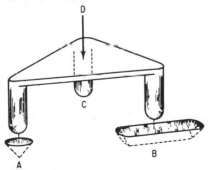

FIG. 14.9. An example of a kinematic locating fixture. The three ball-ended legs of the stool rest in a conical hole at A, a V-groove (aligned with A) at B, and on a flat surface at C.

A perfectly kinematic system is frequently undesirable in practice and semi-kinematic methods are often used. These substitute small area contacts for the point and line contacts of a pure kinematic mount. This is necessary for two reasons. Materials are often not rigid enough to take point contacts without deformation and the wear on a point contact soon reduces it to an area contact in any case.

Thus, in the design of an instrument, optical or otherwise, it is well to start by defining the degrees of freedom to be allowed and the degrees of constraint to be imposed. These can be outlined first by geometrical points and axes and then reduced to practical pads, bearings, and the like. This sort of approach results in a thorough and clear understanding of the effects of manufacturing tolerances on the function of the device and often indicates relatively inexpensive and simple methods by which a high order of precision can be maintained.

Lens Mounts: Optical lens elements are almost always mounted in a close fitting sleeve. A number of methods are used to retain the element in the mount; several are sketched in Fig. 14.10. In sketches A and B the lenses are retained by spring rings. In the left-hand mount A, the spring catches in a V-groove, and if the mount is properly executed, the spring wire (which in its free state assumes a larger diameter) presses against the face of the element and the outer face of the groove. The lens is thus under a light pressure. The flat spring retainer B is less satisfactory, since the retainer will readily slip out unless the spring is strong or has sharp edges which bite into the mount. Other methods suitable for retaining low-precision elements include staking or upsetting ears of metal from the cell which clamp a thin metal washer over the lens element. Condenser systems are often mounted between three rods which are grooved as indicated in Fig. 14.10C. This provides a loose mount which leaves the condenser elements free to expand with the heat from the projection lamp without being constricted by the mount; it also allows cooling air to circulate freely. Both points are especially important in the mounting of a heat absorbing filter.

Where precision is required, the cell is fitted rather closely to the lens. For good quality optics the lens diameter may be toleranced +0.000, −0.001 in. and the inside cell diameter toleranced +0.001, −0.000 in. with 0.001 or 0.0005-in. clearance between the nominal diameters. For small lenses which demand high precision, these tolerances can be halved, at the expense of some difficulty in production. Large diameter optics are usually specified to somewhat looser tolerances. The lenses are most commonly retained by a threaded lock ring, as indicated in Fig. 14.10D or E. Sometimes the lock ring has an unthreaded pilot whose diameter is the same as the lens in order to be certain that the lens will ride on the bored seat and not on the threads. A separate spacer may be substituted for the pilot.

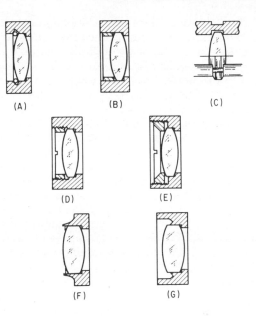

FIG. 14.10. Several methods of retaining optical elements. A) wire spring ring in a V-groove, B) flat spring ring, C) three grooved rods at 120°, D) and E) threaded lock ring, F) spinning shoulder, before burnishing and (dotted) after, G) cemented in place with trough for cement overflow.

A lens may be spun into the mount, as shown in Fig. 14.10F. In this method the mount is made with a thin spinning shoulder which protrudes past the edge of the lens (which is preferably beveled). This shoulder is a few thousandths thick at the outside edge and has an included angle of 10 or 20 degrees. The lens is inserted and the thin lip is turned over, usually by rotating the cell while the lip is bent over. Care and skill are required, but there are a number of advantages to this technique. The pressure of the spinning shoulder tends to center the lens in the mount. In assemblies requiring extreme precision, the seat can be bored to fit the lens diameter and the lens can be spun in place without removing the piece from the lathe; the result is concentricity of an order which is difficult to duplicate by any other means.

Another technique which results in both economy and precision is to cement the lens into its seat. The cement has a modest centering action and with a good plastic cement the lens is securely retained. Care should be taken to provide an overflow groove (Fig. 14.10G) so that excess cement is kept away from the surface of the lens.

In an assembly where several lenses and spacers are retained by a single lockring, care must be taken that the thickness tolerances on the lenses and spacers are not allowed to build up to a point where the outside lens a) extends beyond its seat and is not constrained by the seat diameter, or b) is down into the mount so far that the lockring cannot seat down on it. Another point to watch is that the mouth of a long inside diameter bore is frequently bell shaped and a lens located near the mouth may have several thousandths of an inch more lateral freedom than intended. In critical assemblies, it frequently pays to locate the lens well inside the mouth of the bore.

When elements of different diameters are to be mounted together, the mount should be designed so that the lens seats can all be bored in one operation. This not only tends to reduce the cost of the mount but eliminates a possible source of decentration of each element with respect to the others, which can occur when the lens seats are bored in two or more separate operations.

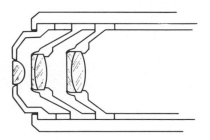

FIG. 14.11. Mounting detail, microscope objective.

The microscope style of element mounting shown in Fig. 14.11 illustrates a number of valuable devices. The lens seat and the outside support diameter of each cell can be turned in the same operation; indeed, in a critical system, the optical element may be spun in place without removing the piece from the lathe.(Cementing the lenses in place can be substituted for spinning.) All the cells are seated in the same bore of the main mount and they are isolated from the lockring threads by a long spacer. All these techniques contribute to maintaining the exquisite concentricity necessary in a first-class microscope objective.

In mounting any type of optical element, it is important to avoid any warping or twisting. In the case of lens elements (which are in effect clamped between a shoulder and a lockring, or their equivalents), this is not too difficult, since the pressure points are opposite each other and result in compression of the lens. More care is necessary in mounting mirrors and prisms, however, since it is quite easy to make the mistake of restraining a mirror in such a way that its surface is warped out of shape. One way to avoid this is to be sure that for each point at which pressure is exerted on a mirror, there is a pad directly opposite so that no twisting moment is introduced.

Figure 14.12 serves as an indication of how few constraints are necessary to kinematically define the location of a piece. This illustration might apply to a piece of cubical shape. The three points in the X-Z plane define a plane on which the lower face of the piece rests; these points take up one translational and two rotational degrees of freedom. The two points in the Y-Z plane take up one translational and one rotational freedom. Note that if there were three (non-aligned) points in this plane, they would then define an angle between the X-Z and Y-Z faces of the piece; if the piece had a different angle, then there would be *two* ways in which the piece could be seated. The single point in the X-Y plane eliminates the last remaining of the six available degrees of freedom. A flexible pressure on the near corner of the piece will now uniquely locate the piece in this mount.

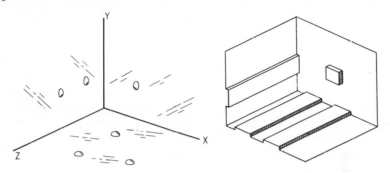

FIG. 14.12. Kinematic and semi-kinematic position defining mount for a rectangular piece.

The sketch on the right illustrates one way of putting this type of mount into practice. The points are replaced by pads or rails. As shown, the two rails in the X-Z plane must be carefully machined in the same operation to assure that they are exactly co-planar: this is not difficult, but if it were, the substitution of a short pad for one rail would eliminate any difficulty on this score.

Prisms and mirrors are usually clamped or bonded to their mounts. In clamp mounts the pressure is usually exerted by a screw on a metal pressure pad. A piece of cork or compressible composition material is placed between the glass and the metal pad to distribute the pressure evenly over the glass; this prevents the pressure from being exerted at a single point. There are a number of excellent cements available for bonding glass pieces to metal mounts. Some care is necessary in designing the mount when bonding a thin mirror, since the cement may warp the mirror (toward the shape of the mount) if the cemented area is large.

14.4 Optical Laboratory Practice

The Lens Bench: An optical bench or lens bench consists, in essence, of a collimator which produces an infinitely distant image of a test target, a device for holding the optical system under test, a microscope for the examination of the image formed by the system, and a means for supporting these components. Each of the components may take various forms, depending on the usage for which it is primarily designed.

The collimator consists of a well-corrected objective and an illuminated target at the focus of the objective. For visual work, the objective is usually a well-corrected achromat; for infrared work, a paraboloidal mirror is used, usually in an "off-axis" or Herschel configuration. The target may be a simple pinhole (for star tests or energy distribution studies), a resolution target, or a calibrated scale if a "focal" collimator is desired.

The lens holder can range in complexity from a simple platform with wax to stick the lens in place, to a T-bar nodal slide which generates a flat image surface. The microscope is usually equipped with at least one micrometer slide, and frequently with two or three orthogonal slides so that accurate measurements may be made.

In subsequent paragraphs, we will discuss some of the applications of the lens bench and will describe the components of the bench more fully in the context of their applications.

The Measurement of Focal Length: There are two basic lens bench techniques for the routine measurement of effective focal length: the nodal slide method and the focal collimator. Both schemes are sketched in Fig. 14.13.

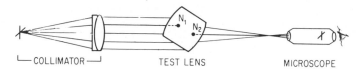

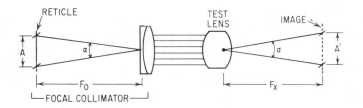

FIG. 14.13. Illustrating the nodal slide (upper) and the focal collimator (lower) methods of measuring focal length on the optical bench.

The nodal slide is a pivoted lens holder equipped with a slide which allows the lens to be shifted axially with respect to the pivotal axis. Thus, by moving the lens forward or backward, the lens can be made to rotate about any desired point. Now note that, if the lens is pivoted about its second nodal point (as indicated in Fig. 14.13), the ray emerging from this point (which by definition emerges from the system parallel to its incoming direction) will coincide with the bench axis (through the nodal point). Thus there will be no lateral motion of the image when the lens is rotated about the second nodal point. Once the nodal point has been located in this manner, the lens is then realigned with the collimator axis and the location of the focal point is determined. Since the nodal points and principal points are coincident when a lens is in air, the distance from the nodal point to the focal point is the equivalent focal length.

This technique is basic and applicable to a wide variety of systems. Its limitations are primarily in the location of the nodal point. The operation of swinging the lens, shifting its position, swinging again, and so on, is tedious, and since it is discontinuous, it is difficult to make an exact setting. If the axis of the test lens is not accurately centered over the axis of rotation of the nodal slide, there will be no position at which the image stands still. Lastly, the measurement of the distance from the axis of rotation to the position of the aerial image is subject to error unless the equipment is carefully calibrated.

A focal collimator consists of an objective with a calibrated reticle at its focal point. The focal length of the objective and the size of the reticle must be accurately known. The test lens is set up and the size of the image formed by the lens is accurately measured with the measuring microscope. From Fig. 14.13 it is apparent that the focal length of the test lens is given by

$$F_x = A' \left(\frac{F_0}{A} \right) \tag{14.10}$$

where A' is the measured size of the image, A is the size of the reticle, and F_0 is the focal length of the collimator objective. Note that the focal collimator may be used to measure negative focal lengths as well as positive; one simply uses a microscope objective with a working distance longer than the (negative) back focus of the lens under test.

It is apparent from Eq. 14.10 that any inaccuracies in the values of A', A, or F_0 are reflected directly in the resultant value of the focal length. Further, any error in setting the longitudinal position of the measuring microscope will be reflected in F_x. Note that both the nodal slide and focal collimator methods assume that the test lens is free of distortion. If an appreciable amount of distortion exists, the measurements must be made over a small angle; this, of course, will limit the accuracy possible.

In setting up a focal collimator, it is necessary to determine the collimator constant (F_0/A) to as high a degree of accuracy as possible. The value of A, the reticle spacing, can be readily measured with a measuring microscope. The focal length of the collimating lens can be determined to a high degree of accuracy by a finite conjugate version of the focal collimator technique. An accurate scale (or glass plate with a pair of lines) is set up 20 to 50 ft from the collimator lens, as shown in Fig. 14.14. The measuring microscope is used to measure the size of the image of the target accurately, and the distance from object to image is measured. The value of p, the distance between the principal points is estimated, either from the design data of the lens or by assuming it to be about one-third the lens thickness. (As long as p is small compared to D, the error introduced by an inaccurate value of p is small.) Now since $D = s + s' + p$ and $A : s = A' : s'$, s and s' can be determined and substituted (with due regard for the sign convention) into

$$\frac{1}{s'} = \frac{1}{f} + \frac{1}{s} \tag{14.11}$$

and the value of the effective focal length determined. The necessity of estimating a value for p can be eliminated, if desired, by measuring the front focal length and applying the Newtonian equation for magnification (Eq. 2.6) or, alternatively, by measuring front and back focal lengths, determining $p = \text{ffl} + \text{bfl} + t - 2f$, and repeating the original calculation; after a few iterations the calculation will converge to the exact p and f.

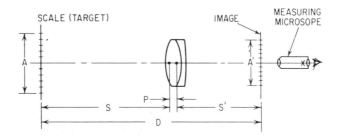

FIG. 14.14. Set up for basic measurement of focal length.

Collimation, and the Measurement of Front and Back Focal Lengths: A basic method of locating the focal points is by auto-collimation. As indicated in Fig. 14.15, an illuminated target is placed near the focus of the lens under test and a plane mirror is placed in front of the lens so as to reflect the light back into the lens. When the reflected image is focused on a screen in the same plane as the target, both screen and target lie in the focal plane. For accurate work an auto-collimating microscope, shown in Fig. 14.16, produces excellent results. The lamp and condenser

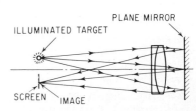

FIG. 14.15. Auto-collimation as a method of locating the focal points. When the object and reflected image are in the same plane (the focal plane), the system is auto-collimated.

illuminate the reticle, which may consist of clear lines scribed through an aluminized mirror. The reticle is then imaged at the focus of the objective. The eyepiece of the microscope is positioned so that its focal plane is conjugate with the reticle. Thus when the microscope is focused on the focal plane of the test lens, the reticle image is auto-collimated by the test lens-plane mirror combination and is seen in sharp focus at the eyepiece. The microscope is then moved in to focus on the rear surface of the test lens; the distance travelled by the microscope is equal to the back focus of the lens.

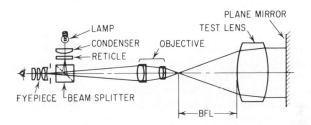

FIG. 14.16. The auto-collimating microscope is used to measure back focal length by focusing first on the surface of the test lens and then on the auto-collimated image at the focal point.

The lens bench collimator itself may be adjusted for exact collimation using this technique. When the collimator reticle and the reflected image of the microscope reticle are simultaneously in focus, then the collimator is in exact adjustment. Note that the mirror must be a precise plano surface if accurate results are expected.

For routine measurements of back focus the bench collimator is substituted for the plane mirror, and, if no auto-collimating microscope is available, a little powder or a grease pencil mark on the rear surface of the test lens can be used as an aid in focusing on the lens surface.

In the absence of many of the usual laboratory trappings, it is still possible to make reasonably accurate determinations of focal lengths and focal points. A lens may be collimated simply enough by focusing it on a distant object. The error in collimation can be determined by the Newtonian equation, $x' = -f^2/x$, where x is the object distance less one focal length and x' is the error in the determination of the focal position. A set of distant targets, such as building edges, smoke stacks, and the like, whose angular separations are known can often be substituted for a focal collimator in determining focal lengths.

Measurement of Telescopic Power: The power of a telescopic system can be measured in three different ways. If the focal lengths of the objective and eyepiece (including any erectors) can be measured, their quotient equals the magnification. The ratio of the diameters of the entrance and exit pupils will also yield the magnifying power. Occasionally the multiplicity of stops in a telescope will introduce some confusion as to whether the pupils measured are indeed conjugates; in this case the image of a transparent scale laid across the objective can be measured at (or near) the exit pupil to determine the ratio. When the field of view is clearly defined, the magnification can be determined by taking the ratio of the tangents of the half field angles at the eyepiece and the objective. Note that the almost inevitable distortion in telescopic eyepieces will usually cause this measurement of power to differ from measurements made by focal lengths or pupil diameters. One should ascertain that the telescope is in afocal adjustment before measuring the power. One way of doing this is to use a low (3 to 5×) power auxiliary telescope or dioptometer (previously focused for infinity) at the eyepiece; this reduces the effect of visual accomodation when the focus is adjusted.

The Measurement of Aberrations: In most instances the aberrations of a test lens can be readily measured on the lens bench by simulating a raytrace. For the measurement of spherical or chromatic aberration, a series of masks, each with a pair of small (to the order of a millimeter in diameter) holes, is useful. As indicated in Fig. 14.17, such a mask, centered over the test lens, simulates the passage of two "rays." When the image is examined with a microscope, a double image of the target is seen except when the microscope is focused at the intersection of the two "rays." By measuring the relative longitudinal position of the ray intersections for masks of various hole spacings, the spherical aberration can be determined. If the measurements are made in red and blue light, the data will yield the chromatic and spherochromatic aberration of the lens.

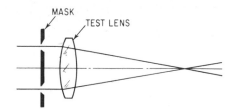

FIG. 14.17. A two hole mask can be used to locate the focus of a particular zone of a lens to determine the spherical aberration.

Figure 14.18 indicates how a similar three-hole mask can be used to measure the tangential coma of a test lens. A multiple hole mask can also be used to measure and plot a ray intercept (rim ray) curve, if desired. The technique for measurement of field curvature is indicated in Fig. 14.19. The bench collimator is

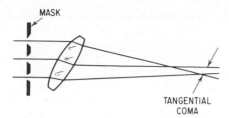

FIG. 14.18. A three hole mask can be used to measure the coma of a test lens.

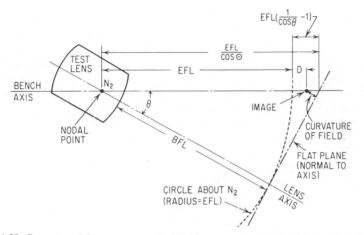

FIG. 14.19. Geometry of the measurement of field curvature using the lens bench nodal slide.

equipped with a reticle consisting of horizontal and vertical lines. The focal length of the test lens is measured. The lens is then adjusted so that its second nodal point is at the center of rotation and the position of the focal point (with the lens axis parallel to the bench axis) is noted. The lens is then rotated through some angle θ. From Fig. 14.19 it is apparent that the intersection of the (flat) focal plane of the lens with the bench axis will shift away from the lens by an amount

$$\text{EFL} \left(\frac{1}{\cos \theta} - 1 \right)$$

as the lens is pivoted through an angle θ. The bench microscope is used to measure D, the amount by which the focus shifts along the axis. Two measurements are necessary, one for the sagittal focus and one for the tangential focus; this is the reason for the orthogonal line pattern of the reticle. Now the departure (along the bench axis) of the image surface from a flat plane is equal to

$$D - \text{EFL} \left(\frac{1}{\cos \theta} - 1 \right)$$

and the curvature of field (parallel to the lens axis) is given by

$$x = \cos\theta \left[D - \text{EFL}\left(\frac{1}{\cos\theta} - 1\right)\right]$$

Much of the numerical work in determining the field curvature by this method can be eliminated by the use of a T-bar attachment to the nodal slide. The cross bar of the T acts as a guide for the bench microscope, causing it to focus on the flat field position as the lens is pivoted. Thus one may measure the value of $x/\cos\theta$ directly; the use of the T-bar eliminates several sources of potential errors inherent in the method described above, although it does complicate the construction of the nodal slide.

Distortion is a difficult aberration to measure. The nodal slide may be used. The lens is adjusted on the slide so that no lateral image shift is produced by a *small* rotation of the lens. Then as the lens is pivoted through larger angles, any lateral displacement of the image is a measure of the distortion. An alternate method is to use the lens to project a rectillinear target and to measure the sag or curvature of the lines in the image, or to measure the magnification of targets of several different angular sizes. The difficulty with any method of measuring distortion is that one invariably winds up basing the work on measurements of magnification (or whatever) vanishingly close to the axis, and the accuracy of such small measurements is usually quite low.

The Star Test: If the object imaged by a lens is effectively a "point," that is, if its image is smaller than the Airy disc, then the image will be a very close approximation to the diffraction pattern. A microscopic examination of the "star" image can indicate a great deal about the lens to the experienced observer. On the axis, the star image of a perfectly symmetrical (about the axis) system is obviously a symmetrical pattern. Therefore, any asymmetry in the pattern is an indication of a lack of symmetry in the system. A flared or coma-shaped pattern on axis generally indicates a decentered element in the system. If the axial pattern is cruciform or shows indications of a dual focus, the cause may be axial astigmatism due either to a non-spherical surface or to a tilted element.

The axial pattern may also be used to determine the state of correction of spherical and chromatic aberration. The outer rings in the diffraction pattern of a well-corrected lens are relatively inconspicuous and the pattern, when defocused, looks about the same both inside and outside the best focus point. In the presence of undercorrected spherical, the pattern will show rings inside the focus and will be blurred outside the focus; the reverse is true of overcorrected spherical. When the spherical aberration is a zonal residual, the ring pattern tends to be heavier and more pronounced.

In the case of undercorrected chromatic, the pattern inside the focus will have a blue center and a red or orange outer flare. As the microscope focus is moved away from the lens, the center of the pattern may turn green, yellow, orange, and will finally become red with a blue halo. The reverse sequence will result from over-corrected chromatic. A chromatically "corrected" lens with a residual secondary spectrum usually shows a pattern with a characteristic yellow-green ("apple" green) center surrounded by a blue or purple halo.

Off-axis star patterns are subject to a much wider range of variations. The classical comet-shaped coma pattern is easily recognized, as is the cross- or onion-shaped pattern due to astigmatism. However, it is rare to find a system with a "pure" pattern off-axis, and it is much more common to encounter a complex mixture of all the aberrations, which are difficult, if not impossible, to sort out.

The star test is a useful diagnostic tool, and, in skilled hands, can be highly effective. The novice should be warned, however, that reliable judgements of quality are difficult, and a considerable amount of experience is necessary before one can safely depend on a star check even for comparative evaluations.

The Foucault Test: The Foucault, or knife-edge, test is performed by moving a knife (or razor-blade) edge laterally into the image of a small point (or line) source. The eye, or a camera, is placed immediately behind the knife and the exit pupil of the system is observed. The arrangement of the Foucault test is shown in Fig. 14.20. If the lens is perfect and the knife is slightly ahead of

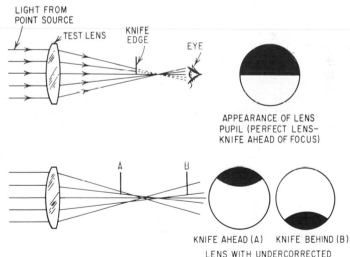

FIG. 14.20. The Foucault Knife Edge Test. Upper: On a perfect lens the knife shadow has a straight edge. Lower: The shadow has a curved edge in the presence of spherical aberration. When the knife cuts through the focus, the pupil (or the zone of the focus) darkens uniformly.

the focus, a straight shadow will move across the exit pupil in the same direction as the knife. When the knife is behind the focus, the direction of the shadow movement is the reverse of the knife direction. When the knife passes exactly through the focus, the entire pupil is seen to darken uniformly.

The same type of analysis can be applied to *zones* of the pupil. If a zone or ring of the pupil darkens suddenly and uniformly as the knife is advanced into the beam, then the knife is cutting the axis at the focus of that particular zone. This is the basis of most of the quantitative measurements made with the Foucault test. The technique generally used is to place a mask over the lens with two symmetrically located apertures to define the zone to be measured. The knife is shifted longitudinally until it cuts off the light through both apertures simultaneously. It is then at the focus for the zone defined by the mask. The process is repeated for other zones and the measured positions of the knife are compared with the desired positions.

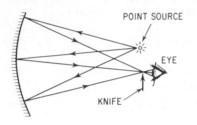

FIG. 14.21. The knife edge test applied to a concave mirror by placing both knife and source at the center of curvature.

This test is extremely useful in the manufacture of large mirrors, which can be tested either at their focus or at their center of curvature. For the center of curvature test, the source is a pinhole adjacent to the knife (Fig. 14.21) and a minimum of space and equipment is required. Obviously if the mirror is a sphere, all zones will have the same focus and a perfect sphere will darken uniformly as the knife passes through the focus. When the surface to be tested is an aspheric, the desired foci for the various zones are computed from the design data and the measurements are compared with the calculated values. It is a relatively simple matter to convert these focus differences into errors in the surface contour; in this way the optician can determine which zones of the lens or mirror require further working to lower the surface.

If the aspheric surface equation is expressed in the form

$$x = f(y)$$

then the equation of the normal to the surface at point (x_1, y_1) is

$$y = y_1 + f(y_1) f'(y_1) - xf'(y_1)$$

and the intersection of the normal with the (optical) axis is then

$$x_0 = x_1 + \frac{y_1}{f'(y_1)}$$

As an example, for a paraboloid represented by

$$x = \frac{y^2}{4f}$$

$$f'(y) = \frac{dx}{dy} = \frac{y}{2f}$$

and the axial intersection of the normal through the point (x_1, y_1) is

$$x_0 = x_1 + \frac{y_1}{\left(\dfrac{y_1}{2f}\right)} = x_1 + 2f = \frac{y_1^2}{4f} + 2f$$

This last equation gives the longitudinal position at which the knife edge should uniformly darken a ring of semi-diameter y_1, when a parabola is tested at the center of curvature (as in Fig. 14.21).

In practice, the knife edge is adjusted longitudinally until the central zone of the mirror darkens uniformly. The distance from the knife to mirror is then equal to $2f$. Then a series of measurements are made using masks with half spacings of y_1, y_2, y_3, etc., each measurement yielding an error e_1, e_2, e_3, etc., where e is the longitudinal distance from the "desired" position for the knife to the actual position.

This data may be readily converted into the difference between the actual slope of the surface and the desired slope by reference to Fig. 14.22. When e is small, we can (to a very good approximation) write

$$\frac{A}{e} = \frac{y}{\sqrt{4f^2 + y^2}}$$

where the term in the right-hand denominator is the distance from the surface to the axis taken along the normal. Now the angle α between the actual normal and the desired normal is equal to

$$\alpha = \frac{A}{\sqrt{4f^2 + y^2}}$$

and substituting for A from the previous expression, we get

$$\alpha = \frac{ye}{4f^2 + y^2}$$

Note that α is also the amount by which the slope of the surface is in error; we can determine the actual departure of the surface from its desired shape by reference to Fig. 14.23. Taking the surface error at the axis as zero, the departure from the desired curve at y_1 is given by

$$d_1 = \frac{-y_1 \alpha_1}{2}$$

At y_2 it is

$$d_2 = d_1 - \frac{1}{2}(y_2 - y_1)(\alpha_1 + \alpha_2)$$

At y_3 it is

$$d_3 = d_2 - \frac{1}{2}(y_3 - y_2)(\alpha_2 + \alpha_3)$$

In general we can write

$$d_n = \frac{1}{2} \sum_{i=1}^{i=n} (y_{i-1} - y_i)(\alpha_{i-1} + \alpha_i)$$

where y_0 and α_0 are assumed zero, and the sign of d is positive if the actual surface is above (to the right in Figs. 14.22 and 23) the desired surface.

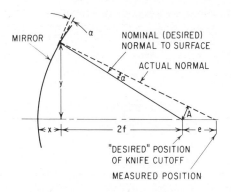

FIG. 14.22. Geometry of knife edge test used to determine the surface contour of a concave (paraboloidal) mirror.

The method outlined above can be readily applied to any concave aspheric. Since it only checks the aspheric at discreet intervals, it must, of course, be supplemented with an over-all knife-edge check to be certain that the surface contour is smooth and free from ridges or grooves. The testing of convex surfaces is more difficult; they are usually checked in conjunction with another mirror chosen so that the combination has an accessible "center focus." The computation of the normal is more involved in this case, but the principles involved are exactly the same.

The Schlieren Test: The Schlieren test is actually a modification of the Foucault test in which the knife blade is replaced by a small pinhole. Thus any ray which misses the pinhole causes a darkened region in the aperture of the optical system. The Schlieren test is especially useful in detecting small variations in index of refraction, either in the optical

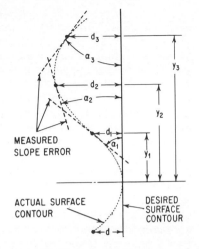

FIG. 14.23. Conversion of measured errors of surface slope (α) into the departure (d) of the actual surface from the desired surface.

system or in the medium (air) surrounding it. In wind-tunnel applications, the tunnel is set up between a collimating optical system and a matching system which focuses the image on the pinhole. When the test is recorded photographically, it is possible to derive quantitative data on the air flow from density measurements on the film.

Resolution Tests: Resolution is usually measured by examining the image of a pattern of alternating bright and dark lines or bars. Conventionally the bright and dark bars are of equal width. A target consisting of several sets of bar patterns of graded spacing is used, and the finest pattern in which the bars can be distinguished (and in which the number of bars in the image is equal to the number in the object) is taken as the limiting resolution of the system under test.

The resolution patterns in use vary in two details of (relatively minor) significance: the number of lines or bars per pattern and the length of the lines relative to their width. The most common practice is to use three bars (and two spaces) per pattern, with a length of five, or more, bar widths. The USAF 1951 target is of this type and the patterns are graded in frequency with a ratio of the sixth root of two between patterns. The National Bureau of Standards Circular No. 533 includes both high- (25:1) and low- (1.6:1) contrast three-bar patterns which are approximately 1-in. long and range in frequency from about one-third line per millimeter to about three lines per millimeter in steps of the fourth root of two. A number of transparent (on film or glass) targets are commercially available; these are, for the most part, based on the USAF target.

The resolution of a photographic system is tested by photographing a suitable target and examining the film under a microscope. In order to obtain optimum results, the photographic processes must be carried out with extreme care, especially with regard to the selection of the best focus, exposure, and development, and the elimination of any vibration in the system. If the microscope used in the examination of the test film has a power approximately equal to the number of lines per millimeter in the pattern, the visual image will have a frequency equal to one line per millimeter, and will be easy to view.

Objective lenses can be tested on an optical bench with a resolution target in the collimator. For lenses with an appreciable angular coverage, a T-bar nodal slide is practically a necessity if reliable off-axis results are to be obtained. Projection of a resolution target is a very convenient means of checking the resolution of lenses designed to cover areas less than a few inches in size. Care must be taken to insure that the illumination system completely fills the aperture of the lens under test; otherwise, the results may be misleading. In all resolution tests, the alignment of the lens axis perpendicular to the target plane is a critical factor. The resolution of telescopic systems can be checked by visual observation of a suitably distant or collimated target. Since the limiting resolution of a telescope is frequently (by design) close to the limiting resolution of the eye, a common practice is to view the image through a low-power auxiliary telescope. Such a telescope serves a dual purpose, in that it reduces the effect of the observer's visual acuity on the measurement and also reduces the effect that his involuntary accomodation (focusing) can have.

The classical criterion for resolution, namely, the ability of a system to separate two point sources of equal intensity, is seldom used (except in astronomy). This is largely because a test using line objects is much easier to make.

Measurement of the Modulation Transfer Function: The measurement of the MTF (frequency response) is, in principle, quite straightforward. The basic elements of the equipment are shown in Fig. 14.24. The test pattern is one in which the brightness varies as a sinusoidal function of one dimension. Such a target is not an easy thing to prepare; fortunately the errors introduced by a target which is not truly sinusoidal are unimportant for most purposes. Some instruments utilize "square wave" targets. The target pattern is imaged by the test lens on a narrow slit whose direction is exactly parallel to the target pattern. The light passing through the slit is measured by a photodetector.

As the target is shifted laterally, the amount of light falling on the detector will vary, and the image modulation is given by

$$M_i = \frac{\text{max} - \text{min}}{\text{max} + \text{min}}$$

where max and min represent the maximum and minimum illumination on the photodetector. The object modulation M_0 is similarly derived from the maximum and minimum brightness levels of the target. The Modulation Transfer Factor (or frequency response, or sine-wave response, or contrast transfer) is then the ratio $M_i : M_0$.

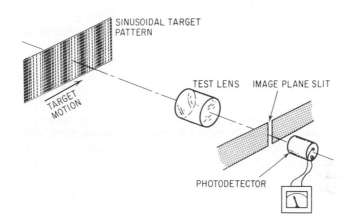

FIG. 14.24. The basic elements of Modulation Transfer (frequency response) measurement equipment. The motion of the target scans its image across the narrow slit, where the maximum and minimum illumination levels are measured. By using targets of different spatial frequency, a plot of the Modulation Transfer Function (vs. frequency) can be obtained.

A provision is usually made to vary the spatial frequency of the target pattern so that the response may be plotted against frequency. The target portion of the system may be as simple as a set of interchangeable targets which are slowly traversed by hand, or it may be a fully automatic device which translates the target and scans a range of frequencies simultaneously.

The image plane slit is almost never just a slit, since the manufacture of a slit of the required narrow dimensions can be fairly difficult. Instead, the image is magnified by a first-class microscope objective; this allows the use of a wider slit.

Obviously any real slit width will have some effect on the measurements, and a slit as narrow as the sensitivity of the photodetector will allow should be used. The effect of the slit width on the response may be readily calculated, since it simply represents a line spread function of rectangular cross section, and the data can be adjusted accordingly where necessary. If the microscope objective is not effectively aberration free, compensation must be made for its effects also.

The source of illumination and the spectral response of the photodetector must, of course, be matched to the application for which the system under test is to be used. Otherwise, serious errors in measurement will result from the unwanted radiation outside the spectral band for which the system has been designed. Usually a set of filters can be found which will provide the proper response.

The situation regarding the measurement of the optical response is somewhat analogous to the measurement of spectral transmission. Relatively simple, rigorous equipment can be used and the measurements can be made on a point-by-point basis; the process is laborious. At the other extreme is the completely automatic equipment which, at the push of a button, plots out the complete response function, with only the operator's initials to be filled in by hand. As in spectrophotometry, however, a poorly designed automatic instrument can, if it produces erroneous results in special cases, be more trouble than it is worth.

The Analysis of "Unknown" Optics: It is frequently necessary to determine the constructional parameters of an optical system. An example might be the analysis of a sample system to determine the reason for its failure to perform to the designer's expectations. For the most part this amounts to the measurement of the radii, thicknesses, spacings, and indices of the system components.

Since the measurements to be made are frequently of a precision barely adequate for the purpose, it is best to provide as many interdependent checks on the process as possible. Thus the first steps should include accurate measurements of effective, back and front focal lengths, as well as the aberrations, so that when the system data is collected, a calculation of the complete (measured) system can provide a final comparison check on the over-all accuracy of the analysis.

The thicknesses and spacings of a system are readily measured. For small systems a micrometer (equipped with ball tips for concave surfaces) is sufficient. A depth gage or an oversize plunger caliper (Nonius gage) is useful for larger systems. If a dimension can be deduced from two different measurements (as a check), the extra time involved is usually a worthwhile investment.

The radius of an optical surface can be measured in many ways. The simplest is probably by use of a thin templet or "brass gage," cut to a known radius and pressed into contact with the surface. Differences between the gage and glass of a few ten-thousandths of an inch can be detected this way, but such a gage is not useful unless it very nearly matches the surface.

The classical instrument for radius measurement is the spherometer, the basic principles of which are outlined in Fig. 14.25. The spherometer measures the sagittal height of the surface over a known diameter; the radius is determined from the formula

$$R = \frac{Y^2 + S^2}{2S}$$

where Y is the semidiameter of the spherometer ring and S is the measured sagittal height. Since the sagittal height is a rather small dimension and thus subject to relatively large errors, the accuracy of a spherometer leaves something to be desired even when extreme precautions are taken. One of the best ways to use a spherometer is as a comparison device, by measuring both the unknown radius and a (nearly equal) carefully calibrated standard radius (e.g., a test glass).

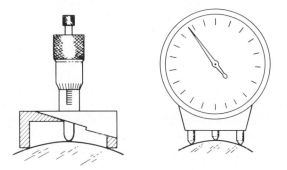

FIG. 14.25. Left: Simple ring spherometer determines the radius of a surface through a measurement of the sagittal height. Right: The diopter gage or lens measure is a spherometer calibrated to read surface curvature in diopters.

The diopter gage, or lens measure, is a handy tool which can provide a quick approximate measure of the surface curvature. As shown in Fig. 14.25, it consists of a dial gage with its plunger between two fixed points. The dial of a diopter gage is calibrated in diopters; the readings may be converted to radii by the formula

$$R = \frac{525}{D} \text{ millimeters}$$

where the 525 is the constant representing 1000 $(N - 1)$ for an "average" opthalmic glass. The accuracy of a typical diopter gage is to the order of 0.1 D at best.

Probably the best way to measure a concave radius is by use of an auto-collimating microscope. The microscope is first focused on the surface and is then focused at the center of curvature (where the reticle image is imaged back on itself by the surface). The distance traveled by the microscope between these two positions is equal to the radius. The precision of this method can be to the order of microns; the accuracy is obviously dependent on the accuracy of the screw or scale used. If the microscope used is of fairly high power (say 150× with NA = 0.3) the quality of the reflected image at the center of curvature is an excellent indication of the sphericity of the surface. Convex surfaces can be measured in this way provided that the working distance of the microscope objective is longer than the radius. A series of long focal length objectives is useful in this regard, although the precision of the method drops as the NA of the objective is lowered (long focal length objectives usually have a small NA) due to the increased depth of focus. If a precise determination of a long convex radius is necessary, a mating concave surface can be made so that it fits perfectly (as tested by interference rings) and the measurement made on the concave glass. Master test plates are measured by this technique.

If a separate piece of glass from which the lens under analysis was made is available, the measurement of its index can be made with considerable precision. The deviation of a test prism may be measured on a laboratory spectrometer, and the prism equations of Chapter 4 used to find the index. Alternatively a refractometer measurement can be made. Either method will readily yield the index value accurate to the fourth decimal place. When one is constrained to measure the lens element itself, without destroying it, the problem is more difficult. A crude determination of the index can be made by measuring the density of the element. A plot of the catalog values of the index against density is then used to determine (very approximately) the corresponding index.

A somewhat more general method is to measure the axial thickness of the element and then to measure the "apparent optical thickness" by focusing a measuring microscope first on one surface and then the other. A simple paraxial calculation, taking into account the refractive properties of the surface through which the second surface is viewed, will yield a value for the index. Depending on the thickness of the element, the index value achieved will probably be almost completely unreliable in the third place, due to the large relative inaccuracy in the measurement of the apparent thickness and to the spherical aberration introduced by the thickness of the glass.

If one measures the radii carefully and makes a good determination of the (paraxial) effective focal length of the element, the thick-lens formula for focal length can be solved for the index of refraction. Although this method requires skilled technique, it is capable of producing results which are accurate to one or two digits in the third place. Note that if care is not taken to eliminate the effects of spherical aberration from the focal length measurement, the resulting index value will tend to err on the high side.

References

Deve, "Optical Workshop Principles," Hilger, London, 1945.
Twyman, "Prism and Lens Making," Hilger, London, 1952.
Elliot and Dickson, "Laboratory Instruments, their Design and Application," Chapman and Hall, London, 1957.

Habell and Cox, "Engineering Optics," Pitman, London, 1948.

Ingalls, "Amateur Telescope Making," Books 1, 2, 3, Scientific American, New York, 1935, 1937, 1953.

Martin, "Optical Measuring Instruments," Van Nostrand, 1924.

Monk, "Optical Instrumentation," McGraw-Hill, N.Y., 1954.

N.B.S. Handbook 77, Vol. III, "Precision Measurement and Calibration," Government Printing Office, Washington, 1961.

Strong, "Procedures in Experimental Physics," Prentice-Hall, 1938.

Taylor, "The Testing and Adjustment of Telescope Objectives," Grubb, Parsons, Newcastle, England, 1945.

Thomson, "A Guide to Instrument Design," Taylor and Francis, London, 1963.

Wright, "The Manufacture of Optical Glass and Optical Systems," Ord. Dept., Doc. 2037, Government Printing Office, Washington, 1921.

Military Specifications:

MIL-O-13830	Optical Components for Fire Control Instruments
MIL-G-174	Optical Glass
MIL-C-675	Coating of Glass Optical Elements (Anti-reflection)
MIL-A-003920	Adhesive; Optical, Thermosetting
MIL-M-13508	Mirrors, Glass, Front Surfaced Aluminized
MIL-G-16592	Glass, Plate (For Optical Instruments)
DD-G-451	Glass, Flat and Corrugated
MIL-STD-150	Photographic Lenses
MIL-P-49	(16 mm Motion Picture Projectors)
JAN-P-245	Projectors (for Slides and Slide Films)
MIL-L-19427	Anamorphic Projection Lenses
MIL-R-6771	Reflectors; Gunsight, Glass
MIL-STD-34	Drawings for Optical Elements and Systems
MIL-STD-1241	Optical Terms and Definitions

Collected Bibliography

Optics — General

Born and Wolf, "Principles of Optics," Macmillan, N. Y., 1964.
Ditchburn, "Light," Wiley-Interscience, N.Y., 1963.
Drude, "The Theory of Optics," Dover, N. Y., 1959.
Francon, "Modern Applications of Physical Optics," Wiley-Interscience, N. Y., 1963.
Hardy and Perrin, "The Principles of Optics," McGraw Hill, N. Y., 1932.
Jenkins and White, "Fundamentals of Optics," McGraw Hill, N. Y., 1957.
Longhurst, "Geometrical and Physical Optics," Longmans, Green, London, 1957.
Menzel, "Fundamental Formulas of Physics," Prentice Hall, N. Y., 1955.
Morgan, "Introduction to Geometrical and Physical Optics," McGraw Hill, N. Y., 1953.
Rayleigh (J. W. Strutt), "The Scientific Papers," Dover, N. Y., 1964.
Rossi, "Optics," Addison Wesley, Cambridge, Mass., 1957.
Sears, "Optics," Addison Wesley, Cambridge, Mass., 1949.
Sommerfeld, "Optics," Academic, N. Y., 1954.
Strong, "Concepts of Classical Optics," Freeman, San Francisco, 1958.
Valasek, "Introduction to Theoretical and Experimental Optics," Wiley, N. Y., 1949.
Wolf, "Progress in Optics," Wiley-Interscience, V.1., 1961; V.2., 1963; V.3., 1964; V.4., 1965.
Wood, "Physical Optics," Macmillan, N. Y., 1934.

Geometrical Optics and Optical Engineering

Bracey, "The Technique of Optical Instrument Design," English Universities Press, London, 1960.
Brouwer, "Matrix Methods in Optical Instrument Design," Benjamin, N. Y., 1964.
Conrady, "Applied Optics and Optical Design," Dover, N. Y., Part I 1957, Part II 1960.
Jacobs, "Fundamentals of Optical Engineering," McGraw Hill, N. Y., 1943.
Kingslake, "Applied Optics and Optical Engineering," Academic, N. Y., 1965.
Martin, "Geometrical Optics," Philosophical Library, N. Y., 1965.
Martin, "Technical Optics," in two volumes, Pitman, London, 1960.
Southall, "Mirrors, Prisms and Lenses," Dover, N. Y., 1964.
Welford, "Geometrical Optics-Optical Instrumentation," North Holland, Amsterdam, 1962.

The Eye

Alder, "Physiology of the Eye-Clinical Application," Mosby, St. Louis, 1959.
Davson, "The Physiology of the Eye," Blakiston, 1950.
Hartrige, "Recent Advances in the Physilogy of Vision," Blakiston, 1950.
Helmholtz, "Treatise on Physiological Optics," Dover, N. Y., 1962.
Wright, "Photometry and the Eye," Hatton, London, 1949.
Zoethout, "Physiological Optics," Professional, 1939.

Color

Gibson, "Spectrophotometry," N.B.S. Circular No. 484 Govt. Printing Office, Washington, 1949.
Judd, "Colorimetry," N.B.S. Circular No. 478 Govt. Printing Office, Washington, 1950.
Judd and Wyszecki, "Color in Business, Science and Industry," Wiley, N. Y., 1963.
M.I.T., "Handbook of Colorimetry," Technology Press, 1936.
Wright, "The Measurement of Color," Van Nostrand, Princeton, N. J., 1964.

Optical Materials

"American Institute of Physics Handbook," McGraw Hill, N. Y., 1963.
Ballard, McCarthy, and Wolfe, "Optical Materials for Infrared Instrumentation," Univ. of Michigan, Ann Arbor, 1959; Supplement 1961.
Morey, "Properties of Glass," Reinhold, N. Y., 1954.

Thin Films

Hass (Ed.), "Physics of Thin Films," Academic, N. Y., Vol. 1, 1963; Vol. 2,1964.
Heavens, "Optical Properties of Thin Films," Butterworth's, London, 1955.
Holland, "Vacuum Deposition of Thin Films," Wiley, N. Y., 1956.
Vasicek, "Optics of Thin Films," North Holland, Amsterdam, 1960.

Infrared and Photometry

Hackforth, "Infrared Radiation," McGraw Hill, N. Y., 1960.
Holter et al., "Fundamentals of Infrared Technology," Macmillan, N. Y., 1962.
Jamieson et al., "Infrared Physics and Engineering," McGraw Hill, N. Y., 1963.
Kruse et al., "Elements of Infrared Technology - Generation, Transmission and Dectection," Wiley, N. Y., 1962.
Smith, Jones and Chasmar, "Detection and Measurement of Infrared Radiation," Clarendon; Oxford, 1957.
Walsh, "Photometry," Dover, N. Y., 1958.

Optical Computation and Evaluation

Buchdahl, "Optical Aberration Coefficients," Oxford, London, 1954.
Conrady, "Applied Optics and Optical Design," Dover, N. Y., Part I, Part II, 1960.
Emsley, "Aberrations of Thin Lenses," Constable, London, 1956.
Herzberger, "Modern Geometrical Optics," Interscience, N. Y., 1958.
Hopkins, "Wave Theory of Aberrations," Oxford, London, 1950.
Linfoot, "Fourier Methods in Optical Design," Focal, N. Y., 1964.
N.B.S. Circular No. 526, "Optical Image Evaluation," Govt. Printing Office, Washington, 1954.
O'Neill, "Introduction to Statistical Optics," Addison-Wesley, Reading, Mass., 1963.

Optical Design and Designs

Bouwers, "Achievements in Optics," Elsevier, N. Y., 1946.
Conrady, "Applied Optics and Optical Design," Dover, N. Y., Part I, 1957; Part II, 1960.
Cox, "A System of Optical Design," Focal, N. Y., 1964.
Cox, "Optics-the Technique of Definition," Focal, N. Y., 1943.
Dimitroff and Baker, "Telescopes and Accessories," Blakiston, Philadelphia, 1945.
Glazebrook, "Dictionary of Applied Physics," Macmillan, London, 1923.
Greenleaf, "Photographic Optics," Macmillan, N. Y., 1950.
Holladay, "Computer Design of Optical Lens Systems," in "Computer Applications," Ed. by Mittman and Unger, Macmillan, N. Y., 1961.
"Handbook of Optical Design," MIL-HDBK-141.
Johnson, "Optical Design and Lens Computation," Hatton, London, 1948.
King, "The History of the Telescope," Sky Pub. Corp., Cambridge, Mass., 1955.
Kingslake, "Lenses in Photography," Barnes, N. Y., 1963.
Kingslake, "Applied Optics and Optical Engineering," Academic, N. Y., 1965.
Linfoot, "Recent Advances in Optics," Clarendon, Oxford, 1955.
Merte, Richter, and VonRohr, "Das Photographische Objektiv," Springer, Vienna, 1932; reprinted by Edwards, Ann Arbor, 1944.
Merte, "Das Photographische Objektiv seit dem Jahre 1929," Springer, Vienna, 1943; reprinted by Edwards, Ann Arbor, 1946.
Merte, "Das Photographische Objektiv," Parts 1 and 2, translation, C.A.D.O., Wright Patterson AFB, Dayton, 1949.
Merte, "The Zeiss Index of Photographic Lenses," Vol. 1 and 2, C.A.D.O., Wright Patterson AFB, Dayton, 1950.
Ordnance Corps Manual, "Design of Fire Control Optics," Vol. 1, 1952; Vol. 2, 1953.
Stavroudis, "Automatic Optical Design," in Vol. 5 of "Advances in Computers," Ed. by Alt and Rubinoff, Academic, N. Y., 1964.

Optical Manufacture

Deve, "Optical Workshop Principles," Hilger, London, 1945.
Ingalls (Ed.), "Amateur Telescope Making," Books 1, 2, 3, Scientific American, N. Y., 1935, 1937, 1953.

"The Optical Industry Directory," Optical Pub. Co., Lenox, Mass., Annually.
Texerau, "How to Make a Telescope," Interscience, N. Y., 1957.
Twyman, "Prism and Lens Making," Hilger and Watts, London, 1952.
Wright, "The Manufacture of Optical Glass and Optical Systems," Ord. Dept. Doc. 2037, Govt. Printing Office, Washington, 1921.

Instruments and Practice

Boutry, "Instrumental Optics," Wiley-Interscience, N. Y., 1962.
Cagnet, Francon, and Thrierr, "Atlas of Optical Phenomena," Prentice Hall, Englewood Cliffs, N. J., 1962.
Candler, "Modern Interferometers," Hilger and Watts, London, 1951.
Elliot and Dickson, "Laboratory Instruments, their Design and Application," Chapman and Hall, London, 1957.
Habell, "Optical Instruments and Techniques, Proceedings of the Conference, London 1961," Chapman and Hall, London, 1962.
Habel and Cox, "Engineering Optics," Pitman, London, 1948.
Johnson, "Optics and Optical Instruments" ("Practical Optics"), Dover, N. Y., 1960.
Martin, "Optical Measuring Instruments," Van Nostrand, 1924.
Monk, "Optical Instrumentation," McGraw Hill, N. Y., 1954.
N.B.S. Handbook 77, Vol. III, "Precision Measurement and Calibration," Selected Papers on Optics, Metrology and Radiation, Govt. Printing Office, Washington, 1961.
"Proceedings of the London Conference on Optical Instruments 1950," Wiley, N. Y., 1952.
Strong, "Proceedures in Experimental Physics," Prentice-Hall, 1938.
Taylor, "The Testing and Adjustment of Telescope Objectives," Grubb, Parsons, Newcastle, England, 1945.
Thomson, "A Guide to Instrument Design," Taylor and Francis, London, 1963.
Brown, "Modern Optics," Reinhold, N. Y., 1965.

Technical Journal References

Abbreviations:

JOSA   Journal of the Optical Society of American
AO   Applied Optics
JSMPTE   Journal of the Society of Motion Pictures and Television Engineers
            (also JSMPE)
PSE   Photographic Science and Engineering
The above journals are published at 20th and Northampton Streets, Easton, Pa.

Chapter 4   Prisms and Mirrors

Walles and Hopkins, "The Orientation of the Image Formed by a Series of Plane Mirrors," AO, v. 3, pp. 1447-1452 (1964).

Pegis and Rao, "Analysis and Design of Plane Mirror Systems," AO, v. 2, pp. 1271-1274 (1963).

Walther, "Comment on . . . Pegis and Rao," AO, v. 3, p. 543 (1964).

Waddell, "Design of Rotating Prisms for High-Speed Cameras," JSMPE, v. 53, pp. 496-501 (1949).

Schwesinger, "Reducing Aberrations in Rotating-Prism Compensators for Non-intermittent Motion Pictures," JOSA, v. 42, pp. 923-930 (1952).

Berkowitz, "Design of Plane Mirror Systems," JOSA, v. 55, pp. 1464-1467 (1965).

Chapter 5   The Eye and Color

"American Standard Methods of Measuring and Specifying Color," JOSA, v. 41, pp. 431-439 (1951).

De Palma and Lowry, "Sine-Wave Response of the Visual System II," JOSA, v. 52, pp. 328-335 (1962).

Krauskopf, "Light Distribution in Human Retinal Images," JOSA, v. 52, pp. 1046-1050 (1962).

Lowry and De Palma, "Sine-Wave Response of the Visual System, I The Mach Phenomenon," JOSA, v. 51, pp. 740-746 (1961).

Schober and Hilz, "Contrast Sensitivity of the Human Eye for Square-Wave Gratings," JOSA, v. 55, pp. 1086-1091 (1965).

Chapter 6   Stops and Apertures

Kingslake, "The Effective Aperture of a Photographic Objective," JOSA, v. 35, pp. 518-520 (1945).

Murty, "On the Theoretical Limit of Resolution," JOSA, v. 47, pp. 667-668 (1945).

Welford, "Use of Annular Apertures to Increase Focal Depth," JOSA, v. 50, pp. 749-753 (1960)

Stimson, "The G-Number: A Photometric Lens-Aperture Designation," JSMPTE, v. 74, pp. 99-101 (1965).

Chapter 7   Optical Materials and Coatings

Pellicori, "Transmittances of Some Optical Materials for Use Between 1900 and 3400 Å," AO, v. 3, pp. 361-366 (1964).

Malitson, "A Redetermination of Some Optical Properties of Calcium Fluoride," AO, v. 2, pp. 1103-1107 (1963).

McCarthy, "The Reflection and Transmission of Infrared Materials," Part 1 "Spectra," AO, v. 2, pp. 591-595; Part 2. "Bibliography," AO, v. 2, pp. 596-603 (1963); Part 3, AO, v. 4, p. 317 (1965); Part 4, AO, v. 4, pp. 507-511, (1965).

Herzberger, "The Dispersion of Optical Glass," JOSA, v. 32, pp. 70-77 (1942).

Murrary, "Effect of Antireflection Films on Color in Optical Instruments," JOSA, v. 46, pp. 790-796 (1956).

Weinstein, "Computations in Thin Film Optics," Vacuum, v. IV, pp. 3-19 (1954).
Hass, "Filmed Surfaces for Reflecting Optics," JOSA, v. 45, pp. 945-952 (1955).
Hass and Erbe, "Method of Producing Replica Mirrors with High Quality Surfaces," JOSA, v. 44, pp. 669-671 (1954).
Hass and Jenness, "Method for Fabricating Paraboloidal Mirrors," JOSA, v. 48, pp. 86-87 (1958).
Bradford, Erbe, and Hass, "Two Step Method for Producing Replica Mirrors with Expoxy Resins," JOSA, v. 49, pp. 990-991 (1959).
Saxton and Kline, "Optical Characteristics and Physical Properties of Filled Epoxy Mirrors," JOSA, v. 50, pp. 1103-1111 (1960).
Kline and Saxton, "Radiation, Moisture and Temperature Effects in Filled Expoxy Replica Mirrors," JOSA, v. 51, pp. 447-451 (1961).
Meyer, "High Accuracy Plastic Replica Optics," JSMPTE, v. 74, pp. 28-32 (1965).

Chapter 8   Radiation and Photometry

Hardy, "The Distribution of Light in Optical Systems," J. Franklin Inst. v. 208, No. 6 (1929).
Hardy, "Flux Calculations in Optical Systems," JOSA, v. 33, pp. 71-74 (1943).
Reiss, "The Cos^4 Law of Illumination," JOSA, v. 35, pp. 283-288 (1945).
Reiss, "Notes on the Cos^4 Law of Illumination," JOSA, v. 38, pp. 980-986 (1948).
Infrared Issue (49 Articles on IR Technology), Proc. of the I.R.E., Sept. 1959.
Canada, "Radiation Slide Rule," General Electric Review, Dec. 1948.

Chapter 9   Basic Optical Devices

Benford, "The Projection of Light," JOSA, v. 35, pp. 149-156 (1945).
Dempster, "The Principles of Microscope Illumination and the Problem of Glare," JOSA, v. 34, pp. 695-710 (1944).
Jones, "Immersed Radiation Detectors," AO, v. 1, pp. 607-613 (1962).
Dreyfus, "Wedge-Immersed Thermistor Bolometer," AO, v. 1, pp. 615-618 (1962).
Greenler, "An Image Reducer Immersed Detector System for the Infrared," AO, v. 1, pp. 674-675 (1962).
Kingslake, "The Optics of the Lenticular Color Film Process," JSMPTE, v. 67, pp. 8-13 (1958).
Williamoon, "Cone Channel Condenser Optics," JOSA, v. 42, pp. 712-715 (1952).
McLeod, "The Axicon: A New Type of Optical Element," JOSA, v. 44, pp. 592-597 (1954).
McLeod, "Axicons and their Uses," JOSA, v. 50, pp. 166-169 (1960).
Krolak and Parker, "The Optical Tunnel-A Versatile Electrooptical Tool," JSMPTE, v. 72, pp. 177-180 (1963).
Potter, Donath, and Tynan, "Light Collecting Properties of a Perfect Circular Optical Fiber," JOSA, v. 53, pp. 256-260 (1963).
Potter, "Transmission Properties of Optical Fibers," JOSA, v. 51, pp. 1079-1089 (1961).
Kapany, "Role of Fiber Optics in Ultra-High-Speed Photography," JSMPTE, v. 71, pp. 75-81 (1962).

Puder and Mortensen, "Xenon Illuminator Systems for 35mm and 70mm Projection," JSMPTE, v. 74, pp. 594-597 (1965).

Courtney-Pratt, "High Speed Photography and Micrography," AO, v. 3, pp. 1201-1209 (1964).

Davis, "Maximizing Exposure-Limited Resolution of Practical Rotating Mirror Cameras," AO, v. 3, pp. 1217-1222 (1964).

Cook, "Recent Developments in Anamorphic Systems," JSMPTE, v. 65, pp. 151-154 (1956).

Bouwers and Blaisse, "Anamorphic Mirror Systems," JSMPTE, v. 65, pp. 146-150 (1956).

Benford, "The Cinemascope Optical System," JSMPTE, v. 62, pp. 64-70 (1954).

Rosin, "Anamorphic Lens System," JSMPTE, v. 66, pp. 407-409 (1957).

Kingslake, "The Development of the Zoom Lens," JSMPTE, v. 69, pp. 534-544 (1960).

Bergstein and Motz, "Third Order Aberration Theory for Varifocal Systems," JOSA, v. 47, pp. 579-593 (1957).

Bergstein, "General Theory of Optically Compensated Varifocal Systems," JOSA, v. 48, pp. 154-171 (1958).

Bergstein and Motz, "Two Component Optically Compensated Varifocal System," JOSA, v. 52, pp. 353-362 (1962).

Bergstein and Motz, "Three Component Optically Compensated Varifocal System," JOSA, v. 52, pp. 362-375 (1962).

Bergstein and Motz, "Four Component Optically Compensated Varifocal System," JOSA, v. 52, pp. 376-388 (1962).

Reiss, "Note on Thin Lens Color Correction for Variable Focal Length Lenses," JOSA, v. 48, pp. 578-579 (1958).

Pegis and Peck, "First-Order Design Theory for Linearly Compensated Zoom Systems," JOSA, v. 52, pp. 905-911 (1962).

Back and Lowen, "The Basic Theory of Varifocal Lenses with Linear Movement and Optical Compensation," JOSA, v. 44, pp. 684-691 (1954).

Back and Lowen, "Generalized Theory of Optically Compensated Varifocal Systems," JOSA, v. 48, pp. 154-171 (1958).

Cook, "Television Zoom Lenses," JSMPTE, v. 68, pp. 25-28 (1958).

Wooters and Silvertooth, "Optically Compensated Zoom Lens," JOSA, v. 55, pp. 347-351 (1965).

Chapter 10   Optical Computation

Feder, "Optical Calculations with Automatic Computing Machinery," JOSA, v. 41, pp. 630-635 (1951).

Spencer and Murty, "Generalized Ray-Tracing Proceedure," JOSA, v. 52, pp. 672-678 (1962).

Allen and Snyder, "Ray Tracing through Uncentered and Aspheric Surfaces," JOSA, v. 42, pp. 243-249 (1952).

Herzberger, "Automatic Ray Tracing," JOSA, v. 47, pp. 736-739 (1957).

Herzberger, "Some Remarks on Ray Tracing," JOSA, v. 41, pp. 805-807 (1951).

Herzberger and Marchand, "Tracing a Normal Congruence thru an Optical System," JOSA, v. 44, pp. 146-154 (1954).

Murray, "A Toric Skew Ray Trace," JOSA, v. 44, pp. 672-676 (1954).

Lessing, "Cylindrical Ray-Tracing Equations for Electronic Computers," JOSA, v. 52, pp. 472-473 (1962).

Stavroudis, "Ray-Tracing Formulas for Uniaxial Crystals," JOSA, v. 52, pp. 187-191 (1962).

Stavroudis and Feder, "Automatic Computation of Spot Diagrams," JOSA, v. 44, pp. 163-170 (1954).

Buchdahl, "Algebraic Theory of the Primary Aberrations of the Symmetrical Optical System," JOSA, v. 38, pp. 14-19 (1948).

Buchdahl, "Optical Aberration Coefficients:

I The Coefficient of Tertiary Spherical Aberration," JOSA, v. 46, pp. 941-943 (1956);

II The Tertiary Intrinsic Coefficients," JOSA, v. 48, pp. 563-567 (1958);

III The Computation of the Tertiary Coefficients," JOSA, v. 48, pp. 747-756 (1958);

IV The Coefficient of Quanternary Spherical Aberration," JOSA, v. 48, pp. 757-759 (1958);

V On the Quality of Predicted Displacements," JOSA, v. 49, pp. 1113-1121 (1959);

VI On the Computations Involving Coordinates Lying Partly in Image Space," JOSA, v. 50, pp. 534-539 (1960);

VII The Primary, Secondary and Tertiary Deformations and Retardation of the Wavefront," JOSA, v. 50, pp. 539-544 (1960);

VIII Coefficient of Spherical Aberration of Order Eleven," JOSA, v. 50, pp. 678-683 (1960);

IX Theory of Reversible Optical Systems," JOSA, v. 51, pp. 608-616 (1961);

X Theory of Concentric Optical Systems," JOSA, v. 52, pp. 1361-1367 (1962);

XI Theory of a Concentric Corrector," JOSA, v. 52, pp. 1367-1372 (1962);

XII Remarks Relating to Aberrations of Any Order," JOSA, v. 55, pp. 641-649 (1965).

Cruickshank, "The Paraxial Differential Transfer Coefficients of a Lens System," JOSA, v. 36, pp. 13-19 (1946).

Cruickshank, "On the Primary Chromatic Coefficients of a Lens System," JOSA, v. 36, pp. 103-107 (1946).

Cruickshank and Hills, "Use of Optical Aberration Coefficients in Optical Design," JOSA, v. 50, pp. 379-387 (1960).

Delano, "A General Contribution Formula for Tangential Rays," JOSA, v. 42, pp. 631-633 (1952).

Herzberger and Marchand, "Image Error Theory for Finite Aperture and Field," JOSA, v. 42, pp. 306-321 (1952).

Wolf, "On a New Aberration Function of Optical Instruments," JOSA, v. 42, pp. 547-552 (1952).

Chapter 11   Image Evaluation

Perrin, "Methods of Appraising Photographic Systems, Part I Historical Review," JSMPTE, v. 69, pp. 151-156 (1960).

Perrin, "Methods of Appraising Photographic Systems, Part II Manipulation and Signifigance of the Sine-Wave Response Function," JSMPTE, v. 69, pp. 239-248 (1960).

Toraldo Di Francia, "On the Image Sharpness in the Central Field of a System Presenting Third- and Fifth-Order Spherical Aberration," JOSA, v. 43, pp. 827-835 (1953).

O'Neill, "Transfer Function for an Annular Aperture," JOSA, v. 46, pp. 258-288 (1956).

Kapany and Burke, "Various Image Assessment Parameters," JOSA, v. 52, pp. 1351-1361 (1962).

Schade, "An Evaluation of Photographic Image Quality and Resolving Power," JSMPTE, v. 73, pp. 81-119 (1964).

Schade, "Modern Image Evaluation and Television," AO, v. 3, pp. 17-21 (1964).

Smith, "Optical Image Evaluation and the Transfer Function," AO, v. 2, pp. 335-350 (1963).

Charman, "Spatial Frequency Spectra and Other Properties of Conventional Resolution Rargets," PSE, v. 8, pp. 253-259 (1964).

Herzberger, "Analysis of Spot Diagrams," JOSA, v. 47, pp. 584-594 (1957).

Herzberger, "Light Distribution in the Optical Image," JOSA, v. 37, pp. 485-493 (1947).

Marechal, "A Coherent Set of Tolerances for Geometrical Aberrations, Including Excentration Effects," JOSA, v. 40, p. 60 (1950).

Hotchkiss, Washer, and Rosberry, "Spurious Resolution of Photographic Lenses," JOSA, v. 41, pp. 600-603 (1951).

Sparrow, Astrophys. J. v. 44, p. 76 (1916) (the Sparrow criterion).

Linfoot, "Information Theory and Optical Images," JOSA, v. 45, pp. 808-819 (1955).

Linfoot, "Transmission Factors and Optical Design," JOSA, v. 46, pp. 740-752 (1956).

Miyamoto, "Image Evaluation by Spot Diagram Using a Computer," AO, v. 2, pp. 1247-1250 (1963).

Miyamoto, "On a Comparison between Wave Optics and Geometrical Optics by Using Fourier Analysis:

     I General Theory," JOSA, v. 48, pp. 57-63 (1958),

     II Astigmatism, Coma, Spherical Aberration," JOSA, v. 48, pp. 567-575 (1958).

     III Image Evaluation by Spot Diagram," JOSA, v. 49, pp. 35-40 (1959).

Barakat, "Total Illumination in a Diffraction Image Containing Spherical Aberration," JOSA, v. 51, pp. 152-157 (1961).

Barakat, "Computation of the Transfer Function of an Optical System from the Design Data for Rotationally Symmetric Aberrations I Theory," JOSA, v. 52, pp. 985-991 (1962).

Barakat and Morello, "Computation of the Transfer Function etc... II Programming and Numerical Results," JOSA, v. 52, pp. 992-997 (1962).

Barakat, "Rayleigh Wavefront Criterion," JOSA, v. 55, pp. 641-649 (1965).

Barakat and Levin, "Application of Apodization to Increase Two-Point Resolution by the Sparrow Criterion:

     I Coherent Illumination," JOSA, v. 52, pp. 276-283 (1962).

     II Incoherent Illumination," JOSA, v. 53, pp. 274-282 (1963).

Barakat and Lev, "Transfer Functions and Total Illuminance of High Numerical Aperture Systems Obeying the Sine Condition," JOSA, v. 53, pp. 324-332 (1963).

Barakat and Houston, "Reciprocity Relations between the Transfer Function and Total Illuminance I," JOSA, v. 53, pp. 1244-1249 (1963).

Barakat and Houston, "Modulation of Squarewave Objects in Incoherent Light," JOSA, v. 53, pp. 1371-1376 (1963).

Barakat and Houston, "Diffraction Effects of Coma," JOSA, v. 54, pp. 1084-1088 (1964).

Barakat, "Numerical Results Concerning the Transfer Functions and Total Illuminance for Optimum Balanced Fifth-Order Spherical Aberration," JOSA, v. 54, pp. 38-44 (1964).

Barakat and Morello, "Computation of the Total Illuminance (Encircled Energy) of an Optical System from the Design Data for Rotationally Symmetric Aberrations," JOSA, v. 54, pp. 235-240 (1964).

Barakat and Houston, "Line Spread Function and Cumulative Line Spread Function for Systems with Rotational Symmetry," JOSA, v. 54, pp. 768-773 (1964).

Barakat and Houston, "Transfer Function of an Annular Aperture in the Presence of Spherical Aberration, "JOSA, v. 55, pp. 538-541 (1965).

Carman and Charman, "Detection, Recognition and Resolution in Photographic Systems," JOSA, v. 54, pp. 1121-1130 (1964).

Coltman, "The Specification of Imaging Properties by Response to a Sine Wave Input," JOSA, v. 44, pp. 468-471 (1954).

Crane, "An Objective Method for Rating Picture Sharpness: SMT Acutance," JSMPTE, v. 73, pp. 643-647 (1964).

DeVelis, "Comparison of Methods for Image Evaluation," JOSA, v. 55, pp. 165-174 (1965).

Toraldo Di Francia, "Resolving Power and Information," JOSA, v. 45, pp. 497-501 (1955).

Drougard, "Optical Transfer Properties of Fiber Bundles," JOSA, v. 54, pp. 907-914 (1964).

Eyer, "Spatial Frequency Response of Certain Photographic Emulsions," JOSA, v. 48, pp. 938-944 (1958).

Hariharan, "Resolution of an Annulus Test Object," JOSA, v. 45, pp. 44-45 (1955).

Hopkins, Oxley, and Eyer, "The Problem of Evaluating a White Light Image," JOSA, v. 44, pp. 692-698 (1954).

Jones, "On the Point and Line Spread Functions of Photographic Images," JOSA, v. 48, pp. 934-937 (1958).

Kuwabara, "Studies on the Image Formed by Lenses:
I On the Characteristics of an Image and their Quantative Representation," JOSA, v. 45, pp. 309-319 (1955).
II The Effect of Spherical Aberration on Optical Images," JOSA, v. 45, pp. 625-636 (1955).

Bray, "Comparative Analysis of Geometric vs Diffraction Heterochromatic Lens Evaluations using Optical Transfer Function Theory," JOSA, v. 55, pp. 1136-1138 (1965).

Lamberts, "Relationship between the Sine Wave Response and the Distribution of Energy in the Optical Image of a Line," JOSA, v. 48, pp. 490-495 (1958).

Lamberts, "Application of Sine-Wave Techniques to Image Forming Systems," JSMPTE, v. 71, pp. 635-640 (1962).

Lucy, "Image Quality Criteria Derived from Skew Traces," JOSA, v. 46, pp. 699-706 (1956).

Marchand, "Derivation of the Point Spread Function from the Line Spread Function," JOSA, v. 54, pp. 915-919 (1964).

Marchand and Phillips, "Calculation of the Optical Transfer Function from Lens Design Data," AO, v. 2, pp. 359-364 (1963).

Murcott and Gottfried, "Interferometer Quality and its Relation to Photographic Resolving Power," JOSA, v. 45, pp. 644-646 (1955).

Higgins, "Methods for Engineering Photographic Systems," AO, v. 3, pp. 1-9 (1964).

Perrin and Altman, "Studies in the Resolving Power of Photographic Emulsions: VI The Effect of the Type of Patterns and the Luminance Ratio in the Test Object," JOSA, v. 43, pp. 780-790 (1953).

O'Connell, "Variation of Photographic Resolving Power with Lateral Chromatic Aberration," JOSA, v. 47, pp. 1018-1020 (1957).

Chapter 12   The Design of Optical Systems: General

Baker, "Planetary Telescopes," AO, v. 2, pp. 111-129 (1963).

Hopkins, "Telescope Doublets," JOSA, v. 49, pp. 200-201 (1959).

Korones and Hopkins, "Some Effects of Glass Choice in Telescope Doublets," JOSA, v. 49, pp. 869-871 (1959).

Rosin, "A New Thin Lens Form," JOSA, v. 42, pp. 451-455 (1952).

Miller, "A Solar Observatory on Long Island," AO, v. 2, pp. 93-103 (1963).

Perrin and Hoadley, "Photographic Sharpness and Resolving Power. I The Design and Performance of an Apochromatic Resolving Power Camera Objective," JOSA, v. 38, pp. 1040-1053 (1948).

Hopkins, "Secondary Color in Optical Relay Systems," JOSA, v. 39, pp. 919-921 (1949).

Herzberger and Jenkins, "Color Correction in Optical Systems and Types of Glass," JOSA, v. 39, pp. 984-989 (1949).

Lessing, "Selection of Optical Glasses in Apochromats," JOSA, v. 47, pp. 955-958 (1957).

Stephens, "Secondary Chromatic Aberration," JOSA, v. 47, p. 1135 (1957).

Lessing, "Further Considerations on the Selection of Optical Glasses in Apochromats," JOSA, v. 48, pp. 269-273 (1958).

Stephens, "Selection of Glasses for Three Color Achromats," JOSA, v. 49, pp. 398-401 (1959).

Smith, "Thin Lens Analysis of Secondary Spectrum," JOSA, v. 49, pp. 640-641 (1959).

Stephens, "Four Color Achromats and Superchromats," JOSA, v. 50, pp. 1016-1019 (1960).

Willey, "Machine-Aided Selection of Optical Glasses for Two-Element, Three-Color Achromats," AO, v. 1, pp. 368-369 (1962).

Herzberger and Salzberg, "Refractive Indices of Infrared Optical Materials and Color Correction of Infrared Lenses," JOSA, v. 52, pp. 420-427 (1962).

Herzberger and McClure, "The Design of Superachromatic Lenses," AO, v. 2, pp. 553-560 (1963).

Herzberger, "Some Recent Ideas in the Field of Geometrical Optics," JOSA, v. 53, pp. 661-671 (1963).

Gelles, "A General Purpose Infrared Lens," AO, v. 1, p. 78 (1962).

Treuting, "An Achromatic Doublet of Silicon and Germanium," JOSA, v. 41, pp. 454-456 (1951).

Stockbarger and Hawley, "On Lithium Fluoride-Quartz Achromatic Lenses," JOSA, v. 29, pp. 29 (1939).

Stockbarger and Hawley, "Lithium-Fluoride Quartz Apochromat," JOSA, v. 29, pp. 350 (1939).

Murray, "An All-Irtran Doublet Objective," AO, v. 4, pp. 254-255 (1965).

Stephens, "Design of Triplet Anastigmat Lenses of the Taylor Type," JOSA, v. 38, pp. 1032-1039 (1948).

Smith, "Comment on Design of Triplet Anastigmat Lenses of the Taylor Type," JOSA, v. 40, pp. 406-407 (1950).

Lessing, "Selection of Optical Glasses in Taylor Triplets (General Method)," JOSA, v. 49, pp. 31-34 (1959).

Lessing, "Selection of Optical Glasses in Taylor Triplets (Special Method)," JOSA, v. 48, pp. 558-562 (1958).

Lessing, "Selection of Optical Glasses in Taylor Triplets with Residual Chromatic Aberration," JOSA, v. 49, pp. 872-877 (1959).

Hopkins, "Third Order and Fifth Order Analysis of the Triplet," JOSA, v. 52, pp. 389-394 (1962).

Wallin, "Design Study of Air Spaced Triplets," AO, v. 3, pp. 421-426 (1964).

Herzberger, "Replacing a Thin Lens by a Thick Lens," JOSA, v. 34, pp. 114-115 (1944).

Glancy, "Differential Correction of Optical Systems," JOSA, v. 41, pp. 389-396 (1951).

Herzberger, "Precalculation of Optical Systems," JOSA, v. 42, pp. 637-640 (1952).

Rumsey, "On the Extension of a System for Differential Correction of Lens Systems to Include Second Order Terms," JOSA, v. 41, pp. 229-234 (1951).

Wallin, "The Control of Petzval Curvature," JOSA, v. 41, pp. 1029-1032 (1951).

Hopkins and Spencer, "Creative Thinking and Computing Machines in Optical Design," JOSA, v. 52, pp. 172-176 (1962).

Hopkins, "Re-Evaluation of the Problem of Optical Design," JOSA, v. 52, pp. 1218-1222 (1962).

Malacara, "Two Lenses to Collimate Red Laser Light," AO, v. 4, pp. 1652-1654 (1965).

Rosen and Eldert, "Least Squares Method of Optical Correction," JOSA, v. 44, pp. 250-252 (1954).

Lucy, "Simultaneous Correction of Meridian Aberrations," JOSA, v. 45, pp. 320-327 (1955) Errata p. 670.

Hopkins, McCarthy, and Walters, "Automatic Correction of Third Order Aberrations," JOSA, v. 45, pp. 363-365 (1955).

McCarthy, "A Note on 'The Automatic Correction of Third Order Aberrations'," JOSA, v. 45, pp. 1087-1088 (1955).

Hopkins and Lauroesh, "Automatic Design of Telescope Doublets," JOSA, v. 45, pp. 992-994 (1955).

Rosen and Chung, "Application of the Least Squares Method," JOSA, v. 46, pp. 223-226 (1956).

Feder, "Automatic Lens Design Methods," JOSA, v. 47, pp. 902-912 (1957).

Feder, "Calculation of an Optical Merit Function and Its Derivative with Respect to System Parameters," JOSA, v. 47, pp. 913-925 (1957).

Merion and Loebenstein, "Automatic Correction of Residual Aberration," JOSA, v. 47, pp. 1104-1109 (1957).

Wynne and Nunn Proc. Phys. Soc. (London) v. 74, pp. 316 (1959), and v. 73, pp. 777 (1959): Articles on Automatic Design by Computer (SLAMS).

Meiron, "Automatic Lens Design by the Least Squares Method," JOSA, v. 49, pp. 293-298 (1959).

Meiron and Volinez, "Parabolic Approximation Method for Automatic Lens Design," JOSA, v. 50, pp. 207-211 (1960).

Meiron, "Damped Least-Squares Method for Automatic Lens Design," JOSA, v. 55, pp. 1105-1109 (1965).

Feder, "Automatic Lens Design with a High-Speed Computer,"JOSA, v. 52, pp. 177-183 (1962).

Grey, "Aberration Theories for Semiautomatic Lens Design by Electronic Computer: I Preliminary Remarks," JOSA, v. 53, pp. 672-676 (1963); II A Specific Computer Program," JOSA, v. 53, pp. 677-680 (1963).

Feder, "Automatic Optical Design," AO, v. 2, pp. 1209-1226 (1963).

Hopkins and Feder, "The Symposium Lens-An Epilogue," AO, v. 2, pp. 1227-1231 (1963).

Wynne and Wormell, "Lens Design by Computer," AO, v. 2, pp. 1233-1238 (1963).

Spencer, "A Flexible Automatic Lens Correction Proceedure," AO, v. 2, pp. 1257-1264 (1963).

Brixner, "Automatic Lens Design for Nonexperts," AO, v. 2, pp. 1281-1286 (1963).

Brixner, "The Symposium Lens Improved," AO, v. 2, pp. 1331-1332 (1963).

Brixner, "Automatic Lens Design: Further Notes for Optical Engineers," JSMPTE, v. 73, pp. 314-320 (1964).

Brixner, "The Symposium Lenses- A Performance Evaluation," AO, v. 3, pp. 780-781 (1964).

Hopkins and Viswanathan, "The Symposium Lens Continued," AO, v. 3, pp. 787-788 (1964).

Brixner, "Automatic Lens Design Illustrated by a 600mm f/2.0, 24° Field Lens," JSMPTE, v. 73, pp. 654-657 (1964).

Grey, "RE: Automatic Lens Design," JSMPTE, v. 74, pp. 799-800 (1965).

O'Brien, "Automatic Optical Design of Desired Image Distributations Using Orthagonal Constraints," JOSA, v. 54, pp. 1252-1255 (1964).

Chapter 13   The Design of Optical Systems: Particular

Stempel, "An Emprirical Approach to Lens Design- The Huygens Eyepiece,"JOSA, v. 33, pp. 278-292 (1943).

Matter, "Eyepieces in Photography," PSE, v. 4, pp. 234-236 (1960).

Reiss, "The Design of a Telephoto-Magnifer," JOSA, v. 33, pp. 641-651 (1943).

Hopkins, "Aspheric Corrector Plates for Magnifiers," JOSA, v. 36, pp. 604-610 (1946).

Coulman and Petrie, "Some Notes on the Designing of Aspherical Magnifiers for Binocular Vision," JOSA, v. 39, pp. 612-613 (1949).

Bennett, "The Development of the Microscope Objective," JOSA, v. 33, pp. 123-128 (1943).

Foster, "Microscope Optics," JOSA, v. 40, pp. 275-282 (1950).

Benford and Butterfield, "A New Series of Photomicrographic Lenses," JOSA, v. 44, pp. 598-600 (1954).

Claussen, "Microscope Objectives with Plano-Correction," AO, v. 3, pp. 993-1003 (1964).

Benford, "Recent Microscope Developments at Bausch and Lomb," AO, v. 3, pp. 1044-1045 (1964).

Cruickshank, "The Trigenometrical Correction of Miscroscope Objectives," JOSA, v. 36, pp. 296-298 (1946).

Foster and Thiel, "An Achromatic Ultraviolet Microscope Objective," JOSA, v. 38, pp. 689-692 (1948).

Grey and Lee, "A New Series of Miscroscopes Objectives:
I Catadioptric Newtonian Systems,"
II Preliminary Investigation of Catadioptric Schwarzchild Systems,"
JOSA, v. 39, pp. 719-723 and 723-728 (1949).

Grey, "A New Series of Microscope Objectives: III Ultraviolet Objectives of Intermediate Numerical Aperture," JOSA, v. 40, pp. 283-290 (1950).

Blout, Bird, and Grey, "Infrared Microspectroscopy," JOSA, v. 40, pp. 304-313 (1950).

Rank, "Some High Speed Spectrograph Objectives," JOSA, v. 40, pp. 462-464 (1950).

Thorndike, "A Wide Angle, Underwater Camera Lens," JOSA, v. 40, pp. 823-824 (1950).

Norris, Seeds, and Wilkins, "Reflecting Microscopes with Spherical Mirrors," JOSA, v. 41, pp. 111-119 (1951).

Grey, "Computed Aberrations of Spherical Schwarzchild Reflecting Miscroscope Objectives," JOSA, v. 41, pp. 183-192 (1952).

Miyata, Yamagawa, and Noma, "Reflecting Microscope Objectives with Nonspherical Mirrors," JOSA, v. 42, p. 431 (1952).

Thornburg, "Reflecting Objective for Microscopy," JOSA, v. 45, pp. 740-743 (1955).

Taylor and Lee, "The Development of Photographic Lenses," Proc. Phys. Soc. London, v. 47-3, pp. 502-518 (1935).

Glancy, "A Photographic Lens System," JOSA, v. 35, pp. 307-308 (1945).

Kingslake, "Lenses for Aerial Photography," JOSA, v. 32, pp. 129-134 (1942).

Kingslake, "A Classification of Photographic Lens Types," JOSA, v. 35, pp. 251-255 (1946).

Kingslake, "Recent Developements in Lenses for Aerial Photography," JOSA, v. 37, pp. 1-9 (1947).

Kaprelian, "Recent and Unusual German Lens Designs," JOSA, v. 37, pp. 466-472 (1947).

Gardner and Washer, "Lenses of Extremely Wide Angle for Airplane Mapping," JOSA, v. 38, pp. 421-431 (1948).

Aklin, "The Effect of High Index Glasses on the Field Characteristics of Photographic Objectives," JOSA, v. 38, pp. 841-844 (1948).

Kaprelian, "Objective Lenses of f/1 Aperture and Greater," JSMPE, v. 53, pp. 86-99 (1949).

Smith, "Control of Residual Aberrations in the Design of Anastigmat Objectives," JOSA, v. 48, pp. 98-105 (1958).

Kingslake, "New Series of Lenses for 16mm Cameras," JSMPE, v. 52, pp. 509-521 (1949)

Pestrecov and Hayes, "Animar Series of Photographic Lenses," JSMPTE, v. 54, pp. 183-198 (1950).

Schade, "A New f/1.5 Lens for Professional 16mm Projectors," JSMPTE, v. 54, pp. 337-344 (1950).

Havlicek, "On Simple Sky Lenses," JOSA, v. 41, pp. 1058-1059 (1951).

Foote and Miesse, "Military-Type Lenses for 35mm Motion Picture Cameras," JSMPTE, v. 59, pp. 219-232 (1952).

Lotman and Stettler, "A New Projection Lens," JSMPTE, v. 64, p. 259 (1955).

Kazamaki and Kondo, "New Series of Distortionless Telephoto Lenses," JOSA, v. 46, pp. 22-31 (1956).

Cook, "Modern Cine Camera Lenses," JSMPTE, v. 65, pp. 155-161 (1956).

Cook, "35mm Camera Lenses," JSMPTE, v. 67, pp. 534-536 (1958).

Hayes, "A New Series of Lenses for Vidicon Type Cameras," JSMPTE, v. 67, pp. 593/595 (1958).

Cook, "Vidicon Camera Lenses," JSMPTE, v. 67, pp. 862-864 (1959).

Rosin, "Concentric Lens," JOSA, v. 49, pp. 862-864 (1959).

Heimer, "The Systematic Design of a High Acuity Long Focal Length Reconnaissance Objective for Aerospace Photography," PSE, v. 7, pp. 157-161 (1963).

Miyamoto, "Fish Eye Lens," JOSA, v. 54, pp. 1060-1061 (1964).

Miyamoto, "The Four- and Five-Element Lens Designed by Computer," AO, v. 4. pp. 81-83 (1965).

Offner and Decker, "An f:1:0 Camera for Astronomical Spectroscopy," JOSA, v. 41, pp. 169-172 (1951).

Lucy, "Exact and Approximate Computation of Schmidt Cameras
I Classical Arrangement," JOSA, v. 30, pp. 251 (1940);
II Some Modified Arrangements," JOSA, v. 30, pp. 358 (1941).

Synge, "The Theory of the Schmidt Telescope," JOSA, v. 33, pp. 129-136 (1943).

Benford, "Design Method for a Schmidt Camera with a Finite Source," JOSA, v. 34, pp. 595-596 (1944).

Herzberger and Hoadley, "The Calculation of Aspherical Correcting Surfaces," JOSA, v. 36, pp. 334-340 (1946).

Glancy, "On the Theory and Computation of an Aspherical Surface," JOSA, v. 36, pp. 416-423 (1946).

Friedman, "Method of Computing Correction Plate for Schmidt System for Near Projection...." JOSA, v. 37, pp. 480-484 (1947).

Linfoot and Wolf, "On the Corrector Plates of Schmidt Cameras," JOSA, v. 39, pp. 752-756 (1949).

Wormser, "On the Design of Wide Angle Schmidt Optical Systems," JOSA, v. 40, pp. 412-415 (1950).

Meiron, "On the Design of Optical Systems Containing Aspheric Surfaces," JOSA, v. 46, pp. 288-292 (1956).

Miyamoto, "On the Design of Optical Systems with an Aspheric Surface," JOSA, v. 51, pp. 21-22 (1961).

Schulte, "Auxiliary Optical Systems for the Kitt Peak Telescopes," AO, v. 2, pp. 141-151 (1963).

DeVany, "Optical Design for Two Telescopes," AO, v. 2, pp. 201-204 (1963).

Nielsen, "Aberrations in Ellipsoidal Mirrors Used in Infrared Spectrometers," JOSA, v. 39, pp. 59-63 (1949).

Yoder, Patrick, and Gee, "Analysis of Cassegrain-Type Telescopic Systems," JOSA, v. 43, pp. 1200-1204 (1953).

Jones, "Coma of a Modified Gregorian and Cassegrainian Mirror System," JOSA, v. 44, pp. 630-633 (1954).

Rosin, "Mirror Condenser for Spectrographs," JOSA, v. 45, p. 398 (1955).

Erdos, "Mirror Anastigmat with Two Concentric Spherical Surfaces," JOSA, v. 49, pp. 877-886 (1959).

Rosin, "Corrected Cassegrain System," AO, v. 3, pp. 151-152 (1964).

Maksutov, "New Catadioptic Meniscus Systems," JOSA, v. 34, pp. 270-284 (1944).

Stephens, "Reduction of Sphero-Chromatic Aberration in Catadioptric Systems," JOSA, v. 38, pp. 733-735 (1948).

Lauroesch and Wing, "Bouwers Concentric Systems for Materials of High Refractive Index," JOSA, v. 49, pp. 410-411 (1959).

Rosin, "Optical Systems for Large Telescopes," JOSA, v. 51, pp. 331-335 (1961).

Waland, "Flat Field Maksutov-Cassegrain Optical Systems," JOSA, v. 51, pp. 359-366 (1961).

Gelles, "A New Family of Flat Field Cameras," AO, v. 2, pp. 1081-1084 (1963).

Churilovskii and Goldis, "An Apochromatic Catadioptric System Equivalent to a Parabolic Mirror," AO, v. 3, pp. 843-846 (1964).

DeVany, "Schmidt-Cassegrain Telescope System with a Flat Field," AO, v. 4, p. 1353 (1965).

Wynne, "Field Correctors for Large Telescopes," AO, v. 4, pp. 1185-1192 (1965).

Schulte, "Anastigmatic Cassegrain Type Telescopes," AO, v. 5, pp. 309-311 (1966).

Schulte, "Prime Focus Correctors Involving Aspherics," AO, v. 5, pp. 313-317 (1966).

Chapter 14 Optics in Practice

McLeod and Sherwood, "A Proposed Method of Specifying Appearance Defects of Optical Parts," JOSA, v. 35, pp. 136-138 (1945).

Military Specifications:

MIL-O-13830	Optical Components for Fire Control Instruments
MIL-G-174	Optical Glass
MIL-C-675	Coating of Glass Optical Elements (Anti-reflection)
MIL-STD-150	Photographic Lenses
MIL-A-003920	Adhesive; Optical, Thermosetting
MIL-P-49	16mm Motion Picture Projectors
JAN-P-245	Projectors (for Slides and Slide Films)
MIL-L-19427	Anamorphic Projection Lenses
MIL-R-6771	Reflectors; Gunsight, Glass

DD-G-451   Glass, Flat and Corrugated
MIL-STD-34   Drawings for Optical Elements
MIL-STD-1241 Optical Terms and Definitions
MIL-M-13508 Mirrors, Glass, Front Surfaced Aluminized
MIL-G-16592 Glass, Plate (for Optical Instruments).

American Standard PH 22.90-1964, "Aperture Calibration of Motion Picture Lenses," JSMPTE, v. 73, pp. 496-499 (1964).

Yoder, Patrick, and Gee, "Permitted Tolerance on Percent Correction of Paraboloidal Mirrors," JOSA, v. 43, pp. 702-703 (1953).

Stevens, "Analysis of Errors in a Reflecting Prism," JOSA, v. 33, pp. 74-78 (1943).

Motz, "On the Nature of a Polished Surface," JOSA, v. 32, pp. 147-148 (1942).

Morey, "The Flow of Glass at Room Temperature," JOSA, v. 42, pp. 856-857 (1952).

Ray, "On the Mechanism of Optical Polishing," JOSA, v. 39, p. 92 (1949).

Schulz, "Preparation of Aspherical Refracting Optical Surfaces by an Evaporation Technique," JOSA, v. 38, pp. 432-441 (1948).

Schulz, "Making Fresnel Off-Axis Parabolic Mirrors by the Evaporation Technique," JOSA, v. 37, pp. 349-354 (1947).

Lewis, "Graphical Ray-Trace and Surface Generation Methods for Aspheric Surfaces," JOSA, v. 41, pp. 456-459 (1951).

Cooke, "Lightweight Mirror Blank," AO, v. 2, pp. 623-624 (1963).

Bowen, "Optical Problems at the Palomar Observatory," JOSA, v. 42, pp. 795-800 (1952).

Sakurai and Shishido, "Study on the Fabrication of Aspherical Surfaces," AO, v. 2, pp. 1181-1190 (1963).

Sakurai and Shishido, "Fabrication of Paraboloidal Mirror Segments for a Large Solar Furnace," AO, v. 3, pp. 813-816 (1964).

Cooke, "Making a Concave f/0.4 Parabola," AO, v. 3, pp. 1148-1149 (1964).

Rank, "Correction of Imperfect Weak Cylindrical Lenses," JOSA, v. 36, pp. 172-175 (1946).

Cooke, "Grinding Aspheric Surfaces," AO, v. 4, pp. 764-765 (1965).

Cooke, "Making a 1.22m Diameter Aluminum Mirror," AO, v. 4, pp. 1210-1213 (1965).

Cooke, "The Making of a Cylindrical Ellipse," AO, v. 5, pp. 209-210 (1966).

Miller, McLeod, and Sherwood, "Thin Sheet Plastic Fresnel Lenses of High Aperture," JOSA, v. 41, pp. 807-814 (1951).

Boettner and Barnett, "Design and Construction of Fresnel Optics for Photoelectric Receivers," JOSA, v. 41, pp. 849-857 (1951).

Epstein, "The Aberrations of Slightly Decentered Optical Systems," JOSA, v. 39, pp. 847-853 (1949).

Ruben, "Aberrations Arising from Decentrations and Tilts," JOSA, v. 54, pp. 45-52 (1964).

Wooters, "Lens Centering in Microscope Objectives," JOSA, v. 40, pp. 521-523 (1950).

Grey, "Athermalization of Optical Systems," JOSA, v. 38, pp. 542-546 (1948).

Neumer, "New Series of Lenses for Professional 16mm Projection," JSMPTE, v. 52, pp. 501-508 (1949).

Strong, "New Johns Hopkins Ruling Engine," JOSA, v. 41, pp. 3-15 (1951).

Brooks, "Adjustable Instrument Mount," JOSA, v. 44, p. 87 (1954).

Schwesinger, "Optical Effect on Flexure in Vertically Mounted Precision Mirrors," JOSA, v. 44, pp. 417-424 (1954).

Embleton, "Improved Mirror Mount for Optical Interferometers," JOSA, v. 45, pp. 152-153 (1955).

Scott, "Optical Engineering," AO, v. 1, pp. 387-397 (1962).

Deterding, "High Precision, Strain-Free Mounting of Large Lens Elements," AO, v. 1, pp. 403-406 (1962).

Chin, "Optical Mirror-Mount Design Philosophy," AO, v. 3, pp. 895-901 (1964).

Phelps, "Making of Spacers for Fabry-Perot Etalons," JOSA, v. 55, pp. 293-295 (1965).

Washer and Williams, "Precision of Telescope Pointing for Outdoor Targets," JOSA, v. 36, pp. 400-411 (1946).

Lovins, "High-Precision Pointing Interferometer," AO, v. 3, pp. 883-887 (1964).

Dyson, "Precise Measurement by Image Splitting," JOSA, v. 50, pp. 754-757 (1960).

Gardner and Bennett, "A Modified Hartman Test Based on Interference," JOSA, v. 11, pp. 441 (1952).

Kingslake, "A New Bench for Testing Photographic Objectives," JOSA, v. 22, pp. 207 (1932).

Kinglake, "Measurement of the Aberrations of a Miscroscope Objective," JOSA, v. 26, pp. 251 (1963).

Leistner, Marcus, and Wheeler, "Lens Testing Bench," JOSA, v. 43, pp. 44-48 (1953).

Saunders, "An Improved Optical Test for Spherical Aberration," JOSA, v. 44, p. 664 (1954).

Washer, "Instrument for Measuring the Marginal Power of Spectacle Lenses," JOSA, v. 45, pp. 719-726 (1955).

Washer and Darling, "Factors Affecting the Accuracy of Distortion Measurements Made on the Nodal Slide Optical Bench," JOSA, v. 49, pp. 517-534 (1959).

Welford, "On the Limiting Sensitivity of the Star Test for Optical Instruments," JOSA, v. 50, pp. 21-23 (1960).

Zimmerman, "A Method of Measuring the Distortion of Photographic Objectives," AO, v. 2, pp. 759-760 (1963).

Schulz, "Quantitative Tests for Off-Axis Parabolic Mirrors," JOSA, v. 36, pp. 588-594 (1946).

Goldberg and Benson, "Apparatus for Determining Spherical Aberration of Convex Surface of Paraboloid of Revolution," JOSA, v. 37, pp. 186-191 (1947).

Yoder, Patrick, and Gee, "Application of the Rayleigh Criterion to the Null Test of Dall-Kirkham Primary Mirrors," JOSA, v. 45, pp. 881-883 (1955).

Offner, "A Null Corrector for Paraboloidal Mirrors," AO, v. 2, pp. 153-155 (1963).

Cooke, "Testing a Schmidt Corrector Aspheric Surface by Autocollimation," AO, v. 2, p. 328 (1963).

McRae, "The Measurement of Transmission and Contrast in Optical Instruments," JOSA, v. 33, pp. 229-243 (1943).

Coleman, Coleman, and Fridge, "Theory and Use of the Dioptometer," JOSA, v. 41, pp. 94-97 (1951).

Rank, "Measurement of the Radius of Curvature of Concave Spheres," JOSA, v. 36, pp. 108-110 (1946).

Cooke, "The Bar Spherometer," AO, v. 3, pp. 87-88 (1964).

Hugo and Lessing, "Determination of Long Radii of Curvature of Positive Lenses," AO, v. 3, pp. 483-485 (1964).

Townsley and Foote, "An Autocollimator for Precise Measurement of Flange Focal Distance of Photographic Lenses," JOSA, v. 37, pp. 42-46 (1947).

Jacobsohn, Beloian, and Hess, "A Rapid Non-Destructive Method of Determining Refractive Index of Lens Components, Using a Spectrometer," JOSA, v. 37, pp. 941-943 (1947).

Rank, "Calibration of a Set of Master Wedges," JOSA, v. 36, pp. 116-119 (1946).

Barnes and Bellinger, "Schlieren and Shadowgraph Equipment for Air Flow Analysis," JOSA, v. 35, pp. 497-509 (1945).

Washer and Gardner, "Method for Determining the Resolving Power of Photographic Lenses," N.B.S. Circular 533, Government Printing Office, Washington, 1953.

American Standards Assoc. 70 E. 45th St. N.Y.:

    Z 52.55   Meth. of Det. Resol. Power, Slide Film Proj. Lenses

    Z 22.53   Meth. of Det. Resol. Power, 16mm Mot. Picture Lenses

    Z 38.7.16 Meth. of Det. Resol. Power, 35mm and 2x2 Slide Lenses

Perrin and Altman, "Photographic Sharpness and Resolving Power II The Resolving Power Cameras in the Kodak Research Laboratory," JOSA, v. 41, pp. 265-272 (1951).

Hutto, "Equipment for Evaluating Lenses of Television Systems," JSMPTE, v. 64, pp. 133-136 (1955).

Murcott and Gottfried, "Photographic Resolving Power and Aberrations of Lenses," JOSA, v. 45, pp. 434-440 (1955).

Carpenter and Hopkins, "Comparison of Three Photoelectric Methods of Image Evaluation," JOSA, v. 46, pp. 764-767 (1956).

Lamberts, Higgins, and Wolfe, "Measurement and Analysis of the Distribution of Energy in Optical Images," JOSA, v. 48, pp. 487-490 (1958).

Taylor and Thompson, "Attempt to Investigate Experimentally the Intensity Distribution near the Focus in the Error-Free Diffraction Pattern of Circular and Annular Apertures," JOSA, v. 48, pp. 844-850 (1958).

Herriot, "Recording Electronic Lens Bench," JOSA, v. 48, pp. 968-971 (1958).

De and SenGupta, "Measurement of Wave Aberrations of Microscope and Other Objectives," JOSA, v. 51, pp. 158-164 (1961).

Kelly, "Systems Analysis of the Photographic Process II Transfer Function Measurements," JOSA, v. 51, pp. 319-330 (1960).

Lamberts, "The Production and Use of Variable Transmittance Test Objects," AO, v. 2, pp. 273-276 (1963).

Shannon and Newman, "An Instrument for the Measurement of the Optical Transfer Function," AO, v. 2, pp. 365-369 (1963).

Shack, "The Influence of Image Motion and Shutter Operation on the Photographic Transfer Function," AO, v. 3, pp. 1171-1181 (1964).

Barakat, Newman, and Humphreys, "Measurement of the Total Illuminance in a Diffraction Image, Part II. Line Sources," JOSA, v. 54, pp. 1256-1260 (1964).

Index